TAB ELECTRONICS GUIDE TO
UNDERSTANDING ELECTRICITY
AND ELECTRONICS

TAB Electronics Guide to Understanding Electricity and Electronics

G. Randy Slone

Second Edition

McGraw-Hill

New York · San Francisco · Washington, D.C. · Auckland · Bogotá
Caracas · Lisbon · London · Madrid · Mexico City · Milan
Montreal · New Delhi · San Juan · Singapore
Sydney · Tokyo · Toronto

Library of Congress Cataloging-in-Publication Data

Slone, G. Randy.
   TAB electronics guide to understanding electricity and electronics / G. Randy Slone.—
2nd ed.
     p. cm.
   Includes index.
   ISBN 0-07-136057-3
   1. Electronics. 2. Electricity. I. Title.

TK7816 S57 2000
621.381—dc21   00-032905

McGraw-Hill

A Division of The McGraw·Hill Companies

ISBN 0-07-136057-3

*The sponsoring editor for this book was Scott Grillo, the editing supervisor was Caroline
Levine, and the production supervisor was Sherri Souffrance. It was set in Vendome ICG by
Joanne Morbit of McGraw-Hill's Professional Book Group composition unit, Hightstown, N.J.*

Printed and bound by R. R. Donnelley & Sons Company.

To my Lo[...] Jesus Christ, from whom
all good things originate, to the glory of our
Heavenly Father.

CONTENTS

Contents

Contents

PREFACE

When I was 10 years old, I purchased an introductory electronics book at a local pharmacy. I cherished this book because I knew it contained marvelous and fascinating knowledge about *electronics*—an awesome, mysterious, and exciting field that was to become the literal foundation of the next step in human progress. Unfortunately, after carrying my beloved electronics book back and forth to school every day for months and spending countless study-hall hours trying to decipher the strange terms and symbolism, I finally became discouraged and concluded that it was beyond my comprehension.

In reflecting on that experience, I recognize now that it would have been almost impossible for me to understand that book. As with so many other introductory electronics textbooks, too many assumptions were made regarding the reader's educational background, especially in regard to mathematics. In part because of my childhood experience, I became very excited (and grateful) when the project of writing this book was offered to me by the people at McGraw-Hill. For years, I have wanted to write the book that should have been available in that small town drugstore. In a sense, you could say this text was written through the eyes of a 10-year-old; the awe, mystery, and excitement are still there!

Understanding Electricity and Electronics has been designed to be the best "first book" for anyone interested in becoming proficient in the electrical/electronics fields. The beginning chapters provide guidance in establishing an inexpensive electronics workplace and in acquiring all of the informational and mechanical tools needed. The succeeding chapters combine electronics projects with the appropriate text, both of which advance at an easy-to-digest pace. Expensive electronics text equipment is not required for building, testing, or utilizing the projects; the only essential piece of equipment needed is a volt-ohm meter. The appendices provide a handy reference for commonly used equations, symbols, and supply sources. In short, *Understanding Electricity and Electronics* contains everything the novice needs to know, to have, and to do to become a competent electronics hobbyist and experimenter.

ACKNOWLEDGMENTS

I would like to thank all of my readers who have provided me with invaluable feedback throughout the past five years—without their support, this second edition would be deficient of many improvements. A special debt of gratitude goes out to my wife, Mary Ann, who continually supports and encourages me, especially when my schedule is hectic and the hours are late. I want to thank the excellent support group at Electronics Workbench (especially Joe Koenig and Luis Alves) for their invaluable support in the production of this work. I'm also greatly indebted to David T. Ronan, Curriculum Developer with the Professional Career Development Institute, for his friendship and technical support. All of my efforts and support would be meaningless without the professionalism and expertise provided by the incomparable group at McGraw-Hill. So a heartfelt thanks goes out to Scott Grillo and Caroline Levine. A good textbook is always a group effort!

G. RANDY SLONE

1

Getting Started

As the old proverb states, "Every long journey starts with the first step." This chapter deals with the first steps in developing a new and fascinating activity in your life. My goal is to make the *journey* a comfortable, entertaining, and rewarding experience. If you will supply a little time, effort, and patience, you can accomplish your goal.

Establishing Reasonable Goals

All successful individuals achieve their varying degrees of success by establishing and accomplishing goals. Many people establish goals without even realizing it. In many cases, these goals can be classified as dreams, aspirations, or concepts. Reflect back to the last time that you were really pleased with some accomplishment in your life. This accomplishment required personal motivation, planning, effort, success, and the satisfaction brought on by that success. The actual goal was determined during the planning stage, but it was not achieved until the success stage. Between these stages occurred all of the work and effort.

Take these concepts one step further. Assume that an unobtainable goal has been established. An indefinite amount of work and effort might be invested in this elusive goal, but the end result will be discouragement and defeat. To make matters worse, the individual who established this unreasonable goal might become hesitant to set new goals, because of the fear of a similarly wasted effort and failure.

In my opinion, the establishment of unreasonable goals is the most significant obstacle that you must face in your journey toward becoming proficient in the field of electronics.

What Is Unreasonable?

For the sake of illustration, assume for the moment that this book is entitled *Building Your Own Automobiles*. Few people would buy such a book—only those motivated by the desire to build their own car from scratch. Such an idea is ludicrous for several reasons. For one, it isn't practical. The cost of buying the individual pieces and parts to build a finished car would cost 5 to 10 times as much as a new car that has been factory built and tested. The builder would have to be very knowledgeable, and experienced in a wide variety of the specialized skills existing within the automobile industry, such as diagnostic alignment, automobile electrical systems, and bodywork and finishing techniques. In addition, it staggers the mind to imagine the vast array of the specialized shop tools that would be required for such a project!

There are similar circumstances in the electronics field. Even electronic geniuses don't build their own television sets from thousands of tiny parts that they pick up at their local electronics dealer. It is also helpful to understand that most electronic systems are not invented, designed, and built by single individuals. For example, the consumer electronic products that we all enjoy every day (TVs, radios, CD players, VCRs, etc.) are actually evolutionary products. They have been redesigned and improved over a period of many years, by many different design engineers.

Generally, goals involving the *from-scratch* building of complex electronic systems are usually unreasonable. Also, from a conceptual point of view, the field of electronics is vast, and divided into many specialized fields. If your goal is to understand everything about electronics, I wish you a very long life. I believe you may need more than one lifetime to accomplish a goal of that magnitude!

What Is Reasonable?

This is the exciting part in regard to the electrical and electronics fields, because the possibilities are limited only by your ingenuity and imagination. I have known many people who got started in electronics as a rewarding hobby, only to find themselves in a high-paying career before they knew it! For example, a good friend of mine became interested in home computers as a hobby. As he continued to expand his computer system, his personal financial situation forced him to locate the most inexpensive places to purchase the pieces for his system. As he began to impress his friends and relatives with his computer system, they decided to buy their own systems. One day, almost by accident, he discovered that he could supply them with systems identical to his own for substantially less money than the local computer store; even after he added in a healthy profit for himself. As a result, he changed his career in the midstream of life (he was an investment counselor), and opened a very successful computer store. I know this story very well, because I bought my first computer from him!

I have had many friends and acquaintances who began tinkering with electronics in their homes as a hobby. Eventually, they found themselves deluged with friends, neighbors, and relatives bringing them everything from portable television sets to computer monitors, all harping the same request, "When you get a few minutes, would you please take a look at this. I think it's a simple problem because it worked just fine yesterday." This leaves the besieged electronics tinkerer with one of two choices: either begin charging for repair services, or become a candidate for the "Good Samaritan of the Year Award." Of those who began to charge for their services, many have found lucrative and rewarding careers.

Many people have the mistaken belief that a career in electronics is not possible without a formal degree from an accredited college or technical institution. A formal degree will certainly enhance and accelerate your career progress, but there are many career pathways for non-degreed individuals as well. For example, many electronic salespersons do not have a deep, intricate knowledge of the products they sell; a functional and applicative understanding is all that is required. The consumer electronics repair field is loaded with people who were called *tube jockeys* back in the 1950s. As the field of electronics evolved from vacuum tubes to solid-state technology, they advanced right along with it by reading the various electronic periodicals published within their field. Many younger people who are successful in these same fields received their education from correspondence or vocational schools.

Nondegreed electronic hobbyists often find careers within their hobby. For example, the hobbyist who collects a large parts inventory for personal use may begin to sell these parts at substantial profits. Many local electronics stores have had their beginnings in this manner; not to mention some large national parts distribution chains.

Your personal interests will play a big role in discovering and opening doors to possible career opportunities. A hobbyist who likes to tinker with automotive sound systems may start a part-time business installing these systems in the local community. A little effort, perseverance, and dedication to performing quality work can convert it to a full-time lucrative career.

Nondegreed electrical and/or electronic career opportunities are common in most industrial manufacturing facilities. The majority of electrical maintenance personnel I have trained over the years have had little or no formal classroom training in the electrical or electronics fields. Within the industrial manufacturing community, any prior experience in this area (even at the hobbyist level) is usually given weighty consideration in hiring and job promotions.

The list of possibilities goes on and on. Don't expect to go around designing sentient robots, or building laboratories that look like they're out of an old Boris Karloff movie. However, don't stumble over the diamonds, while you are looking for the gold! It is always reasonable to expect to go as far as your effort and perseverance will take you.

Obtaining the Informational Tools

The informational tools you will need for a successful hobby or career in the various electrical and electronics fields can be broken down into four categories: textbooks, data books, periodicals, and catalogs.

Textbooks

Textbooks are generally self-explanatory as to their usefulness to any specific individual. As your experience grows, you will probably collect a reasonable library, according to your needs and interests. For the novice, I would recommend an *electronics dictionary* and a beginning-level *electronics math* book.

Data Books

The manufacturers and distributors of electronic components publish data books, containing cross-referencing information and individual component specifications. A few examples of such books are *NTE Semiconductors, The GE Semiconductor Replacement Guide,* and *SK Replacement Cross-Reference Directory.* As you can see, the data book titles are self-explanatory.

Your first project in the field of electronics is to obtain all of the electronics data books that you can get your hands on. The reason, for having included this section at the beginning of this book, is to give you ample time to order and receive a fair quantity of data books before you start your first projects. They are *that* essential. Many electronics manufacturers will supply their data books free of charge, if you simply call and ask them. This is especially true if you have started a small part-time or full-time business. Try this approach first; then, if you are not successful, they can be purchased from many different electronics supply companies.

Manufacturers' data books can be general or specific in nature. Try to acquire the general or broad-based data books in the beginning. As your interests begin to lean toward certain specific areas, you can obtain what you need at a later date.

Data books provide needed information in two critical areas: cross-referencing and parts specifications. Back in the days of vacuum tubes, the tube manufacturers would identify their tubes with certain generic numbers. In other words, a 12AU7 tube would always be labeled *12AU7*, regardless of the manufacturer. Unfortunately, this tradition did not carry over into the solid-state field. Although there are generic numbers for solid-state components, they are only used occasionally. Instead, you must rely on the manufacturer's cross-references, which are supplied in their data books. For example, suppose you needed to replace a defective transistor labeled *NTE 130*. If you had some NTE 130s in your personal stock, or if your local electronics shop carried the NTE line of components, you would simply use another NTE 130 as a replacement. But, if you didn't have one and the local parts store only carried the SK line of components, you would have to cross-reference the NTE 130 to its SK equivalent. In this case, it would be an SK3027. When you consider that there are dozens of major parts manufacturers, each using their own unique part numbers, you begin to appreciate the usefulness of an exhaustive cross-reference library.

Cross-referencing will also play an important role in acquiring a respectable parts inventory. All of the large surplus and wholesale electronics houses offer many electronic components at buy-out or wholesale prices. In most cases, you will have to cross-reference these parts to know if you can use them. In addition, if you salvage parts from used equipment to place in an inventory, you will have to cross-reference the used parts to know what they are.

One final word on cross-referencing electronic parts—it is not as difficult as it may seem from just reading this book. On receipt of your first few data books, spend a few minutes scanning through the cross-reference section. You should easily recognize how the parts are arranged according to sequential numbers and letters.

Electronics data books also provide the detailed specifications for electronic parts. When you begin to build or design electronics projects, you will need a working knowledge of the specifics of the various parts you intend to use. For example, the device parameters will define the electrical conditions for reliable operation (breakdown voltage, power dissipation, maximum current, etc.), and the pictorial diagrams will provide the necessary mechanical information (case style, lead designation, pin definitions, etc.).

If you have not understood some of the terms I have used in describing and explaining data books, don't worry. When you have a chance to skim through one, much of what is written here will become clear. The rest will be understood as you begin to build a few of the projects covered in the following chapters. The important thing to do right now is to *get them!* A list of the sources, from which to obtain electronic data books, is provided in Appendix B.

Periodicals

Periodicals are very important to anyone involved in the electrical or electronics fields for a variety of reasons. First, they keep the electronics enthusiast current and up-to-date in the latest technology. It has been said that a college-degreed electrical engineer will become obsolete in 5 years without a strong personal effort to stay current with technological advances. I do not necessarily agree with that statement; I believe it may take less than 5 years! Consider this; the IRS allows a business owner to totally depreciate a computer system in 3 years. The advancements in digital technology are so rapid that a typical computer system could be outdated in only 1 or 2 years after it has been purchased. The majority of electronic textbooks are either revised, or taken off the market in 3 to 5 years. Although this may seem to imply that a proficient electronics person must become a chronic bookworm, you actually stay current by reading periodicals.

Most periodicals and electronics magazines are constructed in such a way that they are enjoyable and entertaining to read. The general-interest periodicals will always contain something of interest to almost everyone and they will motivate the reader with new ideas and perspectives. The newest innovations in the industry are covered, along with their practical aspects. In many cases, the reader is provided with home-brew projects to utilize these innovations. Periodicals also make great wish books, with their large variety of product advertisements. The publishers of these magazines also recognize that all of their readers are not at the same technical level; thus, the home projects will vary from easy to difficult, and from practical to just plain fun. Without even realizing it, you can stay up-to-date and have a great time doing it. Periodicals also keep you up-to-date on the most current book releases.

I recommend that you subscribe to several of the general-interest electronic periodicals. By the time you finish this textbook, you should be able to understand and build most of the projects provided in these magazines, especially those that are of special interest to you. In addition, you will pick up little bits of information here and there that will help your progress, and spur your interest. One word of caution: Do not become discouraged if you experience some degree of confusion as you read through these magazines for the first time. Electricity and electronics are, generally, not that difficult to understand. Simply be content with what you do understand, and the rest will follow in time as you progress toward your goal. A list of several of the best general-interest electronics periodicals is included in Appendix B.

Setting up a Lab

While you are waiting for your first few data books and electronics periodicals to arrive in the mail, it would be prudent to turn your attention to setting up an electronics lab. The *lab* is the room or area in your home or business where you will build, test, or repair electronic equipment. It is also the place where you will probably spend considerable time studying, experimenting, sorting parts, jumping for joy, and stewing in frustration. The lab is a *dangerous* place for novices and children, an eyesore to visiting guests, a probable aggravation to your spouse, a collection area for a large volume of "yet to be salvaged" electronic junk, and a secure area for all of your expensive tools and test equipment. The environment in the lab should be quiet, comfortable, and well lighted. With all of these considerations in mind, it is wise to put a little forethought into the best location to set up shop.

I highly recommend choosing a room with a door that can be locked to keep out children, pets, and the overly curious. In addition to the possibility of personal injury to the uninvited, you will probably be working with equipment or projects that are easily damaged or tampered with. Frequently, you will want to leave your work in progress overnight, or even for days at a time, so it must not be disturbed.

Garages are usually not the best place for a lab because of the difficulty in keeping a controlled environment. In addition to your personal discomfort from temperature extremes, most electronic equipment is very sensitive to the moisture that will condense on it in a cold, damp environment. A damp environment is hard on tools and test equipment because of corrosion problems. If you plan on setting up your lab in a basement, a dehumidifier would be a wise investment. A spare bedroom or den is an excellent choice for a lab. Keep in mind that a lab does not have to be an enormously large place. It need only contain a workbench (about 3 × 6 feet), a three- or four-level bookshelf, a bare wall or shelving unit for small-parts cabinets, and a closet or floor area for storage. A nice luxury would be a desk for studying, reading, drawing schematics, and miscellaneous paperwork.

The Workbench

A good size for an electronics lab workbench is about 3×6 feet. Of course, this can greatly vary according to your needs and what you might already have available. In most instances, it should be sturdy enough to hold about 80 pounds, stable enough to not be easily shaken (very annoying when trying to solder small or intricate parts), devoid of any cracks (small parts have a way of finding them), and of the correct height for comfortable use.

Various community organizations and churches often sell heavy-duty, fold-up tables, used for group meetings and meals, at very affordable prices. These same tables can be purchased at office supply stores. They make excellent electronics workbenches, and I personally use two of them for my office and lab. The Formica tops are very durable, resistant to heat, and aesthetically pleasing.

If you prefer, building a lab workbench is a simple project. Particle board or plywood makes a good top, and 2×4s are adequate for legs. An old hollow-core (or solid-core) door also makes an excellent top with 2×4s or saw horses for legs. Commercial electronics workbenches, providing many convenient features, are also available, but they are very expensive.

Hand Tools

If you are any kind of do-it-yourselfer, you probably already have the majority of hand tools you will need for working in electronics. However, in the event that your forté has been working on diesel trucks, keep in mind that the majority of your work will be with small items. A good electronics toolbox will consist of small to medium sizes of the following common tools: needle-nose pliers, side cutters, wire strippers, screwdrivers (both flat and phillips), nut drivers, socket sets, wrenches, tack hammers, files, and hack saws. In regard to powered hand tools, a $1/_{4}$-inch drill and a scroll saw are a good beginning for most work. A high-speed, hand-held grinder, with a variety of attachments, (such as the *Dremel MotoTool*) will be extremely handy for any fabrication project. A *nibbling tool* does a great job of cutting printed circuit board material, chassis material, and various types of metal or plastic enclosure boxes. Of course, there are specialized tools intended exclusively for use in the electronics industry. Most electronics parts suppliers will carry a fair selection of these.

It is best not to go overboard spending a small fortune on a great variety of hand-held tools in the beginning. The tools mentioned so far are only suggestions. You will probably save a great deal of money by adding tools only as you need them.

One of the most important tools to the electronics enthusiast is the soldering iron. Do not get a soldering iron confused with a soldering gun. Soldering guns have a pistol grip, and are intended primarily for heavy-duty soldering applications. Their usefulness in the electronics field is very limited because they can easily damage printed circuit boards and electronic components. A soldering iron is straight like a pencil, with a very small point or tip. A typical soldering iron with stand is shown in Fig. 1-1.

The critical variable, with a soldering iron, is its tip temperature. If the tip temperature is too high, it could destroy heat-sensitive electronic components. If the tip temperature is too low, the solder might not flow or adhere properly to the joint and a poor electrical connection might result.

If the *load* on a soldering iron always remained constant, it would be fairly easy to maintain a constant tip temperature. But in actual use, the load will vary depending on the size of the joint and the amount of solder used. In other words, a large solder joint will conduct more heat away from the soldering iron tip, than will a smaller joint, causing the tip temperature to fall to a lower level. If the wattage (or heating power) to the tip were increased to compensate for this temperature drop, the tip might then become too hot when the iron is not in use, and

Figure 1-1
A typical soldering iron with stand and extra tips.

potential destruction of components could result. Because of performance-versus-price reasons, there are four commonly available types of soldering irons on the marketplace:

1. *Nonadjustable*—Specified by wattage.
2. *Regulated*—Specified by wattage and tip temperature.
3. *Adjustable*—Wattage (power) control.
4. *Adjustable*—Temperature control.

The most common and least expensive type of soldering iron is the first type listed. It is nonadjustable, and is rated (or specified) by its wattage. This is an acceptable type with which to start out, or to throw into the tool box for emergency repairs when away from the lab. The disadvantage with this type of iron is the variation in tip temperature relative to the load. For general-purpose electronic work, try to find one that is rated at about 30 watts.

The second type of soldering iron listed is the type that I use most of the time. The tip of this iron contains a special thermostatic switch that will maintain the tip temperature reasonably close to a specified point.

The third type of soldering iron incorporates a *light-dimmer* circuit to vary the amount of wattage the soldering iron is allowed to dissipate. For bigger jobs, the wattage can be increased to facilitate easier soldering. For smaller jobs, the wattage can be decreased for a lower tip temperature. The wattage control is manual with this type of iron, and the user must use caution not to get the tip temperature too hot when soldering smaller components.

The fourth type of soldering iron is the most versatile, and it is the most expensive. A heat-sensing device is implanted close to the tip, so that the tip temperature can be continuously monitored and controlled by an electronic controller located in the holding stand for the soldering iron. These *soldering stations* have an adjustment control in their bases; thus the exact tip temperature can be set and maintained, regardless of the load placed on the iron itself.

In addition to the soldering iron, you will need some accessories to perform quality soldering jobs. If your soldering iron doesn't come with a stand, buy one! I ended up burning my table, an expensive pair of pants, and my hand before I finally learned this lesson. Also, be sure the stand has a sponge holder. This essential tip-cleaning convenience is well worth the small additional expense. Pick up a few extra sponges, too.

A quality soldering iron will have a variety of different-size tips available for it. If it doesn't, look for another soldering iron. You should

Figure 1-2
A common type of desoldering tool with an extra tip.

choose a couple of small tips for intricate work, and some medium-size tips for general-purpose work.

When purchasing solder, buy only the *60/40 rosin-core* type. Acid-core solder should never be used on electronic equipment.

If the need arises to remove a soldered-in component (or to correct a mistake), you will need a desoldering tool. Most desoldering tools consist of a spring-loaded plunger-in-a-tube housing with a hollow tip at one end. To remove unwanted solder, melt the solder with a soldering iron, place the desoldering tool tip close to the molten solder, and press the trigger. The trigger releases the spring-loaded plunger, thus creating a vacuum in the tube and causing the molten solder to be sucked up into the tube. When purchasing the desoldering tool, be sure that replacement tips are available. The Teflon tips will flare out with use, decreasing the effectiveness of the desoldering tool. A typical desoldering tool is shown in Fig. 1-2.

Soldering is a learned skill, but certainly not a difficult one to master. A little practice, coupled with a conscientious attitude, is all that is required. The actual techniques to soldering are covered in Chapter 3 of this book.

Miscellaneous Supplies

For the most part, the miscellaneous supplies that you will accumulate over a period of time will be a matter of common sense. This section lists a few items to help you get started.

For making temporary connections and setting up certain tests, you will need a number of *alligator clip* leads. These consist of a short length of insulated wire (usually 10 to 15 inches) with a spring-loaded alligator clip at each end. I recommend purchasing (or fabricating) about a dozen.

A variety of sizes and colors of wire will be needed to build various projects. Some electronics supply stores offer a *variety pack*, which would be ideal for starting a new lab. The minimum you will probably need is a spool of medium hook-up wire (stranded, insulated, about 18 to 22 gauge), a spool of small hook-up wire (stranded, insulated, about 26 to 28 gauge), and a small spool of shielded, coaxial cable.

A few helpful cleaning supplies would be cotton swabs, isopropyl alcohol, some very fine sandpaper or emory paper, a can of flux remover, and a variety of small brushes (save your old toothbrushes for the lab).

Electrical Lab Power

The area you have chosen for your lab will hopefully have a sufficient number of 120-volt AC wall outlets to accommodate your needs. If, by any chance, you live in an older home that does not have grounded outlets (three holes for each plug: hot, neutral, and ground), they should be upgraded to the grounded type in your lab. If you have any questions on how to do this, it would be wise to have a professional do it for you. Ungrounded or incorrectly wired outlets are dangerous!

You will find it very convenient to install an outlet strip (a rectangular enclosure with multiple outlets, as in Fig. 1-3) somewhere within easy access on your workbench. Try to find one with a lighted on-off switch, a plastic (or nonconductive) housing, and a 15-amp circuit breaker. It is best to find some way of mounting it securely to the workbench so that you can easily remove plugs with only one hand.

Whenever the need arises to service or repair line-powered electronic equipment (the term *line-powered* means that the equipment must be plugged into a standard 120-volt AC outlet), you will probably want to purchase an isolation transformer to power it (Fig. 1-4). This will

Figure 1-3
Multiple-outlet AC
power strip.

Figure 1-4
An isolation trans-
former specifically
designed for an elec-
tronics test bench.

Figure 1-4
An isolation trans-
former specifically
designed for an elec-
tronics test bench.

become critically important if you are also using line-powered test equipment to perform tests and measurements. It will not be necessary to purchase an isolation transformer for any of the projects in this book, but you might want to keep your eyes open for a good bargain for future needs. The theory and purpose of isolation transformers are covered in a later chapter.

If your interests lie in the industrial electronics field, you might find a need to provide 220-volt AC power to your lab. This is easily accomplished with a step-up power transformer. As in the case of isolation transformers, an appropriate step-up transformer can be added on at any time in the future if the need arises.

Basic Test Equipment

The most frequently used (and most important) piece of test equipment to the electrical or electronics enthusiast is commonly called the *digital voltmeter* (DVM), which is used to measure and display voltage, current, and resistance. A typical DVM is shown in Fig. 1-5. In regard to function, a DVM is basically the same instrument as a *volt-ohm-milliammeter*

Figure 1-5

A versatile, multifunction digital voltmeter (DVM) *(Photograph Courtesy of Elenco Electronics Inc., Wheeling, IL.)*

(VOM), a *digital multimeter* (DMM), or a *vacuum-tube voltmeter* (VTVM). The term *DVM* is most frequently used today, and it is the term I will use through the remainder of this book. When talking with experienced electrical or electronics personnel, they will probably refer to a DVM as a "voltmeter," or simply "meter."

Modern DVMs can typically perform a variety of functions other than simple voltage, current, and resistance measurements. Some might come in very handy; others are somewhat *gimmicky*. The only important additional feature to look for, when shopping for a DVM, is a "diode test" function. The purpose for this will be explained later in this book. The other functions are between you and your pocketbook.

A DVM is the only piece of test equipment you will need to accomplish the goals of this textbook. If you continue to pursue the electrical/electronic field, you will use this instrument for many years to come, so try to find one from a reputable company that is durable and well proven. A well-established electronics dealership might provide some valuable guidance in this area. As you progress through this book, you will discover many uses for a DVM. But at this point, I offer the following words of caution: *Don't try to use a DVM without reading and under-*

standing the operator's manual! This caution also applies to other types of test equipment. Electricity is dangerous!

The remainder of this section covers the additional types of commonly used electronic test equipment. This is not a suggestion that you should go out and buy all of this equipment. Depending on your interests, you might never need some of these; some pieces you might decide to buy as you progress through this book, and fully understand their function and purpose. You might even decide to build some of it yourself for the fun, satisfaction, and great savings. In any event, read through the remainder of this section, get some basic concepts (don't get upset if you don't understand it all right now), and consider it for future reference as needed.

The *oscilloscope* (Fig. 1-6) can provide a visual representation of voltage and current variations (commonly called *waveforms*) within an operating circuit. In addition to displaying these waveforms, the oscilloscope can also be used to measure their amplitude and frequency.

In many ways, an oscilloscope is similar to a small television set. The cathode-ray tube (CRT; picture tube) is used for displaying the voltage or current waveforms. The waveform amplitude is calculated by measuring the vertical height of the waveform, and comparing it to the vertical sensitivity adjustment on the front panel of the oscilloscope.

The frequency of the waveform is calculated by measuring the horizontal length of one complete waveform (one complete cycle), and comparing it to the horizontal-sweep frequency adjustment on the front panel of the oscilloscope.

In addition to amplitude and frequency measurements, an oscilloscope is used for *waveform analysis.* Simply stated, this means that the person performing a test should have a good idea of what the waveform

Figure 1-6
A modern dual-trace oscilloscope. *(Photograph courtesy of Elenco Electronics Inc., Wheeling, IL.)*

should look like at the point being checked. If the waveform is not correct, the defect in the waveform can often identify the problem.

Logic probes and *logic pulsers* are used by personnel involved with digital electronics. Logic probes give a visual (and sometimes aural) indication of the logic state of the checkpoint in question; either high, low, or pulsing. A logic pulser "injects" a continuous train of highs and lows (called *pulses*) into a digital circuit, so that its operation might be observed.

Power supplies are used to externally power circuits that are being built, serviced, or tested. For example, if you wanted to functionally test an automobile radio, it would be necessary to connect it to a +12-volt DC power supply to simulate the automobile battery. Lab power supplies are usually line-powered, and adjustable over a wide range of voltages and currents. Most electronics enthusiasts will collect a variety of power supplies (some purchased, some home-built, some salvaged from used equipment) for maximum lab versatility. Figure 1-7 shows an example of power supplies.

Frequency counters are instruments used to count the frequency of any periodic waveform. They are more convenient, and more precise for measuring frequency, than an oscilloscope.

Signal generators produce a test signal to be injected into an electronic circuit for testing and design purposes (Fig. 1-8). Signal generators typically produce a selectable "sine" wave or "square" wave test signal with adjustable amplitude and frequency. Signal generators that produce a greater variety of test waveforms (such as triangular waves) are called *function generators*. A special type of signal generator, which automatically varies the output frequency within preselected limits, is called a *sweep generator.*

Figure 1-7
A triple-output lab power supply. *(Photograph courtesy of Interplex Electronics Inc., New Haven, CT.)*

Figure 1-8
An Audiofrequency
signal generator and
function generator.
(Photographs cour-
tesy of Elenco Elec-
tronics Inc.,
Wheeling, IL and
Interplex Electronics
Inc., New Haven,
CT.)

Current-transformer ammeters (commonly known by a manufacturer's trade name *Amprobe*) are most often used by electricians for measuring high values of AC current. (This type of ammeter will not measure DC currents.) This instrument measures the strength of the moving magnetic field created around any wire through which AC current is flowing. The circuit does not have to be opened, and no physical contact needs to be made to the wire. It then converts this field strength reading to a proportional current reading.

Starting a Parts and Materials Inventory

Collecting and organizing a good parts inventory is one of the more enjoyable and profitable aspects of the electrical-electronics field. But, like most other endeavors in life, there is a right way and a wrong way to do it. The wrong way will cost you plenty. The right way will open up a fascinating hobby within a hobby (or career within a career) that is both entertaining and educational. If you follow many of the suggestions I have outlined in this section, you can easily acquire a parts inventory that will rival most local electronic parts stores, at very little expense.

Salvaging

The cheapest way to acquire electronic parts is to remove them from electronic "junk" that someone has thrown away, or given to you. The key to success here is to be able to differentiate the good junk from the bad junk. Unless your interest lies exclusively in the TV repair business, old defective television sets should probably remain in the bad-junk category for several reasons. First, they're big, bulky, and hard to move, and may take up too much space. Second, they can be dangerous to tinker around with, unless you know what you're doing. The CRT (picture tube) is subject to implosion (the opposite of explosion, but with the same catastrophic results) if accidentally fractured. Also, the second anode of the CRT can retain a very nasty high voltage for months. Whereas the shock probably wouldn't do you any physical harm, the involuntary response from your muscles could! Third, the quantity of useful parts that can be removed from TV sets is usually small, unless you have a need for high-voltage components. Fourth, they're hard to get rid of after you're through with them.

The good-junk category would include radios, VCRs, CD players, stereo systems, computers, tape players, electronic musical instruments, automobile stereo systems, and some types of commercial or industrial electronic equipment.

Good junk can be acquired in a variety of ways. Most electronics repair shops get stuck with a large volume of junk from customers who don't want to pick up their electronic equipment when they find out how expensive it will be to repair. Many shop owners will give it away for the asking. The local garbage dump is also a good place to find scrap electronics. Don't forget to spread the word around (to all of your friends, neighbors, and relatives) that you would like to have any of their electronic junk destined for the trash can. If you're really ambitious, contact the maintenance superintendents or plant engineers at the local manufacturing plants in your area. Manufacturers often upgrade to new electronic systems, and will discard their old systems for salvage prices. (I once purchased three truckloads of extremely valuable electronic equipment from a large manufacturer for only a penny a pound!)

Keep in mind that an electronic bargain is in the eye of the *knowledgeable* beholder. For illustration, let's assume your particular interest is in the audio electronics field (speaker systems, amplifiers, CD players, etc.). If you receive catalogs from the various electronics suppliers listed in Appendix B, you'll be amazed at the low prices on top-quality internal components for audio systems. (These are the same companies that many of your local electronics dealerships probably buy from.) Unfortunately, the biggest problem you will run into is obtaining suitable cabinets, housings, or enclosures into which to mount your internal components for a finished product. High-quality speaker cabinets might cost $100 or more. Professional-quality *project enclosures* are very expensive, or unavailable in the size needed. This is one of the areas where the junk market really pays off. Junked CD players, VCRs, and computers often have beautiful cases that might require little or no modification to install your parts. If an expensive speaker system goes bad (usually from cone dry rot or misuse), most people simply throw them away; cabinets and all! The speakers can easily be replaced for a third (or less) the cost of a new speaker system of equal quality. Of course, these are just a couple of examples of the cost-effectiveness of collecting the right kinds of electronic junk. Try to be somewhat selective, according to your interests, or you might wind up with more junk than living space in your home.

A good place to find materials for cabinet or enclosure fabrication is the local scrap metals yard. You will be particularly interested in sheet aluminum, aluminum plate, and extruded-aluminum stock (both channel and angle stock). Most scrap yards will sell this stuff for about 30 to 50 cents a pound. If you attempt to buy this material new, you'll appreciate how much of a savings this is. It is also handy to know that most road signs are made from plate aluminum. When the local highway department replaces old road signs, they will typically sell them to scrap metal companies. Look for road signs while you're at the scrap yard, or you might try contacting your local highway department to ask if they will sell them to you directly.

If your interests happen to lie in robotics, automation, or car sound systems, don't forget your local automobile junkyard. Old automobiles are a good source for powerful electric motors (used in windshield wipers, automatic seat adjusters, etc.), lamps and fixtures, car radios and stereo systems, and miscellaneous hardware.

One of the best ways to be selective and to obtain all of the salvageable equipment you need, is to advertise in the local "shopper" magazines (often called "advertisers"). For noncommercial individuals, these ads are usually placed free of charge. If your interests happen to be in the digital electronics field (computers and accessories), a typical ad might read as follows:

> Electronics experimenter interested in purchasing obsolete or nonfunctional computers or computer equipment. Call (your telephone number).

Hypothetically, if you placed an ad such as this in a local shopper magazine, and someone called you in response to that ad, try to keep a few points in mind. There are many obsolete and useless computers in homes and businesses today. In many cases, obsolete or defective computers are not even supportable by the companies that manufactured them. The person calling you probably has only two options available: sell the computer stuff to you, or throw it away. You shouldn't pay any more than "scrap" value for equipment of this sort. This is just an example, but the technique should work equally well in any personal-electronics interest you might have. But I have one word of caution; do a little research, and know your market before you try this technique.

What to Salvage

Many junked electronic items will contain *subassemblies* that are valuable to the electronics hobbyist and experimenter. Old CD players, computers, and VCRs often contain good power supplies, usable for other projects, or even for lab power supplies. (Two of the power supplies that I use the most often in my lab came from junked equipment.) Junked stereo systems might contain good audio power amplifier subassemblies. These are only a few examples. The point is, always check out the value of equipment subassemblies before tearing everything apart to salvage components.

Any person interested in robotics will find a wealth of electromechanical items in old VCRs, including motors, gears, pulleys, belts, limit switches, and optical sensors. Junked CD players often contain functional laser diodes for experimentation or making laser pointers (be very careful to protect your eyes if you experiment with any kind of laser, or laser diode). Junked computers might contain good floppy drives, hard drives, or memory chips. Commonly used electronic hardware (fuse holders, line cords, switches, etc.) is found in almost all electronic equipment.

Salvaging Electronic Components

If you are not familiar with the names or appearance of commonly used electronic components, it would be advisable to read the section entitled *Electronic Components* in Chapter 2 of this book before proceeding.

Now a few words of common sense. An inventory must be organized to be of value to the user. If you have limited inventory space, be practical and selective according to your needs. Every part you salvage will cost you time to remove, and time to enter into your inventory. Although small-parts cabinets are not extremely expensive, their cost will add up in time. Try to be more conscious of variety than quantity. Compare the cost of "grab bag" specials from surplus dealers versus your time spent in acquiring the same items by salvaging.

It is usually not a good idea to mix salvaged parts with new "assumed to be good" parts. Salvaged parts are used and could be defective. Some salvaged parts might be destroyed in trying to remove them. Unless you want to go through exhaustive functional testing procedures, simply keep your salvaged stock separate from your new stock. Then, if the need arises for a salvaged component, you will want to check out that component before using it in a circuit.

The remainder of this section provides some basic guidelines in salvaging electronic components for inventory purposes. These are only suggestions. You might want to do things a little differently to meet your specific needs.

RESISTORS   Common $1/4$- and $1/2$-watt resistors are probably going to be more trouble than they're worth to salvage. Electronics surplus dealers sell mixed resistors in this size range for as low as a dollar a pound. On the other hand, power resistors (1 watt or higher) are more expensive, and often easier to remove. They are usually a good salvage component.

POTENTIOMETERS   If easily removed, potentiometers are good salvage components. There is a high risk that salvaged potentiometers will function erratically, or contain "dead" spots. They should be thoroughly checked before using in any valuable or critical circuit function.

CAPACITORS   Small capacitors typically fall into the same category as small resistors; they're not practical to salvage. Large electrolytic capacitors are practical to salvage, if they're not too old. The functional characteristics of electrolytic capacitors deteriorate with age. Old oil-filled capacitors should not even be brought home; they might contain polychlorinated biphonyls (PCBs)! Large nonpolarized capacitors are good salvage items.

TRANSFORMERS   Step-down power transformers almost always make good salvage items because of their versatility. Other types of transformers become a matter of choice, depending on your interests.

DIODES   Small *signal* or *switching* diodes are not practical to salvage because of their low value and ease of availability. Large high-current diodes, high-power zener diodes, and high-current bridge rectifiers are good components to salvage.

TRANSISTORS   Transistors fall into a "gray" area in regard to salvaging. Even small-signal transistors can be practical to salvage if they are marked with generic or easily cross-referenced part numbers. Generally speaking, any transistor marked with a part number that you cannot cross-reference (many manufacturers use "in house" part numbers) will probably be more trouble than it's worth, considering the time required to analyze it. (In a personal lab, certain types of parameter analysis would not even be possible without damaging the transistor.)

LEDS   Light-emitting diodes are easily tested and make good salvage items, if their leads are long enough for future use.

INTEGRATED CIRCUITS   There are many snags to salvaging ICs. Soldered-in ICs are difficult to remove (especially the 40-pin varieties); and, without a professional desoldering station, you stand a good chance of destroying it in the removal process. Once removed, a salvaged IC is difficult to functionally test without very expensive and specialized test equipment. Many ICs are in-house marked. You might run into special cases where it is practical to salvage an IC, such as computer memory chips, but this is not the general rule. If your electronic involvement is very specific, you might find it practical to keep a big box in your lab as a storage place for junk printed circuit boards containing a large quantity of ICs on them. Then, if the need arises, you might find it practical to remove and test the IC.

Buying from Surplus Dealers

Before proceeding, let's define the word *surplus* as it applies to electronic parts. *Surplus* means "extra stock." Surplus electronic components and equipment are not substandard or defective, they're just extra stuff that has to be moved out to make room for new. To understand the surplus market, here's how a hypothetical electronic item might get there. Most modern electronic printed circuit boards are manufactured in mass quantities by automated processes. Small "runs" of electronic equipment are not cost-competitive in today's market. If a medium-to-large-size electronic manufacturing firm wants to market a new product, they might start out by making 10,000 pieces. If the marketing idea goes sour, or if the product is improved and redesigned, or if the company goes out of business, thousands of these pieces might be left over and sold to a surplus dealership at below manufacturing cost. The surplus dealership can then sell these items far below retail cost and still make a profit.

A typical surplus dealership will sell more than just manufacturers' overruns and excess stock. Because all of the branches of the armed services use considerable electronic equipment (much of it specialized), many surplus dealerships sell *government surplus* equipment as well. Usually, government surplus equipment is used and obsolete, but that doesn't mean it's not valuable. Many surplus dealerships sell factory returns or factory-refurbished items. *Factory returns* are defective items sent back to the manufacturer for repair or replacement. *Factory-refurbished* items are factory returns that have been repaired by the manufacturer. Surplus dealerships often buy industrial salvage for resale. The surplus dealership might sell industrial subassemblies (control panels, circuit board assemblies, etc.) by listing all of the "goodies" the buyer can get out of it, or they might salvage the subassemblies themselves, and sell the individual components for a greater profit.

Electronic surplus dealerships are great places to buy electronic components. With few exceptions, the components will be new and in "prime" condition. Besides offering low prices on specific components, the grab-bag specials are an excellent way to stock up your general inventory.

If you are lucky, you might have one or more electronics surplus stores in your local area. Because these stores do not cater to the general public, you might have to do a little investigative work to find the ones nearest you, but it will be worth the effort. If you live in a rural area, mail-order surplus is a good alternative. A list of some good mail-order electronics surplus dealerships is included in Appendix B.

Electronics surplus is an excellent way to round out your parts and materials inventory, but there are a few cautions and considerations. Don't automatically assume that every item offered for sale by a surplus dealer must be far below retail cost. In some cases, it is not! Before buying or ordering equipment, be aware of its status. It might be new, used, government surplus, factory-returned, factory-refurbished, or sold "as is" (no guarantee of anything). Grab-bag specials will consume a lot of tedious sorting time. Be sure that your eyes and nerves are up to it.

A few additional cautions are relative to mail-order surplus. Most surplus dealerships require a minimum order. Take care to meet this minimum before placing an order. Also, there are *hidden* costs associated with insurance, postage, COD fees, and shipping/handling. It's also wise to verify that the surplus dealership has a lenient return policy, if you are not satisfied for any reason. A good, ethical firm shouldn't have any problem in this area.

2

Basic Electrical Concepts

If you are a novice in the electrical or electronics fields, you will soon discover that this chapter is not light reading. I'm not suggesting that it will be difficult to understand, but I am providing an advanced warning that it will be concentrated. This chapter was designed to maximize reading efficiency. Therefore, there are some basic terms used in conjunction with component descriptions that you might be a little confused about. Don't panic! All unexplained terms are covered later in the chapter as it becomes more appropriate to do so.

Much forethought has gone into the structure of this and successive chapters, in order to provide you with the easiest possible way of comprehending the fundamentals of the electrical and electronics fields. It is not always prudent to define every detail of a topic under discussion because the most important aspects might be obscured by issues that could be more clearly explained in a different context. Therefore, I suggest that you read this chapter twice. You will find the confusing terms, of the first reading, to be much easier to understand during the second reading. If, after the second reading, you are still a little shaky in a few areas, don't become frustrated. The basics explained in this chapter will be utilized throughout the book. I have tried to be thorough in placing reminders throughout the text as an additional aid to understanding. Also, because we all learn by doing, these same fundamentals will be used in the construction of many practical circuits. In this way, concepts are removed from the realm of theory and put into actual practice. One last suggestion—remember, this is a textbook. Textbooks are not designed or intended to be read in the same way as a good novel; they are designed to be studied. You might have to go back and reread many chapters as your progress continues.

Electronic Components

The purpose of this section is to provide the reader with a basic concept of the physical structure of individual electronic components, and how to determine their values.

Resistors

A *resistor,* as the name implies, resists (or opposes) current flow. As you will soon discover, the characteristic of opposing current flow can be used for many purposes. Hence, resistors are the most common *discrete* components found in electronic equipment (the term *discrete* is used in electronics to mean *nonintegrated,* or standing alone).

Resistors can be divided into two broad categories: fixed and adjustable. *Fixed resistors* are by far the most common. *Adjustable resistors* are called *potentiometers* or *rheostats.*

Most fixed resistors are of carbon composition. They utilize the poor conductivity characteristic of carbon to provide resistance. Other types of commonly manufactured resistors are *carbon film, metal film, molded composition, thick film, and vitreous enamel*. These different types possess various advantages, or disadvantages, relating to such parameters as temperature stability, tolerance, power dissipation, noise characteristics, and cost.

The two most critical resistor parameters are *value* (measured in *ohms*) and *power* (measured in *watts*). Resistors that are larger in physical size can dissipate (handle) more power than can smaller resistors. If a resistor becomes too hot, it can change value or burn up. Consequently, it is very important to use a resistor with adequate power-handling capability, as determined by the circuit in which it is placed. Unfortunately, there is not a good, standard way of looking at most resistors and determining their power rating by some standardized mark or code. Comparative size can be deceiving depending on the resistor construction. For example, a 10-watt wire-wound resistor might be about the same size as a 2-watt carbon composition resistor. Common power ratings of the vast majority of resistors are $\frac{1}{8}$, $\frac{1}{4}$, $\frac{1}{2}$, 1, and 2 watts. Don't worry about being able to determine resistor power ratings at this point. It is largely an experience-oriented talent that you will acquire in time.

The value of most resistors is identified by a series of colored bands that encircle the body of the resistor. Each color represents a number, a multiplier, or a tolerance value. The first digit will always be the colored band closest to one end.

There are two *color-coded systems* in common use today: the four-band system and the five-band system. In the *four-band* system, the first band represents the first digit of the resistance value, the second band is the second digit, the third band is the multiplier, and the fourth band is the tolerance (resistors with a tolerance of 20% will not have a fourth band). The *five-band* system is the same as the four-band one, except for the addition of a third band representing a third significant digit. The five-band system is often used for precision resistors requiring the third digit for a higher level of accuracy. There are also a few additional colors used for tolerance identification in the five-band system.

You will need to memorize the following color-code table:

Resistor Color Codes

Black = 0	Green = 5
Brown = 1	Blue = 6
Red = 2	Violet = 7

Orange = 3	Gray = 8
Yellow = 4	White = 9
Brown = 1% tolerance	Yellow = 4% tolerance
Red = 2% tolerance	Silver = 10% tolerance
Orange = 3% tolerance	Gold = 5% tolerance

The following are a few examples of how these color-code systems work. As stated previously, the color band closest to one end of the resistor body is the first digit. Suppose that you had a resistor marked blue-gray-orange-gold. Because there are only four bands, you know it uses the four-band system. Therefore:

Blue = 6 gray = 8 orange = 3 gold = 5% tolerance

The first two digits of the resistor value are defined by the first two bands. Therefore, the two most significant digits will be 68. The third band (multiplier band) indicates how many zeros will be added to the first two digits; in this case it's 3. So, the full value of the resistor is 68,000 ohms. The tolerance band describes how far the actual value can vary. A 5% tolerance means that this resistor can vary as much as 5% more, or 5% less, than the value indicated by the color bands. 5% of 68,000 is 3400. Therefore, the actual value of this resistor might be as high as 71,400 ohms, or as low as 64,600 ohms.

Another example of the four-band system could be red-violet-brown-silver. Therefore:

Red = 2 violet = 7 brown = 1 silver = 10% tolerance

The value of this resistor is 27 with 1 zero added to the end; or 270 ohms. The 10% tolerance indicates it can vary either way by as much as 27 ohms.

Try an example of the five-band system: brown-black-green-brown-red. The first three bands are the first three digits: brown-black-green = 105. The fourth brown band indicates that 1 zero should be added to the end of the first three digits resulting in 1050 ohms. The fifth band is tolerance; red = 2%.

Just to make things a little more confusing, some manufacturers add a *temperature coefficient* band to the four-band system resulting in a resistor appearing to be coded in the five-band system. The easy way to recognize this is the fourth band will always be either gold or silver. If this is the case, simply disregard the fifth band. Also, as stated previously, a complete absence of a fourth band indicates a 20% tolerance.

On some precision resistors or larger power resistors, the actual value and tolerance rating might be printed on the resistor body. A variety of typical resistors is illustrated in Fig. 2-1.

A good way to become very proficient at reading resistor color codes and estimating power ratings (an extremely handy talent to have) is to order one of the large grab-bag resistor specials from one of the mail-order surplus stores listed in Appendix B. Use your DVM (set to measure ohms) to verify that your color-code interpretations are correct. (Don't hold onto the DVM test lead ends with both hands when measuring; the DVM will include your body resistance in the reading.) You'll wind up with a good assortment of cheap sorted resistors. The experience gained in using your DVM will be a side benefit!

Potentiometers, Rheostats, and Resistive Devices

Technically speaking, a *potentiometer* is a three-lead continuously variable resistive device. In contrast, a *rheostat* is a two-lead continuously variable resistor. Unfortunately, these two terms are often confused with each other, even by many professionals in the field.

A few good examples of potentiometers can be found in the typical consumer electronic products around the home. Controls labeled as "volume," "level," "balance," "bass," "treble," and "loudness" are almost always potentiometers.

Figure 2-1
A sampling of typical resistor shapes and styles.

A potentiometer has a round body, about $\frac{1}{2}$ to 1 inch in diameter, a rotating shaft extending from the body, and three terminals (or leads) for circuit connection. It consists of a fixed resistor, connected to the two outside terminals, and a *wiper* connected to the center terminal. The wiper is mechanically connected to the rotating shaft, and can be moved to any point along the fixed resistance by rotating the shaft. A rheostat can be made from a potentiometer by connecting either outside terminal to the wiper terminal.

Potentiometers are specified according to their power rating, fixed resistance value, taper, and mechanical design. The fixed resistance value is usually indicated somewhere on the potentiometer body (it can be determined by measuring the resistance value between the two outside terminals with a DVM). If the power rating is not marked on the body, you will probably have to estimate it (if you cannot cross the manufacturer's part number to a parts catalog). Standard-size potentiometers (about 1 inch in diameter) will typically be rated at 1 or 2 watts. *Taper* refers to the way the resistance between the three terminals will change in respect to a rotational change of the shaft. A *linear taper* potentiometer will produce proportional resistance changes. For example, rotating the shaft by 50% of its total travel will result in 50% resistance changes between the terminals. *Logarithmic taper* potentiometers are nonlinear, or nonproportional, in their operation. They are often called *audio taper* potentiometers because they are commonly used for audio applications. The sensitivity-to-volume level of the human ear is nonlinear. Logarithmic potentiometers closely approximate this nonlinear sensitivity and are used for most audio volume-level controls, as well as many other nonlinear applications. The mechanical construction of potentiometers will vary according to the intended application. Most are *single-turn,* meaning that the shaft will rotate only about 260 degrees. For some precision applications, multiple-turn potentiometers are used.

The term *trim-pot* is used to describe small, single-turn potentiometers intended to be mounted on printed circuit boards and adjusted once, or very infrequently. Figure 2-2 illustrates some common types of potentiometers and trim-pots.

If you work with various types of commercial or industrial electrical or electronic equipment, you might run across some *dedicated rheostats.* These are rheostats that were not made from potentiometers. They are usually intended for high-power applications, and are specially designed to dissipate large amounts of heat.

Adjustable resistors should not be confused with potentiometers or rheostats. Potentiometers and rheostats are commonly used in applica-

Figure 2-2
Common poten-
tiometer types.

tions requiring frequent or precise adjustment. Adjustable resistors are meant to be adjusted once (usually to obtain some hard-to-find resistance value). The body of an adjustable resistor is similar to that of a typical power resistor, with the addition of a metal ring that can be moved back and forth across the resistor body. The position of this metal ring determines the actual resistance value. Once the desired resistance is set, the metal ring is permanently clamped in place.

Capacitors

Next to resistors, capacitors are probably the most common component in electronics. *Capacitors* are manufactured in a variety of shapes and sizes. In most cases, they are either flat, disk-shaped components; or they are tubular in shape. They vary in size from almost microscopic to about the same diameter, and twice the height, of a large coffee cup. Capacitors are two-lead devices, sometimes resembling fixed resistors without the color bands.

Virtually all tube-shaped capacitors will have a capacitance value and a voltage rating marked on the body. Identification of this type is easy. The small disk- or rectangle-shaped capacitors will usually be marked with the following code (where μF=microfarad):

Capacitor Markings

01–99 = Actual Value in picofarads

101–0.0001 μF	331–0.00033 μF
102–0.001 μF	332–0.0033 μF
103–0.01 μF	333–0.033 μF
104–0.1 μF	334–0.33 μF
221–0.00022 μF	471–0.00047 μF

222–0.0022 μF	472–0.0047 μF
223–0.022 μF	473–0.047 μF
224–0.22 μF	474–0.47 μF

With the previous coding system, a letter will usually follow the numeric code to define the tolerance rating. *Tolerance*, in the case of capacitors, is the allowable variance in the capacitance value. For example, a 10-μF capacitor with a 10% tolerance can vary by plus or minus (±) 1 μF. The tolerance coding is as follows:

Capacitor Tolerance-Markings

B = ±0.1 pF	J = ± 5%
C = ±0.25 pF	K = ±10%
D = ±0.5 pF	M = ±20%
F = ±1%	Z = +80%, −20%
G = ±2%	

Variable capacitors are, as the name implies, adjustable within a narrow range of capacitance value. The most common type is called the *air dielectric* type and is typically found on the tuning control of an AM (amplitude modulation) radio. Another type of variable capacitor is called a *trimmer* capacitor. These are usually in the range of 5 to 30 pF and are used for high-frequency applications.

Capacitor characteristics, functions, and additional parameter information will be discussed in Chapter 5. Some common shapes and sizes of capacitors are illustrated in Fig. 2-3.

Inductors

Inductors are broken down into two main classifications: *transformers* and *chokes* (or *coils*). Transformers designed for power conversion applications, called *power transformers* or *filament transformers*, are usually heavy, block-shaped components consisting of two or more coils of wire wound around rectangular wafers of iron. Most of these types of transformers are intended to be mounted on a sturdy chassis because of their weight,

Figure 2-3
Some examples of
capacitors. The two
larger capacitors are
electrolytics.

but some newer types of *split-bobbin* power transformers are designed for printed circuit board mounting. Transformers designed for signal handling applications are smaller, and are usually mounted on printed circuit boards. Figure 2-4 illustrates several common variations of transformers. Chokes are coils of wire either wound on a metallic core or a nonferrous form (to hold the shape of the coil).

Some types of small transformers and chokes are manufactured with an adjustable *slug* in the center. The slug is made of a *ferrite* material, and has threads on the outside of it like a screw. Turning the slug will cause it to move further into (or out of) the coil, causing a change in the inductance value of the coil or transformer. (Inductance is discussed in Chapter 3.) Depending on their application, these small adjustable inductors are called chokes, traps, IF transformers, and variable core transformers.

Diodes

Diodes are two-lead semiconductor devices with tubular bodies similar to fixed resistors. Their bodies are usually either black or clear, with a single-colored band close to one end.

Diode bridges are actually four diodes encased in square or round housings. They have four leads, or connection terminals, two of which will be marked with a horizontal S symbol (sine-wave symbol), one will be marked "+" and the other "−." Several types of diodes and a diode bridge are illustrated in Fig. 2-5a.

Figure 2-4
A few examples of transformers. The small transformer in front is an adjustable IF transformer.

Stud-mount diodes are intended for high-power rectification applications. They have one connection terminal at the top, a body shaped in the form of a hex-head (hexagonal-, i.e., six-sided-head) bolt (for tightening purposes), and a base with a threaded shank for mounting into a heatsink. (*Heatsinks* are devices intended to conduct heat away from semiconductor power devices. Most are made from aluminum and are covered with extensions, called *fins*, to improve their thermal convection properties.)

Special types of diodes, called *LEDs* (the abbreviation for light-emitting diodes), are designed to produce light, and are used for indicators and displays. They are manufactured in a variety of shapes, sizes, and colors, making them attractive in appearance. Figure 2-5*b* illustrates a sampling of common LEDs. A specific type of LED configuration, called "seven-segment" LEDs, are used for displaying numbers and alphanumeric symbols. The front of their display surface is arranged in a block "8" pattern.

Transistors

Transistors are three-lead semiconductor devices manufactured in a variety of shapes and sizes to accommodate such design parameters as power dissipation, breakdown voltages, and cost. A good sampling of transistor shapes and styles is illustrated in Fig. 2-6.

Figure 2-5
(*a*) Two common diode types, a dual-diode (three-lead device) and a diode bridge (with the heatsink); (*b*) Common styles of light-emitting diodes (LEDs).

(*a*)   (*b*)

Figure 2-6
Common transistor types.

Some power transistors have an oval-shaped body with two mounting holes on either side (this is called a *TO-3* or *TO-5* case style). At first glance, it appears to only have two leads, but the body itself is the third (collector) lead.

Integrated Circuits

Integrated circuits, or ICs, are multicomponent semiconductor devices. A single IC might contain thousands of individual components manufactured through a series of special processes. They have multiple leads, called *pins,* which can vary in number from 8 to 40. ICs can be soldered directly into printed circuit boards, or plugged into specially designed IC sockets. Figure 2-7 illustrates several varieties of integrated circuits.

Although the vast majority of integrated circuits have flat, rectangular bodies, a few types of ICs have round "can type" bodies similar to a common transistor case style. The difference is readily detectable by simply counting the number of lead wires coming from the can; a transistor will have three, but an IC of this style will have eight or more.

Figure 2-7
Examples of integrated circuits. These are "dual in-line" (DIP) styles.

A Final Note on Parts Identification

If you think this is all rather complicated, you're right! This section covers only the most common types of components, and the easiest ways of identifying them and ascertaining their values. There are many more styles, types, and configurations than I have time or space to cover. A good way to build on this information is to open up some electronics supply catalogs and skim through them, paying special attention to the pictorial diagrams and dimensions relating to the various components. It will also be a great help to visit your local electronics supply store, and spend some time browsing through the component sales section. And, if it's any consolation, I still get stuck on a "what the heck is it" from time to time!

Characteristics of Electricity

At some point in your life, you have probably felt electricity. If this experience involved a mishap with 120 volts AC (common household power from an outlet), you already know that a physical force is associated with electricity. You constantly see the effects of electricity when you use electrical devices such as electric heaters, television sets, electric fans, and other household appliances. You also see the effects of it when you pay the electric bill! But even with the almost constant usage of electricity, it often possesses an aura of mystery because it cannot be seen. In many ways, electricity is similar to other things that you do understand, and to which you can relate. Electricity follows the basic laws of physics, as do all things in our physical universe.

The electrical and electronics fields depend on the manipulation of subatomic particles called *electrons*. Electrons are negatively charged particles that move in orbital patterns, called *shells*, around the nucleus of an atom. Magnetic fields, certain chemical reactions, electrostatic fields, and the conductive properties of various materials affect the movement and behavior of electrons. As you progress through this book, you will learn how to use electron movement and its associated effects to perform a myriad of useful functions in our lives.

As a beginning exercise in understanding electricity, you are going to compare it to something you can easily visualize and understand. Examine the simple water-flow system shown in Fig. 2-8. This hypothetical system consists of a water pump, a valve to adjust the water flow, and a water pipe to connect the whole system together. Assuming that the pump runs continuously and at a constant speed, it will produce a continuous and constant pressure to try to force water through the pipe.

If the valve is closed, no water will flow through the pipe, but water pressure from the running pump will still be present. The valve is totally "resisting" the flow of water.

If the valve is opened approximately half way, a "current" of water will begin to flow in the pipe. This will not be the maximum water current possible because the valve is "resisting" about half of the water flow.

If the valve is opened all the way, it will pose no resistance to the flow of water. The maximum water flow for this system will occur, limited by the capacity of the pump and the size (diameter) of the pipe.

The electrical circuit shown in Fig. 2-9 is very much like the water system shown in Fig. 2-8. Note the symbols used to represent a battery, resistor, switch, and wire. The battery is analogous to the pump. It provides an electrical *pressure* that produces an electrical flow through the wire. The electrical pressure is called *voltage*. The electrical flow is called *current*.

If the switch in Fig. 2-9 is open (in the "off" position), it will not allow current to flow because a continuous electrical path will not exist. This condition is analogous to the water valve (Fig. 2-8) being completely

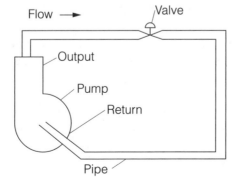

Figure 2-8
Fluid analogy of an electric circuit.

Figure 2-9
Basic electrical circuit.

closed. Because a continuous electrical path from one side of the battery to the other does not exist with the switch open, this circuit is without *continuity.*

If the switch is closed (in the "on" position), an electrical current will begin to flow through the circuit. This current will start at the negative (−) side of the battery, flow through the resistor (R1), through the closed switch (S1), and return to the positive (+) side of the battery. The wire used to connect the components together is analogous to the pipe (Fig. 2-8). The resistor poses some opposition to current flow, and is analogous to the water valve (Fig. 2-8) being partially closed. The actual amount of current flow that will exist in this circuit cannot be determined because I have not assigned absolute values to the components.

If the resistor was removed, and replaced with a piece of wire (assuming that the switch is left in the closed, or "on" position), there would be nothing remaining in the circuit to limit (or resist) the maximum possible current flow. This condition would be analogous to the water valve (Fig. 2-8) being fully opened. A maximum current would flow limited only by the capacity of the battery, and the size (diameter) of the wire.

Notice that in these two comparative illustrations, there is a device producing "pressure" (water pump or battery), a pathway through which a medium could flow (water pipe or wire), a device or devices offering "resistance" to this flow (water valve or resistor/switch combination), and a resultant flow, or "current," limited by certain variables (water pipe or wire diameter, water valve or resistor opposition, and water pump or battery capacity).

Simply stated, in any operational electrical circuit, there will always be three variables to consider: voltage (electrical pressure), current (electrical flow), and resistance (the opposition to current flow). These variables can now be examined in greater detail.

Voltage

Voltage does not move; it is applied. Going back to the illustration in Fig. 2-8, even with the water valve completely closed blocking all water flow, the water pressure still remained. If the water valve were suddenly opened, a water flow would begin and the water pressure produced by the pump would still exist. In other words, the water valve does not control the water pressure; only the water flow. The water pressure is an applied force promoting movement of the water. The water pressure does not move, only the water moves. The basic concept is the same for the electrical circuit of Fig. 2-9. The voltage (electrical pressure) produced by the battery is essentially independent of the current flow.

To help clarify the previous statements, consider the following comparison. Imagine a one-mile-long plastic tube with an inside diameter just large enough to insert a ping-pong ball. If this tube were to be completely filled from end to end with ping-pong balls and you inserted one additional ping-pong ball into one end, a ping-pong ball would immediately fall out of the other end. The force, or pressure, used to insert the one additional ping-pong ball was applied to every ping-pong ball within the tube at the same time. The movement of the ping-pong balls is analogous to electrical current flow. The pressure causing them to move is analogous to voltage. The pressure did not move; it was applied. Only the ping-pong balls moved.

Voltage is often referred to as *electrical potential,* and its proper name is *electromotive force.* It is measured in units called *volts.* Its electrical symbol, as used in formulas and expressions, is E.

The level, or *amplitude,* of voltage is usually defined in respect to some common point. For example, the battery in most automobiles provides approximately 12 volts of electrical potential. The negative side of the battery is normally connected to the main body of the automobile, causing all of the metal parts connected to the body and frame to become the common point of reference. When measuring the positive terminal of the battery with respect to the body, a positive 12-volt potential will be seen.

The terms *positive* and *negative* are used to define the polarity of the voltage. *Voltage polarity* determines the direction in which electrical current will flow. Referring to Fig. 2-8, note how the direction of the water flow is dependent on the direction, or orientation, of the water pump. If the pump had been turned around, so that the output and return were reversed, the direction of water flow would also have been reversed.

Another way of looking at this condition is to consider the output side of the pump as having a positive pressure associated with it; and the return, a negative pressure (suction) associated with it. The water will be pushed out of the output and be sucked toward the vacuum.

A similar condition exists with the electrical circuit of Fig. 2-9. The negative terminal of the battery "pushes out" an excess of electrons into an electrical conductor, and the positive terminal "attracts," or sucks in, the same number of electrons as the negative terminal has pushed out. Thus, if the battery terminals are reversed, the direction of current flow will also reverse. (Note the symbol for a battery in Fig. 2-9. A battery symbol will always be drawn with a short line on one end and a longer line on the other. The short line represents the negative terminal; the long line represents the positive terminal.)

Current

Electrical current is the movement, or flow, of electrons through a conductive material. The fluid analogy of Fig. 2-8 showed some common principles of water flow. In the United States, the flow of water (and most other fluids) is measured in units as *gallons per hour* (gph). In the electrical and electronics fields, you need a similar standard to measure the flow of electrons. The *gallon* of electrons is called a *coulomb*. A coulomb is equal to 6,280,000,000,000,000,000 electrons, give or take a few! In scientific notation, that number is written 6.28×10^{18} (6.28 with the decimal point moved 18 places to the right). Whenever one coulomb of electrons flows past a given point in one second of time, the current flow is equal to one *ampere*.

While on the topic of standards, let me provide you with the standard relationship between voltage, current, and resistance. *One volt is the electrical pressure required to push one coulomb of electrons through one ohm of resistance in one second of time.* In other words, 1 volt will cause a 1 ampere current flow in a closed circuit with 1 ohm of resistance. You will understand this relationship more clearly as you begin to work with *Ohm's law*.

The direction of current flow will always be from negative to positive. This is referred to as the *electron flow*. As stated previously, electron flow rate is measured in units called *amperes*, or simply *amps*. Its electrical symbol, as used in formulas and expressions is *I*.

As illustrated in Fig. 2.10, current can flow only in a *closed circuit*; that is, a circuit that provides a continuous conductive path from the negative potential to the positive potential. If there is a break in this continuous path, such as the open switch (S1) in Fig. 2-11, current cannot flow, and the circuit is said to be *open.* Another way of stating the same principle is to say the circuit of Fig. 2-10 has *continuity* (a continuous path through which current can flow). The circuit in Fig. 2-11 is without continuity.

Resistance

Resistance is the opposition to current flow. The open switch in Fig. 2-11 actually presents an infinite resistance. In other words, its resistance is so high, it doesn't allow any current flow at all. The closed switch shown in Fig. 2-10 is an example of the opposite extreme. For all practical purposes, it doesn't present any resistance to current flow and, therefore, has no effect on the circuit, as long as it remains closed.

The circuit illustrated in Fig. 2-10 shows the symbol for a resistor (R1) and the current flow (*I*) through the circuit. A resistor will present some resistance to current flow that falls somewhere between the two extremes presented by an open or closed switch. This resistance will normally be much higher than the resistance of the wire used to connect the circuit together. Therefore, under most circumstances, this wire resistance is

Figure 2-10
Example of a closed circuit.

Figure 2-11
Example of an open circuit.

considered to be negligible. Resistance is measured in units called *ohms*. Its electrical symbol, as used in formulas and expressions, is *R*.

Alternating Current (AC) and Direct Current (DC)

The periodic reversal of current flow is called *alternating current* (AC). In AC-powered circuits, the polarity of the voltage changes perpetually at a specified rate, or *frequency*. Because the polarity of the voltage is what determines the direction of the current flow, the current flow changes directions at the same rate. The symbol for an AC voltage is shown in Fig. 2-12.

The frequency of current alternations is measured in units called *hertz* (Hz). A synonym for hertz is *cycles per second* (cps). Both of these terms define how many times the current reverses direction in a one-second time period. Common power of the average home in the United States has been standardized at 60 Hz. This means the voltage polarity and current flow reverse direction 60 times every second. (See Fig. 2-12.)

In contrast to ac power sources (Fig. 2-12), a *direct-current* (DC) power source never changes its voltage polarity. Consequently, the current in a DC powered circuit will never change direction of flow. The battery power sources shown in Figs. 2-9 through 2-11 are examples of DC power sources, as are all batteries.

Conductance

Sometimes it is more convenient to consider the amount of current allowed to flow, rather than the amount of current opposed. In these cases, the term *conductance* is used. *Conductance* is simply the reciprocal of resistance. The unit of conductance is the *mho* (ohm spelled backward), and its electrical symbol is *G*. The following equations show the

Figure 2-12
Basic AC circuit illustrating current flow.

ac Source

Current (*I*) flows
in both directions

relationship between resistance and conductance (for example, 2 ohms of resistance would equal 0.5 mho of conductance, and vice versa):

$$\frac{1}{G} = R \qquad\qquad\qquad\qquad \frac{1}{R} = G$$

The reciprocal of conductance is resistance

The reciprocal of resistance is conductance.

In the 1990s, the accepted term of conductance, mho, and its associated symbol *G* were replaced with the term *siemen* and its associated symbol *S*. These two terms, with their associated symbols, mean exactly the same thing. However, throughout the remainder of this book, you will continue to use the older term *mho*, which is still the most commonly used method of expressing conductance.

Power

The amount of energy dissipated (used) in a circuit is called *power*. Power is measured in units called *watts*. Its electrical symbol is *P*.

In circuits such as the one shown in Fig. 2-10, the resistance is often referred to as the *load*. The battery is called the *source*. This is simply a means of explaining the origin and destination of the electrical power used up, or dissipated, by the circuit. For example, in Fig. 2-10, the electrical power comes from the battery. Therefore, the battery is the source of the power. All of this power is being dissipated by the resistor (R1), which is appropriately called the *load*.

Laws of Electricity

As with all other physical forces, physical laws govern electrical energy. Highly complex electrical engineering projects might require the use of several types of high-level mathematics. The electrical design engineer must be well acquainted with physics, geometry, calculus, and algebra.

However, if you do not possess a high degree of proficiency with high-level mathematics, this does not mean that you must abandon your goal of becoming proficient in the electrical or electronics fields. Many books are available on specific subjects of interest that simplify the complexities of design into rule-of-thumb calculations. Also, if you own a computer system, you can purchase computer programs (software) to perform virtually any kind of design calculations that you will probably ever need. Unfortunately, this doesn't mean that you can ignore electronics math altogether. The electronics math covered in this chapter is necessary for establishing and understanding the definite physical relationships between voltage, current, resistance, and power.

Ohm's Law

The most basic mathematical form for defining electrical relationships is called *Ohm's law*. In the electrical and electronics fields, Ohm's law is a basic tool for comprehending electrical circuits and analyzing problems. Therefore, it is important to memorize, and become familiar with, the proper use of Ohm's law, just as a carpenter must learn how to properly use a saw or hammer. Ohm's law is expressed as follows:

$$E = IR \qquad (2.1)$$

This equation tells us that voltage (E) is equal to current (I) multiplied by resistance (R). The simple circuit of Fig. 2-13 illustrates how this equation might be used. The value of R1 is given as 10 ohms (note the capital Greek *omega* symbol is used to abbreviate the word ohm), and the current flow is given as 1 amp, but the battery voltage (E) is unknown. By substituting the electrical symbols in Eq. (2.1) with the actual values, E can easily be calculated.

$$E = IR$$
$$E = (1 \text{ amp})(10 \text{ ohms})$$
$$E = 10 \text{ volts}$$

Now you can go through a few more examples of using Ohm's law to really get the hang of it. Note the circuit illustrated in Fig. 2-14. I have

Figure 2-13
Basic DC circuit illustrating Ohm's law.

Figure 2-14
Basic DC circuit.

not assigned any values to this circuit, so you can use the same basic circuit format to go through several exercises. If R is 6.8 ohms, and I is 0.5 amp, what is the source voltage?

$$E = IR$$

$$E = (0.5)(6.8)$$

$$E = 3.4 \text{ volts}$$

If R is 22 ohms, and I is 2 amps, what would E be?

$$E = IR$$

$$E = (2)(22)$$

$$E = 44 \text{ volts}$$

If R is 10,000 ohms, and I is 0.001 amp, what would E be?

$$E = IR$$

$$E = (0.001)(10,000)$$

$$E = 10 \text{ volts}$$

Now consider some different ways you can use this same equation. An important rule in algebra is that you might do whatever you want to one side of an equation, as long as you do the same to the other side of the equation. The principle is the same as with a balanced set of scales. As long as you add or subtract equal amounts of weight on both sides of the scales, the scales will remain balanced. Likewise, you can add, subtract, multiply, or divide by equal amounts to both sides of Eq. (2-1) without destroying the validity of the equation. For example, divide both sides of Eq. (2-1) by R:

$$\frac{E}{R} = \frac{IR}{R}$$

The two Rs on the right side of the equation cancel each other out. Therefore:

$$\frac{E}{R} = I \text{ or } I = \frac{E}{R} \qquad (2.2)$$

Equation (2-2) shows that if you know the voltage (E) and resistance (R) values in a circuit, you can calculate the current flow (I). Referring back to Fig. 2-13, you were given the resistance value of 10 ohms and the current flow of 1 amp. From these values you calculated the source voltage to be 10 volts. Assume, for the moment, that you don't know the current flow in this circuit. Equation (2-2) will allow you to calculate it:

$$I = \frac{E}{R} = \frac{10 \text{ volts}}{10 \text{ ohms}} = 1 \text{ amp}$$

Now go back to the previous three exercises where you calculated voltage using Fig. 2-14 as the basic circuit. Assume the current flow value to be unknown, and using the known voltage and resistance, solve for current. In each exercise, the calculated current flow value should be the same as the given value.

Similarly, you could rearrange Eq. (2-1) to solve for resistance, if you knew the values for current flow and voltage. Note that each side of the equation must be divided by I:

$$\frac{E}{I} = \frac{IR}{I}$$

The two Is on the right side of the equation cancel out each other, leaving

$$\frac{E}{I} = R \quad \text{or} \quad R = \frac{E}{I} \qquad (2\text{-}3)$$

Referring back to Fig. 2-13, assume that the resistance value of R1 is not given. Using Eq. (2-3), you can solve for R:

$$R = \frac{E}{I} = \frac{10 \text{ volts}}{1 \text{ amp}} = 10 \text{ ohms}$$

Once again, refer back to the exercises associated with Fig. 2-14, and assume that the value of resistance is not given in each case. Using Eq. (2-3) and the known values for current flow and voltage, solve for R. In each exercise, the calculated value of R should equal the previously given value.

At this point, you should be starting to understand what is meant by a *proportional relationship* between voltage, current, and resistance. If one of the values is changed, one of the other values must also change. For example, if the source voltage is increased, the current flow must also increase (assuming that the resistance remains constant). If the resistance is increased, the current flow must decrease (assuming that the source voltage remains constant), and so forth. For practice, you might try substituting other values for current, voltage, and resistance, using Fig. 2-14 as the basic circuit. It is vitally important to become very familiar with how this relationship works.

Parallel-Circuit Analysis

If Fig. 2-14 was an example of the most complex circuit you would ever have to analyze, you wouldn't need to go any further in circuit analysis. Unfortunately, electrical circuits become much more complicated in real-world applications. But don't become discouraged; even extremely complicated circuits can usually be broken down into simpler equivalent circuits for analysis purposes.

Figure 2-15 is an example of a simple parallel circuit. A *parallel circuit* is one in which two or more electrical components are electrically connected across each other. Note that resistors R1 and R2 are wired across each other, or in parallel. In a parallel circuit, the total current supplied by the source will divide between the parallel components. Note that there are two parallel paths through which the current might flow.

Before covering the calculations for determining current flow in parallel circuits, consider these functional aspects of Fig. 2-15.

Figure 2-15
Simple parallel DC circuit.

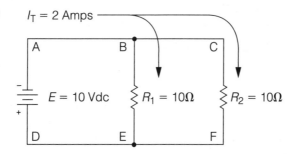

Electrically speaking, points *A, B,* and *C* are all the same point. Like-wise, points *D, E,* and *F* are also at the same *electrical* point in the circuit. Although this might seem confusing at first, remember that the wire shown in an electrical circuit diagram (electrical circuit diagrams are called *schematics*) is considered to have negligible resistance as compared to the rest of the components in the circuit. In other words, for convenience sake, consider the wire to be a perfect conductor. This means that point *A* is the exact same electrical point as point *B* because there is only wire between them. Point *B* is also the same electrical point as point *C*. From a circuit analysis perspective, the top of R1 is connected directly to the negative terminal of the battery. Likewise, the top of R2 is also connected directly to the negative terminal of the battery. This means that Fig. 2-15 can be redrawn into the form shown in Fig. 2-16. It is important to recognize that Fig. 2-15 and Fig. 2-16 are exactly the same circuit. (Because this concept is a little abstract, you might need to reread this paragraph and study Figs. 2-15 and 2-16 several times.)

Figure 2-16 makes it easy to see that both resistors are actually connected directly across the battery. Because the battery voltage is 10 volts DC, the voltage across both resistors will also be 10 volts DC. Now that two circuit variables relating to each resistor are known (the voltage across each resistor and its resistance value); Ohm's law can be used to calculate the current flow through R1:

$$I = \frac{E}{R} = \frac{10 \text{ volts}}{10 \text{ ohms}} = 1 \text{ amp}$$

The previous calculation was for determining the current flow through R1 only. R2 is also drawing a current flow from the battery. It should be obvious that, because R2's resistance is the same as R1, and they both have 10 volts DC across them, the calculation for determining R2's current flow would be the same as R1. Therefore, you can conclude that 1 amp of current is flowing through R1, and 1 amp of current is flowing through R2. The battery in this circuit is a single source that must pro-

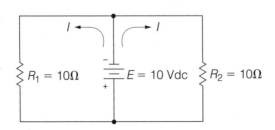

Figure 2-16
The equivalent circuit of Fig. 2-15.

vide the electrical pressure to promote the total electron flow for both resistors. Therefore, 2 amps of total current must leave the negative terminal of the battery, divide into two separate flows of 1 amp through each resistor, and finally recombine into the 2-amp total current flow before entering the positive terminal of the battery. This 2-amp total current flow is illustrated in Fig. 2-15.

Try another example. Referring back to Fig. 2-16, assume R1 is 5 ohms instead of 10 ohms. Note that the voltage across R1 will not change (regardless of what value R1 happens to be), because it is still connected directly across the battery. The current flow through R1 would be

$$I = \frac{E}{R} = \frac{10 \text{ volts}}{5 \text{ ohms}} = 2 \text{ amps}$$

Now you have 2 amps flowing through R1, plus the 1 amp flowing through R2 (assuming the value of R2 remained at 10 ohms). The total circuit current being drawn from the source must now be 3 amps, because 2 amps are flowing through R1, and 1 amp is still flowing through R2.

As the previous examples illustrate, a few general rules apply to parallel circuits:

In a closed parallel circuit, the applied voltage will be equal across all parallel legs (or *loops,* as you think of them)

In a closed parallel circuit, the current will divide *inversely proportionally* to the resistance of the individual loops (the term *inversely proportional* simply means that the current will increase as the resistance decreases, and vice versa)

In a closed parallel circuit, the total current flow will equal the sum of the individual loop currents

When analyzing parallel circuits, it is often necessary to know the combined, or *equivalent,* effect of the circuit instead of the individual variables in each loop. Referring back to Fig. 2-15, it is possible to consider the combined resistance of R1 and R2 by calculating the value of an imaginary resistor, R_{equiv}. The following equation can be used to calculate the equivalent resistance of any two parallel resistances:

$$R_{equiv} = \frac{(R_1)(R_2)}{(R_1) + (R_2)} \tag{2-4}$$

By inserting the values given in Fig. 2-15, we obtain

$$R_{equiv} = \frac{(R_1)(R_2)}{(R_1) + (R_2)} = \frac{(10)(10)}{(10) + (10)} = \frac{100}{20} = 5 \text{ ohms}$$

The 5 ohm (R_{equiv}) represents the total combined resistive effect of both R1 and R2. Because you already know the total circuit current is 2 amps, you can use Ohm's law to prove the calculation for R_{equiv} is correct. Using Eq. (2-1), we have

$$E = IR = (2 \text{ amps})(5 \text{ ohms}) = 10 \text{ volts}$$

Because it is true that the source voltage is 10 volts DC, you know that the calculation for R_{equiv} is correct.

In circuits where there are three or more parallel resistances to consider, the calculation for R_{equiv} becomes a little more involved. In these circumstances, a calculator is a handy little tool to have! You must calculate the individual conductance values of each resistor in the parallel network, add the conductance values together, and then calculate the reciprocal of the total conductance. This sounds more complicated than it really is. Referring to the complex parallel circuit illustrated in Fig. 2-17, note that there are four resistors in parallel. To calculate R_{equiv}, the first step is to find the individual conductance values for each resistor. For example, to find the conductance of R1:

$$G = \frac{1}{R_1} = \frac{1}{10} = 0.1 \text{ mho}$$

The conductance of the remaining resistors should be calculated in the same way. If you perform these calculations correctly, your conductance values should be

$$G(R1) = 0.1 \text{ (previous calculation)}$$

$$G(R2) = 0.2$$

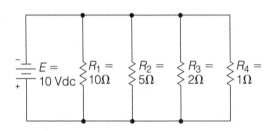

Figure 2-17
Complex parallel circuit.

$$G(R3) = 0.5$$

$$G(R4) = 1$$

By adding all of the conductance values together, the total conductance value comes out to 1.8 mhos. The final step is to calculate the reciprocal of the total conductance, which is

$$\frac{1}{1.8} = 0.5555 \text{ ohm}$$

R_{equiv} for Fig. 2-17 is approximately 0.5555 ohms. Note that in the previous examples, R_{equiv} is less than the lowest value of any single resistance loop in the parallel network. This will always be true for any parallel network.

The procedure for calculating R_{equiv} in parallel circuits having three or more resistances can be put in an equation form:

$$R_{equiv} = \frac{1}{\dfrac{1}{R_1} + \dfrac{1}{R_2} + \dfrac{1}{R_3} \cdots \dfrac{1}{R(n)}} \qquad (2\text{-}5)$$

When using Eq. (2-5), remember to perform the steps in the proper order as previously shown in the example with Fig. 2-17.

Series-Circuit Analysis

The circuit in Fig. 2-18 is an example of a simple *series* circuit. Notice the difference between this circuit and the circuit shown in Fig. 2-15. In Fig. 2-18, the current must first flow through R1, and then through R2, in order to return to the positive side of the battery. There can only be one current flow, and this flow must pass through both resistors.

Figure 2-18
Simple series DC
circuit.

$I = 0.5A$

$R_1 = 10\Omega$

$E = 10 \text{ Vdc}$

$R_2 = 10\Omega$

In a series circuit, the combined effect of the components in series is simply the sum of the components. For example, in Fig. 2-18, the combined resistive effect of R1 and R2 is the sum of their individual resistive values: 10 ohms + 10 ohms = 20 ohms. In series circuits, this combined resistive effect is usually called R_{total}, although it might be correctly referred to as R_{equiv}.

Knowing the total circuit resistance to be 20 ohms, and the source voltage to be 10 volts DC, Ohm's law can be used to calculate the circuit current (I_{total}):

$$I = \frac{E}{R} = \frac{10 \text{ volts}}{20 \text{ ohms}} = 0.5 \text{ amp}$$

In a series circuit, the current is the same through all components, but the division of the voltage is *proportional to the resistance*. This rule can be proved through the application of Ohm's law. The resistance of R1 (10 ohms) and the current flowing through it (I_{total} = 0.5 amp) give us two known variables relating to R1. Therefore, R1's voltage would be

$$E = IR = (0.5 \text{ amp})(10 \text{ ohms}) = 5 \text{ volts}$$

The 5 volts that appear across R1 in this circuit is commonly called its voltage *drop*. R2's voltage drop can be calculated in the same manner as R1's:

$$E = IR = (0.5 \text{ amp})(10 \text{ ohms}) = 5 \text{ volts}$$

The fact that the two voltage drops are equal should come as no surprise, considering that the same current must flow through both resistors, and they both have the same resistive value.

If the voltage dropped across R1 is added to the voltage dropped across R2, *the sum is equal to the source voltage*. This condition is applicable to all series circuits.

A few general rules can now be stated regarding series circuits:

In a closed series circuit, the sum of the individual voltage drops must equal the source voltage.

In a closed series circuit, the current will be the same through all of the series components, but the voltage will divide proportionally to the resistance.

As has been noted previously, the resistors in Fig. 2-18 are of equal resistive value. However, in most cases, components in a series circuit will not present equal resistances. To calculate the voltage drops across the individual resistors, you might use Ohm's law as in the previous example. Another way of calculating the same drops is by using the *ratio method*. To use the ratio method, calculate R_{total} by adding all of the individual resistive values. Make a division problem with R_{total} the divisor, and the dividend the resistive value of the resistor for which the voltage drop is being calculated. Perform this division, and then multiply the quotient by the value of the source voltage. The answer is the value of the unknown voltage drop.

Referring to Fig. 2-19, assume you desire to calculate the voltage drop across R1 using the ratio method. Start by finding R_{total}:

$$R_{total} = R_1 + R_2 + R_3 = 2 \text{ ohms} + 3 \text{ ohms} + 5 \text{ ohms} = 10 \text{ ohms}$$

Next, make a division problem with R_{total} as the divisor and R_1 as the dividend and perform the division:

$$\frac{2 \text{ ohms}}{10 \text{ ohms}} = 0.2$$

Finally, multiply this quotient by the value of the source voltage:

$$(10 \text{ volts})(0.2) = 2 \text{ volts}$$

The voltage dropped across R1 in Fig. 2-19 will be 2 volts. As an exercise, calculate the voltage drop across R1 using Ohm's law. You should come up with the same answer.

Series-Parallel Circuits

In many situations, series and parallel circuits are combined to form *series-parallel* circuits (Fig. 2-20). In this circuit, R2 and R3 are in parallel,

Figure 2-19
Series circuit with
unequal resistances.

Figure 2-20
Example of a series-
parallel circuit.

but R1 is in series with the parallel network of R2 and R3. If this is confusing, notice that the total circuit current must flow through R1, indicating that R1 is in series with the remaining circuit components. However, in the case of R2 and R3, the total circuit current can branch; with part of it flowing through R2, and part of it flowing through R3. This is the indication that these two components are in parallel.

To analyze the circuit of Fig. 2-20, you would start by simplifying the circuit into a form that's easier to work with. As previously explained regarding parallel circuits, resistors in parallel can be converted to an R_{equiv} value. In this case, R2 and R3 are in parallel, so Eq. (2-4) might be used to calculate R_{equiv}:

$$R_{equiv} = \frac{(R_1)(R_2)}{(R_1) + (R_2)} = \frac{(10)(10)}{(10) + (10)} = \frac{100}{20} = 5 \text{ ohms}$$

Once the value of R_{equiv} is known, the circuit of Fig. 2-20 can be redrawn into the form illustrated in Fig. 2-21. You should recognize this as a simple series circuit. R_{total} can be calculated by adding the values of R_1 and R_{equiv}:

$$R_{total} = R_1 + R_{equiv} = 5 \text{ ohms} + 5 \text{ ohms} = 10 \text{ ohms}$$

Using the values of R_{total}, and the source voltage, Ohm's law can be used to calculate the total circuit current (I_{total}):

$$I = \frac{E}{R} = \frac{10 \text{ volts}}{10 \text{ ohms}} = 1 \text{ amp}$$

The voltage drop across R1 can be calculated using the variables associated with R1 (I_{total} and R_1):

$$E = IR = (1 \text{ amp})(5 \text{ ohms}) = 5 \text{ volts}$$

Figure 2-21
Simplified equivalent
circuit of Fig. 2-20.

In this particular circuit, you can take a shortcut in determining the voltage drop across R_{equiv}. Because you know that the individual voltage drops must add up to equal the source voltage in a series circuit, you can simply subtract the voltage drop across R1 from the source voltage, and the remainder must be the voltage drop across R_{equiv}:

$$10 \text{ volts (source)} - 5 \text{ volts (R1)} = 5 \text{ volts (R}_{equiv})$$

As stated previously, the applied voltage (in a parallel circuit) is the same for each loop of the circuit. Because you know that 5 volts is being dropped across R_{equiv}, this actually means that 5 volts is being applied to both R2 and R3 in Fig. 2-20. Now that two variables are known for R2 and R3 (the voltage across them, and their resistive value), the current flow through them can be calculated. Calculating the current flow through R2 first:

$$I = \frac{E}{R} = \frac{5 \text{ volts}}{10 \text{ ohms}} = 0.5 \text{ amp}$$

Obviously, with the same resistive value and the same applied voltage, the current flow through R3 would be the same as R2. Knowing that the individual current flows in a parallel circuit must add up to equal the total current flow, you can add the current flow through R2 and R3 to double-check our previous calculation of I_{total}:

$$0.5 \text{ amp } (R_2) + 0.5 \text{ amp } (R_3) = 1 \text{ amp}$$

Power

An important variable to consider in most electrical and electronics circuits is power. *Power* is the variable which defines a circuit's ability to perform work. Electrical power is dissipated (used up) in the forms of heat and work. For example, the majority of electrical power used by a home stereo is dissipated as heat, but a good percentage is converted to *acoustic energy* (varying pressure waves in the air which is called *sound*).

It is often necessary to calculate the amount of power that must be dissipated in a circuit or component to keep from destroying something. Also, if you are supplying the power to a circuit, you need to know the amount of power to supply. The following three equations are used for power calculations:

$$P = IE \tag{2-6}$$

Power is equal to the current multiplied by the voltage:

$$P = I^2R \tag{2-7}$$

Power is equal to the current squared, multiplied by the resistance:

$$P = \frac{E^2}{R} \tag{2-8}$$

Power is equal to the voltage squared, divided by the resistance.

Examine how these equations could be used to calculate the power dissipated by R1 in Fig. 2-18. (Remember, you calculated the voltage drop across R1 to be 5 volts in an earlier problem.) Using Eq. (2-6) and the known variables for R1:

$$P = IE = (0.5 \text{ amp})(5 \text{ volts}) = 2.5 \text{ watts}$$

Using Eq. (2-7) and the known variables for R1:

$$P = I^2R = (0.5 \text{ amp})(0.5 \text{ amp})(10 \text{ ohm}) = 2.5 \text{ watts}$$

Using Eq. (2-8) and the known variables for R1:

$$P = \frac{E^2}{R} = \frac{(5 \text{ volts})(5 \text{ volts})}{10 \text{ ohms}} = 2.5 \text{ watts}$$

As you can see from the previous calculations, you need to know only two of three common circuit variables (voltage, current, and resistance) to solve for power.

There is a very important point to keep in mind when performing any electrical or electronic calculation; *don't mix up circuit variables with component variables*. For example, in the previous power calculations, the power dissipated by R1 was solved for. To do this, you used the voltage across R1, the current flowing through R1, and the resistance value of R1. In other words, every variable you used was "associated with R1." If you wanted to calculate the power dissipated by the entire circuit of Fig. 2-18, you would have to use the variables associated with the entire circuit. For example, using Eq. (2-8) and the known circuit variables:

$$P = \frac{E^2}{R} = \frac{(10 \text{ volts})(10 \text{ volts})}{20 \text{ ohms}} = 100/20 = 5 \text{ watts}$$

Notice that the voltage used in the previous equation is the source voltage (the total voltage applied to the circuit), and the resistance value used is R_{total} ($R_1 + R_2$; the total circuit resistance). Our answer, therefore, is the power dissipated by the *total circuit*.

Common Electronics Prefixes

Throughout this chapter, the example problems and illustrations have used low value, easy-to-calculate values for the circuit variables. In the real world, however, you will need to work with extremely large and extremely small quantities of these circuit variables. Because it is cumbersome, and both space- and time-consuming, to try to work with numbers that might have 12 or more zeros in them, a system of prefixes has been standardized. These prefixes, together with their associated symbols and values are:

Electronic Prefixes

pico (symbol "p")	= 1/1,000,000,000,000
nano (symbol "n")	= 1/1,000,000,000
micro (symbol "μ")	= 1/1,000,000
milli (symbol "m")	= 1/1,000
kilo (symbol "K")	= 1,000
mega (symbol "M")	= 1,000,000

Notice that prefixes for numbers greater than one are symbolized by capital letters, but symbols for prefixes of less than one are lowercase letters. Also, note that the Greek symbol μ (pronounced "mu") is used as the symbol for micro, instead of "*m*," to avoid confusion with the symbol used for the prefix "milli."

The following list provides some examples of how prefixes would be used in conjunction with common circuit variables. The abbreviation provided with each example is the way that value would actually be written on schematics or other technical information.

EXAMPLES

1 picovolt	= 0.000000000001 volt	(abbrev. 1 pV)
15 nanoamps	= 0.000000015 amp	(abbrev. 15 nA)
200 microvolts	= 0.0002 volt	(abbrev. 200 μV)
78 milliamps	= 0.078 amp	(abbrev. 78 mA)
400 Kilowatts	= 400,000 watts	(abbrev. 400 KW)
3 Megawatts	= 3,000,000 watts	(abbrev. 3 MW)

Using Ohm's Law in Real-World Circumstances

Thus far, you have been exposed to the basic concept of Ohm's law and how circuit variables have a definite and calculable relationship to each other. As you have seen, the mathematics involved with Ohm's law are not difficult, but it takes some practice to become proficient at applying Ohm's law correctly. For example, in the early stages of becoming accustomed to using Ohm's law, it is a common error to mistakenly mix up (or confuse) circuit variables with individual component variables. It is also easy to misplace a decimal point when working with the various standardized prefixes (milli, micro, Mega, etc.).

When electronics professionals draw a complete circuit diagram of a complex electronic system, usually referred to as a *schematic*, they often complicate the easy visualization of circuit subsections in the effort to make the complete schematic fit into a nice, neat, aesthetically pleasing square or rectangle. This means that the electronics enthusiast must become proficient at being able to visualize a parallel circuit, for example, drawn horizontally, vertically, or in any number of possible configurations. In other words, it is necessary to develop the ability to *conceptualize* a schematic, which means merely being capable of mentally simplifying it into familiar subsections that can be easily worked with.

This is a *learned* talent, and the more time you spend at conceptualizing schematics, the more skilled you will become.

The purpose of this section is to familiarize you with using Ohm's law in conjunction with more typical real-world circuits. The following exercises are designed to help you become acquainted with the following areas of difficulty:

1. Using Ohm's law correctly
2. Becoming familiar with commonly used prefixes
3. Simplifying more complex schematic diagrams

In addition, you will examine the following exercises in several ways. First, you will go through each exercise, working out the electrical problems, and developing the aforementioned skills. Next, you will go through the exercises a second time, examining the real-world component values that could be inserted into these circuits to make them practical to construct and test. As a final, optional, exercise, you may want to purchase the materials at a local electronics store to literally construct and test these circuits. The only components required will be a handful of resistors, a 9-volt battery (the small, rectangular type), and a few general-purpose LEDs (i.e., light-emitting diodes). All these circuits can be easily constructed using clip leads. The total cost of the components should be less than $10.00 (it is assumed that you already have a good DVM and an adequate supply of clip leads—these are part of your basic lab supplies, and will be used over and over). If you have not yet purchased a DVM or the majority of your lab supplies, you can always come back to this section and acquire some hands-on experience with building these circuits at a later time.

Electrical Calculations (Ideal) The most common method of analyzing electrical circuits is to consider all of the circuit components and circuit variables to be "ideal." In real life, various circuit variables, such as battery voltages, resistor values, and other component values, are not *exact*. The simple fact that virtually all electronic components are specified as having a tolerance (i.e., an allowable variance from an exact specification) indicates that calculated circuit variables will always contain some error depending on how broad the accepted tolerance values are. In other words, a 10-ohm resistor dropping 10 volts would be expected to conduct 1 amp of current (10 volts divided by 10 ohms = 1 amp). This is an *ideal* calculation. However, if the resistor were specified as being 10 ohms with a 10% tolerance, its *actual* value could range between 9 and 11 ohms, causing the actual, or *real-life* current flow to be between 1.11 and 0.909 amps. Unfortunately, there isn't any method of determining actual circuit variables without literally constructing the circuit and measuring the variables. Even in doing so, the actual circuit variables could be different in two identical circuits, as a result of the unpredictable effects of component tolerances. Consequently, the only practical method of analyzing circuits is to consider all of the circuit conditions to be "ideal." This simply means considering such things as component values and source voltages to be *exactly* as specified. You will also consider all interconnecting wiring and wiring connections to be ideal, or representing "zero" resistance. This is really no different from a carpenter drawing blueprints for a construction project, assuming the available lumber to be straight and dimensionally accurate.

All of the circuit diagrams used for the following exercises contain one or more LEDs. Light-emitting diodes are most commonly used to provide a visual light for indication purposes. You will learn more about the theory and physics of LEDs in Chapter 7, but for now, you will only need to remember a few basic characteristics and approximations.

First, the current flow through an LED must be in the proper direction for it to produce visible light. In other words, if you build any of the following circuits and connect the LED backward in the circuit, it will not light. The low voltages used in these circuits will not harm the LED, so if it doesn't light, simply try reversing the leads. Second, it will be assumed that all of the LEDs in the following exercises will drop 2 volts, and they will require 10 milliamps (i.e., 0.01 amp) of current flow to light. These LED specifications are close enough to be applicable to the majority of general-purpose LEDs.

NOTE *I chose this technique of using a few LEDs, resistors, and 9-volt battery for several reasons. First, we all like to occasionally take a break, get away from the books, and physically work with something related to what we have learned. Ohm's law can be learned academically, and it can be aptly demonstrated using only a power source and a few resistors; however, the circuit doesn't appear to "do" anything. In my experience, it is more fun and memorable if an experimental circuit performs some practical function, even if that function is as simple as lighting an LED.*

Exercise 1   Figure 2-22a illustrates a 9-volt battery connected to a resistor (R1) and an LED (D1) (Note the two small arrows pointing away from D1. This is the typical method of drawing the schematic symbol for an LED—the two arrows symbolizing the light emitted by the LED.) You should recognize this circuit as a simple series circuit, since the same

Figure 2-22
Circuit diagram
examples for practice
problems using
Ohm's law.

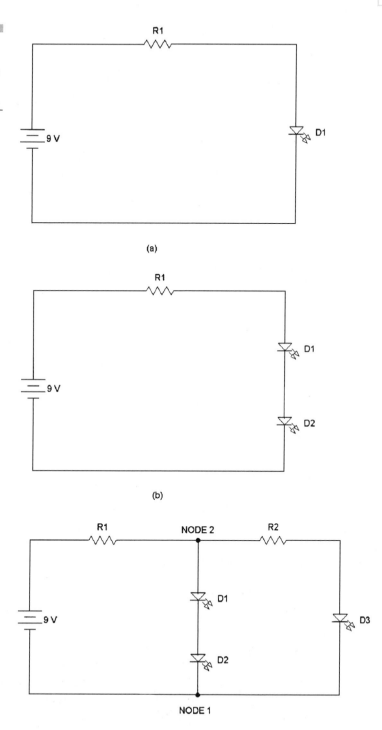

(a)

(b)

(c)

Figure 2-22
(Continued)

(d)

(e)

current flow must flow through all of the components (i.e., the current cannot "branch" into any other circuit legs).

Remembering the aforementioned approximations for all general-purpose LEDs, you know that you want 10 milliamps of current to flow through the LED, and you also know the LED will drop 2 volts. As a beginning exercise, use Ohm's law to calculate the value of R1. (Don't worry if you're a little confused at this point. You'll get the hang of it as you proceed through a few more of these exercises.)

The same current must flow through all of the components in a series circuit. Therefore, if the current flow through the LED is 10 milliamps, the current flow through R1 must be 10 milliamps. All the voltage drops in a series circuit must add up to equal the source voltage. In this example, the source voltage is the 9-volt battery. If the LED is dropping 2 volts, this means that R1 must be dropping 7 volts (i.e., 9 volts − 2 volts = 7 volts). Now you know two electrical variables that *specifically* relate to R1: the current flow through it (10 milliamps) and the voltage drop across it (7 volts). Ohm's law states that the resistance value of R1 will equal the voltage dropped across it divided by the current flow through it. Therefore:

$$R = \frac{E}{I} = \frac{7}{0.01} = 700 \text{ ohms}$$

Of what practical value is this exercise? Suppose you wanted to construct an illuminated Christmas tree ornament using a single LED and a 9-volt battery (such ornaments are commonly sold at exorbitant prices). You now know how to choose an appropriate "limiting" resistor so that the maximum current specifications of the LED are not exceeded.

If you physically construct this circuit for experimental purposes, you can also use it to test an unknown grab-bag assortment of LEDs. R1 can be a common 680-ohm resistor. If some LEDs won't light (assuming that you have already tried reversing the connections), it is possible that they are special-purpose LEDs with internal limiting resistors. Put these aside for future use and categorization after you read and understand more about LEDs in Chapter 7.

Exercise 2   Again referring to Fig. 2-22a, what would be the power dissipation of R1? Refer to Eq. (2-6), (2-7), or (2-8).

Since you previously calculated the resistance of R1 to be 700 ohms, you now know the three most important circuit variables related specifically to R1: its resistance value, the current flow through it, and the voltage drop across it. Therefore, you can use any of the three common power equations to determine the power dissipated by R1. Using Eq. (2-6), we obtain

$$P = IE = (0.01 \text{ amp})(7 \text{ volts}) = 0.07 \text{ watts} \quad \text{or} \quad 70 \text{ mW}$$

Using Eq. (2-7)

$$P = I^2R = (0.01 \text{ amp})(0.01 \text{ amp})(700 \text{ ohms}) = (0.0001)(700 \text{ ohms})$$
$$= 0.07 \text{ watts} \quad \text{or} \quad 70 \text{ mW}$$

Using Eq. (2-8)

$$P = \frac{E^2}{R} = \frac{(7 \text{ volts})(7 \text{ volts})}{700 \text{ ohms}} = \frac{49}{700 \text{ ohms}} = 0.07 \text{ watts} \quad \text{or} \quad 70 \text{ mW}$$

Note how either of the three most commonly used power equations will provide the same answer.

Of what practical value is this exercise? Suppose you wanted to construct the Christmas tree ornament as discussed in Exercise 1 using a resistor for R1 rated at only $\frac{1}{4}$ watt; $\frac{1}{4}$ watt is equal to 0.25 watts, or 250 mW. The actual power dissipated by R1 will be only 70 mW, so you know it will be safe to use a $\frac{1}{4}$-watt resistor for this application.

Exercise 3   Refer to Fig. 2-22b. Again, you should recognize this circuit as a simple series circuit; the same current must flow through all of the circuit components. Using Ohm's law, determine the resistance value of R1 to provide the desired 10 milliamps of current flow through the two LEDs (i.e., D1 and D2).

Since 10 milliamps of current will be flowing through D1 and D2, 10 milliamps must also flow through R1. All of the voltage drops in a series circuit must add up to equal the source voltage. Because D1 and D2 will drop 2 volts each, totaling 4 volts, the remaining source voltage must be dropped across R1. Therefore, R1 must be dropping 5 volts (i.e., 9 volts − 4 volts = 5 volts). Now that the voltage across R1 is known (5 volts) and the current flow through it is known (10 milliamps), Ohm's law can be used to calculate its resistance:

$$R = \frac{E}{I} = \frac{5 \text{ volts}}{10 \text{ milliamps}} = 500 \text{ ohms}$$

Of what practical value is this exercise? Many modern "high-brightness LEDs" can produce light intensities of up to 3000 mcd (i.e., millicandela). Roughly speaking, two LEDs of this type mounted in close proximity can provide about as much light as a typical flashlight bulb. By adding an on-off switch to the circuit of Fig. 2-22b, a very compact *inspection light, key-chain light,* or similar type of illumination project can be easily constructed. In addition, LEDs consume much less power, they have a much longer operational life, and they are virtually shockproof and vibrationproof, as compared to incandescent flashlight bulbs.

Exercise 4   Again referring to Fig. 2-22b, what would be the power dissipation of D1? Equation (2-6) states that power dissipation is equal to the current multiplied by the voltage. The current flow through D1 is 10 milliamps, and the voltage drop across it is assumed to be 2 volts. Therefore:

$$P = IE = (10 \text{ milliamps})(2 \text{ volts}) = 20 \text{ mW} \quad \text{or} \quad 0.02 \text{ watts}$$

Exercise 5   Again referring to Fig. 2-22*b*, what would be the power dissipation of R1? As in the case of Exercise 2, since all three of the common electrical variables are known regarding R1 (i.e., voltage, current, and resistance), either of the three common power equations can be used. In this case, the easiest to use is Eq (2-6):

$$P = IE = (10 \text{ milliamps})(5 \text{ volts}) = 50 \text{ mW} \quad \text{or} \quad 0.05 \text{ watts}$$

(Remember, you calculated the voltage drop across R1 in Exercise 3.)

Exercise 6   Again referring to Fig. 2-22*b*, what is the *total* power dissipation of this circuit? Total power dissipation in any electronic circuit is the summation of all individual power dissipations of all individual components. In Exercise 4, you determined that the power dissipated by D1 is 20 mW. Since the electrical variables for D2 is identical to those of D1, D2 will also dissipate 20 mW. In Exercise 5, you found that the power dissipated by R1 is 50 milliwatts. The summation of all of these individual power dissipations will equal the total circuit power dissipation. Therefore:

$$P_{total} = 20 \text{ mW} + 20 \text{ mW} + 50 \text{ mW} = 90 \text{ mW} \quad \text{or} \quad 0.09 \text{ watts}$$

There is an easier method of calculating the total power dissipation of Fig. 2-22*b*. Since you are calculating the *total* circuit power dissipation, you can use the electrical variables associated with the *total*, or *complete*, circuit. You know the total circuit current (i.e., the current leaving the power source) to be 10 milliamps. You also know the total voltage applied to the circuit (i.e., the source voltage) to be 9 volts. Therefore, you can calculate the total circuit power dissipation using Eq. (2-6):

$$P = IE = (10 \text{ milliamps})(9 \text{ volts}) = 0.09 \text{ watts} \quad \text{or} \quad 90 \text{ mW}$$

Of what practical value are Exercises 4, 5, and 6? Electronic power supplies are often used in place of batteries to provide operational power to electronic circuits. (You will learn more about power supplies in Chapters 3 to 6.) Electronic power supplies are rated, or specified, according to the maximum power they can supply to a load. If, for example, a hypothetical power supply happened to be rated at 5 watts, this simply means that it cannot supply more than 5 watts of total power dissipation to any operational circuit. Therefore, it becomes obvious that it is important to know the total power dissipation of any electronic circuit that you may construct in the future, so that you will be capable of selecting an appropriate power supply to operate it.

Exercise 7 Figure 2-22c illustrates a series-parallel circuit. To visualize this fact, use your pencil point to trace out the following two current paths. Current flows from negative to positive, so start at the negative end of the battery (the bottom end illustrated with a short horizontal line) and follow the current path to the first connection point (labeled "node 1"), continue up through D2, then D1, through the next connection point (labeled "node 2"), through R1, and finally ending at the positive side of the battery. Tracing out the second current path, begin again at the negative side of the battery and follow the current flow to node 1, but this time follow the *other* current path through D3, then through R2, through node 2, through R1, and finally back to the positive side of the battery. As can be seen, there are two current "loops" in this circuit. One consists of D2, D1, and R1. The other consists of D3, R2, and R1. Notice that the total current flow from the battery branches at node 1, with some of it flowing through the D2-D1 leg, while the rest of it flows through the D3-R2 leg. However, at node 2, the two current flows combine again, and the *total* current flows through R1. Since the total circuit current flows through R1, R1 must be in *series* with the remainder of the circuit. However, because the current flow branches, causing only a portion of the total circuit current to flow through the other circuit components, the remainder of the circuit must be in some type of *parallel* configuration. Therefore, this circuit is referred to as a *series-parallel* circuit.

As stated previously, a certain portion of the current branches at node 1 and flows through the D2-D1 leg. Whatever value of current that happens to flow through D2 must also flow through D1, since there are no connections between them wherein the current could branch. Likewise, whatever value of current that happens to flow through D3 must also flow through R2. Therefore, D2 and D1 are in series with each other, and D3 and R2 are also in series with each other. In addition, *the series leg of D2 and D1 is in parallel with the series leg of D3 and R2.*

Refer to Fig. 2-22d and compare it to the circuit of Fig. 2-22c. Note that they are *exactly the same from the electrical perspective.* The only difference is in the way R2 is illustrated. Both methods of drawing the circuit are correct, but the Fig. 2-22d illustration makes it easier to visualize that the D3-R2 series leg is connected directly across from, or parallel with, the D2-D1 series leg.

What would be the desired resistance value of R2 in Fig. 2-22c? Going back to the general rules that govern a parallel circuit, you already know that the voltage across parallel circuit legs must be equal. In other words, the voltage across nodes 1 and 2 is the voltage across the D1-D2 leg, and this same voltage exists across the R2-D3 leg also. Since D1 and D2 are each dropping 2 volts, the total voltage drop from node 1 to node 2 will be 4 volts (remember, voltage drops of components in series are additive). This automatically means that 4 volts must be dropped across the R2-D3 series leg as well.

D3 will drop 2 volts, so the voltage drop across R2 must be the difference between the total voltage drop across the entire leg (i.e., 4 volts) and the 2 volts dropped across D3. Therefore, R2 will drop 2 volts (4 volts − 2 volts = 2 volts). The current flow through D3 should be 10 milliamps, and since it is in series with R2, the current flow through R2 will also be 10 milliamps. Now you know two electrical variables relating specifically to R2: the voltage drop across it (2 volts) and the current flow through it (10 milliamps). Using Eq. (2-3), we obtain

$$R = \frac{E}{I} = \frac{2 \text{ volts}}{10 \text{ milliamps}} = 200 \text{ ohms}$$

Exercise 8 Referring to Fig. 2-22c, what should the resistance value of R1 be? From the previous exercise, you already know that 10 milliamps of current should flow through the D1-D2 leg; 10 milliamps of current should also be flowing through the D3-R2 leg. Since these two parallel currents will sum at node 2, the total circuit current flow, which will be the current flow through R1, will be 20 milliamps.

As you recall, R1 is in series with the two parallel circuit legs (D1 and D2 make up one parallel leg, while R2 and D3 make up the other parallel leg). For simplification purposes, you can think of the parallel circuitry as a single equivalent resistor. This "imaginary" resistor will have a 20-milliamp current flow through it (i.e., the sum of the currents through both parallel legs) and will drop 4 volts (the previously calculated voltage drop across both parallel legs). The imaginary resistor, R_{equiv}, could be substituted for both legs of the parallel circuitry by placing it in the circuit between node 1 and node 2, as illustrated in Fig. 2-22e.

Now the resistance value of R1 is easy to calculate. The current flow through R_{equiv} is 20 milliamps, and because it is in series with R1, the current flow through R1 must be 20 milliamps also (this fact was shown earlier in this exercise). Since all voltage drops in a series circuit must add up to equal the source voltage, remembering that 4 volts is being dropped across R_{equiv}, the voltage drop across R1 must be 5 volts (9 volts

− 4 volts = 5 volts). Now that two electrical variables relating *specifically* to R1 are known, its resistance value can be calculated from Eq. (2-3):

$$R = \frac{E}{I} = \frac{5 \text{ volts}}{20 \text{ milliamps}} = 250 \text{ ohms}$$

Exercise 9   Referring to Fig. 2-22c, what is the power dissipation of R1? Now that you know the current flow through R1 and the voltage drop across R1 (from the previous exercise), you can use Eq. (2-6) to calculate power:

$$P = IE = (20 \text{ milliamps})(5 \text{ volts}) = 100 \text{ mW} \quad \text{or} \quad 0.1 \text{ watt}$$

Of what practical value are Exercises 7, 8, and 9? Hypothetically, suppose you wanted to illuminate 9 LEDs from a single 9-volt battery. If you attempted to place all 9 LEDs in a series circuit with a suitable current limiting resistor, you would probably receive very little, if any, light from the LEDs. Why? Because all of the voltage drops in a series circuit must add up to equal the source voltage. Since most general-purpose LEDs require approximately 2 volts across them to attain the current flow required for nominal brightness, you would have to supply a source voltage of at least 18 volts (i.e., 2 volts × 9 LEDs = 18 volts) to come close to the LED voltage requirements, even if the current-limiting resistor were dropping a negligible percentage of the source voltage. However, by using various series-parallel circuit combinations, you could illuminate dozens of LEDs with a 9-volt battery, provided the maximum *current flow* capability of the battery was high enough.

Exercise 10   Referring to Fig. 2-23a, what resistance value should R1 and R2 be, assuming that they both have the same resistance value? You should recognize that R1 and R2 are in parallel, with D1 in series with the parallel *network* of R1 and R2. (Electronics personnel often use the term *network* to describe a partial circuit or circuit subsystem—it is generally synonymous with *circuit.*) Note that all of the circuit current must flow through D1, but at the node 1 point, it can branch into two different circuit paths; one path is through R1, while the other path is through R2. Also note that it will sum again at node 2, before flowing to the positive terminal of the source.

Now, 10 milliamps of current should flow through D1, and since the resistance value of R1 and R2 are assumed to be equal, the current should branch equally at node 1, causing 5 milliamps to flow through

Figure 2-23
Circuit diagram
examples for practice
problems using
Ohm's law.

(a)

(b)

(c)

Figure 2-23
(*Continued*)

(d)

R1 and 5 milliamps to flow through R2. The voltage drop across D1 is 2 volts, which means that 7 volts must be dropped across the parallel network (9 volts from the source − 2 volts across D1 = 7 volts). Remembering that the voltage is the same across all legs of a parallel network, you now know that the voltage drop across both R1 and R2 is 7 volts. Calculating the resistance of R1 using Eq. (2-3), we obtain

$$R = \frac{E}{I} = \frac{7 \text{ volts}}{5 \text{ milliamps}} = 1400 \text{ ohms} \quad \text{or} \quad 1.4 \text{ Kohm}$$

Since R1 is equal to R2 in resistance value, R2 is 1.4 Kohm also.

Of what practical value is this exercise? There are several important areas of practicality in understanding the principles involved in this exercise. If you use Eq. (2-4) to calculate the R_{equiv} value for the parallel network of R1 and R2, you will discover that R_{equiv} is equal to 700 ohms. As discussed in the following section, 700 ohms is not a *standard* resistance value, and neither is 1.4 Kohm. Therefore, if you needed a 700-ohm current-limiting resistor for a circuit similar to that in Fig. 2-23a, you could use two parallel resistors of unequal resistance value to approximate an R_{equiv} of virtually any needed resistance value. For example, if you wanted to approximate an R_{equiv} value of 700 ohms, you could place a 1.2-Kohm resistor and a 1.8-Kohm resistor in parallel. This parallel combination would produce an R_{equiv} of 720 ohms (not exact, but typically close enough for most applications). Electronics engineers often use various combinations of series and parallel networks to obtain unusual or nonstandard resistance values. Another practical application of the principles in this exercise is to *distribute power dissipation* to multiple components. You will probably run into situations in which the power dissipation requirements for a resistor will exceed its maximum specification. In these cases, you can distribute the total power dissipation to multiple resistors, thus decreasing the individual power dissipation requirements. For example, if you inserted a single 700-ohm resistor in place of the parallel network of R1 and R2 in Fig. 2-23a, the power dissipation requirement on this single resistor would be

$$P = IE = (10 \text{ milliamps})(7 \text{ volts}) = 70 \text{ mW} \quad \text{or} \quad 0.07 \text{ watts}$$

By using two 1.4-Kohm resistors in parallel as originally illustrated in Fig. 2-23a, the power dissipation requirement on "each" resistor would be

$$P = IE = (5 \text{ milliamps})(7 \text{ volts}) = 35 \text{ mW} \quad \text{or} \quad 0.035 \text{ watts}$$

Note how the "total" power dissipation remained at 70 milliwatts (with two resistors dissipating 35 milliwatts, the total dissipation is 35 mW + 35 mW = 70 mW), but by incorporating the parallel network, each individual resistor was required to dissipate only 35 milliwatts.

Exercise 11   Referring to Fig. 2-23b, what should the resistance value of R1 be, assuming both R1 and R2 to have the same resistance value? This is another example of a series-parallel circuit. Note that D1 is in series with R3, and R3 is in series with the *parallel network* of R1 and R2. The total circuit current flow must flow through D1 and R3, but the current branches at node 1, with part of the current flowing through the R1 leg and the remainder of it flowing through the R2 leg. The two currents sum at node 2, with the total circuit current flowing to the positive terminal of the source.

The first step required in this exercise is to determine the voltage drop across R3. Since R3 and D1 are in series, and 10 milliamps should flow through D1, you know that 10 milliamps must flow through R3 as well. Now that you know two electrical variables specifically relating to R3 (i.e., its resistance value and the current flow through it), you can calculate the voltage drop across it using Eq. (2-1):

$$E = IR = (10 \text{ milliamps})(100 \text{ ohms}) = 1 \text{ volt}$$

Since D1 is dropping 2 volts and R3 is dropping 1 volt, the remainder of the source voltage must be dropped across the parallel network of

R1 and R2. Therefore, the voltage across R1 and R2 is 6 volts. Remembering that R1 and R2 are assumed to have equal resistance values, you know that the total circuit current will divide evenly at node 1, meaning that 5 milliamps will flow through R1 and the remaining 5 milliamps will flow through R2. Now you have two electrical variables relating specifically to R1, its current flow (5 milliamps) and the voltage across it (6 volts), using Eq. (2-3), its resistance value will be

$$R = \frac{E}{I} = \frac{6 \text{ volts}}{5 \text{ milliamps}} = 1200 \text{ ohms} \quad \text{or} \quad 1.2 \text{ Kohm}$$

What is the practical value of this exercise? In Exercise 10, you examined how it was possible to use two unequal resistance values connected in parallel to come up with unusual, or "nonstandard" resistance values. By connecting a 1.2-Kohm resistor and a 1.8-Kohm resistor in parallel, it was possible to derive an R_{equiv} resistance of 720 ohms, which was relatively close to the desired target resistance of 700 ohms. This exercise shows how it is possible to use a series-parallel resistor network to obtain "exactly" 700 ohms. If R1 and R2 are each 1.2 Kohms, the R_{equiv} is 600 ohms. Adding this to the value of R3 (100 ohms) provides a total resistance of 700 ohms. Also, the resistance values of 1.2 Kohms and 100 ohms are standard resistor values. Using various combinations of series-parallel circuits, it is possible to obtain virtually any resistance value for virtually any application.

Exercise 12 Referring to Fig. 2-23c, what would be the required resistance value of R1? The first step in solving a problem of this sort is to be certain that you have properly *conceptualized* the circuit. Note that R2 and R3 are in series, with this series network in parallel with R1, and this entire series-parallel network in series with R4, which is in series with D1. If this is confusing, compare Fig. 2-23c to Fig. 2-23d. They are electrically identical, but Fig. 2-23d is redrawn to make it easier to visualize.

Since D1 is in series with R4, remembering that 10 milliamps of current should be flowing through D1, you know that 10 milliamps should also flow through R4. With this information, you can calculate the voltage drop across R4 with Eq. (2-1):

$$E = IR = (10 \text{ milliamps})(100 \text{ ohms}) = 1 \text{ volt}$$

Since D1 is dropping 2 volts and R4 is dropping 1 volt, the remaining source voltage must be dropped across the series-parallel network of R1, R2, and R3. In other words, D1 and R4 are dropping a total of 3 volts, so the remaining 6 volts of the 9-volt source must appear across the two node connection points. Therefore, there is 6 volts across R1, and the same 6 volts also appear across the series combination of R2 and R3.

At this point, the current through R2 and R3 can be calculated. Since they are in series, their resistance values are added:

$$R_{total} = R_1 + R_2 = 1 \text{ Kohm} + 1 \text{ Kohm} = 2 \text{ Kohm}$$

The voltage across the series combination of R2 and R3 was previously determined to be 6 volts. Therefore, the current flow through them can be calculated using Eq. (2-2):

$$I = \frac{E}{R} = \frac{6 \text{ volts}}{2 \text{ Kohm}} = 3 \text{ milliamps} \quad \text{or} \quad 0.003 \text{ amps}$$

(Remember, 2 Kohm = 2000 ohms.)

Now that you know the current flow through the R2-R3 leg to be 3 milliamps, you also know that the remaining 7 milliamps (of the total 10-milliamp current flow) must flow through the R1 leg. Since you previously determined the voltage across R1 to be 6 volts, you now have two electrical variables relating specifically to R1 from which to calculate its resistance. Using Eq. (2-3), we have

$$R = \frac{E}{I} = \frac{6 \text{ volts}}{7 \text{ milliamps}} = 857 \text{ ohms} \quad \text{(rounded off from 857.14 ohms)}$$

Of what practical value is this exercise? This exercise is primarily intended to help you understand and develop the concept of *conceptualizing* a circuit, regardless of how it happens to be illustrated. Take a few moments to compare Fig. 2-23c and *b*, and pay particular attention to how you can prove both circuits to be electrically identical by following the current flow path(s) throughout the circuits. It is a common fault to focus one's mind on the size, position, or proximity of components when attempting to analyze a schematic when, in reality, these factors have little to do with circuit analysis. Until you become more accomplished at conceptualizing circuits, it is helpful to begin at one of the source terminals and methodically follow the current flow through the circuit, paying close attention to connection points, called *nodes*, in which the current will divide or sum.

Exercise 13 Referring to Fig. 2-24a, what should the resistance values of R1 and R2 be? It should be relatively obvious that this circuit consists of two series circuits placed in parallel with each other, with R5 in series with the entire parallel network. In other words, starting at the negative terminal of the source, note that all circuit current must flow through R5. Consequently, you know that R5 is in series *with everything else in the entire circuit.* On reaching node 1, note that the circuit current branches, with part of it flowing through the R4-D2-R2 leg, and the remainder flowing through the R3-D1-R1 leg.

Since the current flow through D2 should be 10 milliamps, this means that 10 milliamps will flow through R4 and R2 as well. The same applies

Figure 2-24
Circuit diagram examples for practice problems using Ohm's law.

(a)

(b)

Figure 2-24
(*Continued*)

(c)

to the other parallel leg; 10 milliamps should flow through D1, so 10 milliamps will flow through R3 and R1. You now know that at node 1 the total circuit current will branch, with 10 milliamps flowing through one parallel leg and 10 milliamps flowing through the other parallel leg. Therefore, the total circuit current must be 20 milliamps, since it is split into two equal current flows of 10 milliamps each at node 1. Likewise, the current will sum at node 2, combining into the total circuit current of 20 milliamps before entering the positive source terminal.

Now that the total circuit current flow is known, you can calculate the voltage drop across R5 with Eq. (2.1):

$$E = IR = (20 \text{ milliamps})(220 \text{ ohms}) = 4.4 \text{ volts}$$

Since 4.4 volts is being dropped across R5, the remainder of the source voltage must be dropped across the parallel network. The source voltage is 9 volts, so after subtracting 4.4 volts, the difference is 4.6 volts. Therefore, 4.6 volts is being dropped across the parallel network (i.e., from node 1 to node 2).

As determined previously, 10 milliamps will flow through the parallel leg of R4, D2, and R2. The voltage drop across R4 can be calculated from Eq. (2.1):

$$E = IR = (10 \text{ milliamps})(100 \text{ ohms}) = 1 \text{ volt}$$

The total voltage drop across the R4-D2-R2 parallel leg is 4.6 volts (as determined previously). Since you know that D2 is dropping 2 volts and

R4 is dropping 1 volt (a total of 3 volts), the remaining portion of the total 4.6 volts must be dropped across R2. Therefore, R2 is dropping 1.6 volts (4.6 volts − 3 volts = 1.6 volts). Now you know two electrical variables pertaining *specifically* to R2: the voltage across it (1.6 volts) and the current flow through it (10 milliamps). Its resistance value can be calculated from Eq. (2-3):

$$R = \frac{E}{I} = \frac{1.6 \text{ volts}}{10 \text{ milliamps}} = 160 \text{ ohms}$$

The resistance value of R1 can be determined in the same manner. Since 10 milliamps is flowing through the R3-D1-R1 parallel leg, the current flow through R3 will be 10 milliamps. The voltage drop across R3 can be calculated using Eq. (2-1):

$$E = IR = (10 \text{ milliamps})(68 \text{ ohms}) = 0.68 \text{ volts} \quad \text{or} \quad 680 \text{ millivolts}$$

The sum of the 2 volts dropped across D1 and the 680-millivolt drop across R3 is 2.68 volts. Remembering that 4.6 volts is dropped across the entire R3-D1-R1 leg, you can find the voltage drop across R1 by subtracting the combined voltage drops of D1 and R3. 4.6 volts minus 2.68 volts provides a difference of 1.92 volts, which is the voltage drop across R1. Now that you know the voltage drop across R1 (1.92 volts) and the current flow through it (10 milliamps), you can calculate its resistance value using Eq. (2-3):

$$R = \frac{E}{I} = \frac{1.92 \text{ volts}}{10 \text{ milliamps}} = 192 \text{ ohms}$$

Now that you have gone through this exercise involving Fig. 2-24*a*, refer to Fig. 2-24*b*. Note that this circuit is electrically identical to the Fig. 2-24*a* circuit, even though its visual appearance is quite different. Prove this to yourself by following the current flow from the negative terminal to the positive terminal of the source. Finally, refer to Fig. 2-24*c*. You'll discover that this circuit is also electrically identical to the Fig. 2-24*a* circuit diagram. Again, prove this to yourself by following the current flow paths.

What is the practical purpose of this exercise? Again, this exercise is primarily directed toward helping you improve your ability of conceptualizing circuits. It may be wise for you to repeat all of the previous exercises several times. Remember, conceptualizing electronic circuitry is a learned process—as you continue through this book, your abilities in this regard will improve greatly.

Exercise 14 You have now examined all of the previous circuit examples from an ideal and theoretical perspective. If you decide to physically construct and test these circuits, there are several reasons why you may run into some problems and discrepancies.

First, you will probably not be able to find resistor values that match many of your previously calculated resistance values. This is because most resistors are manufactured in *standard* values. The following list will provide you with the common, standardized values of resistors that you can use to physically construct any of the circuits illustrated in Figs. 2-22 through 2-24:

Fig. 2-22*a* R1 = 680 ohms

Fig. 2-22*b* R1 = 470 ohms

Fig. 2-22*c* R1 = 270 ohms R2 = 180 ohms

Fig. 2-23*a* R1 = 1.2 Kohm R2 = 1.8 Kohm

Fig. 2-23*b* R1 = 1.2 Kohm R2 = 1.2 Kohm R3 = 100 ohms

Fig. 2-23*c* R1 = 820 ohms R2 = 1 Kohm R3 = 1 Kohm R4 = 100 ohms

Fig. 2-24*a* R1 = 180 ohms R2 = 150 ohms R3 = 68 ohms R4 = 100 ohms
R5 = 220 ohms

The previous list of resistor values will be easy to obtain in any electronics parts store and they can all be inexpensive $\frac{1}{4}$-watt carbon film types.

The second problem you will probably encounter is discrepancies in your measured and calculated values. Several variables contribute to this phenomenon:

1. Your battery voltage will probably not be *exactly* 9 volts.

2. Your actual resistor values will not be *exactly* equal to your ideal calculated values—error due to resistor tolerances will also contribute to this effect.

3. The general-purpose LEDs you use may not drop *exactly* 2 volts (most general-purpose LEDs will drop about 1.8 to 2.2 volts at approximate current flows of 10 milliamps).

An important thing to remember when building and experimenting with these circuits is that *Ohm's law will always remain exactly accurate and predictable*. For example, suppose you build one of the aforementioned circuits and measure the actual voltage across a 100-ohm resistor. You calculated this voltage to be 1 volt, but you measure an actual voltage of

0.92 volts. The discrepancy could be due to the source voltage being a little low, or the actual resistance value of the resistor could be a little lower than its specified 100 ohms, but the discrepancy could *not* be due to any error or tolerance in Ohm's law. In other words, it is normal to encounter measured circuit variables that do not precisely match the calculated variables due to component tolerances and/or power supply variations, but the relationships defined by Ohm's law are a matter of physics, and these relationships cannot vary or suffer from any error— they are always precise and absolute.

Just to provide you with a feel for the expected results you should obtain from constructing and testing the circuits used in these exercises, I constructed the circuit illustrated in Fig. 2-24a, using a common alkaline 9-volt battery, general-purpose LEDs (randomly selected), and the *standard* resistor values for this circuit listed previously. My results are provided in the following list:

Measured voltage across R1 = 1.95 volts	Calculated value was 1.92 volts
Measured voltage across R2 = 1.61 volts	Calculated value was 1.6 volts
Measured voltage across R3 = 0.73 volts	Calculated value was 0.68 volts
Measured voltage across R4 = 1.06 volts	Calculated value was 1 volt
Measured voltage across R5 = 4.75 volts	Calculated value was 4.4 volts
Measured voltage across D1 = 1.87 volts	Assumed value was 2 volts
Measured voltage across D2 = 1.88 volts	Assumed value was 2 volts
Measured battery voltage = 9.3 volts	Ideal value was 9 volts

As you can see, my 9-volt battery was actually a little higher than 9 volts, and the LEDs I randomly chose dropped a little less than 2 volts. However, the circuit action was predictable and the measured voltages were reasonably close to the calculated ones. If I would have used a power supply set to precisely 9 volts, estimated my LED voltage drops at 1.9 volts instead of 2 volts, and incorporated 1% tolerance resistors, the voltage measurements would have been almost exactly the same as calculated.

You may want to build and experiment with a few circuits of your own creation. This is great practice in driving home all of the points brought out in this chapter. Just keep in mind that LEDs can be damaged if the current flow through them is too high. Until you learn a little more about the technicalities of LEDs, calculate your resistance values to allow them about 10 milliamps of current flow. Have fun!

3

The Transformer and AC Power

Described briefly in Chapter 2, the term *alternating current* (abbreviated AC) refers to current flow that periodically changes direction. This condition is related to the periodic changes in the polarity of the applied voltage. Alternating voltage and current are very important because of a phenomenon known as *transformer action*. Transformer action enables the efficient transmission of large quantities of power over long distances. This is why all common household current is AC.

AC Waveshapes

AC voltages and currents can be thought of as having waveshapes. *Waveshapes* are the visual, or graphical, representations of AC voltage or current amplitudes (or levels) relating to time. Common household AC has a type of waveshape called *sinusoidal* (*sine wave* for short). An illustration of a sine wave is shown in Fig. 3-1. Note that it is similar to a circle that is split in half, and joined together at opposite ends on the zero reference line. The sine wave is the most common type of AC waveshape you will be working with in the electrical and electronics fields.

Any repetitive, cyclic condition can be represented in the same manner as the illustration of the sine wave in Fig. 3-1. On the horizontal plane (that is, reading from left to right), a certain period of time is represented. On the vertical plane (from top to bottom), levels or amplitudes are indicated. The amplitudes above the zero reference usually denote positive variations, and the amplitudes below are usually negative. A *cycle* is defined as one complete periodic variation. In other words, if the illustration in Fig. 3-1 were to be carried out any further, it would begin to repeat itself. A *half-cycle* is either the positive half, or the negative half of a full cycle. Notice how the positive half-cycle would be identical to the negative half-cycle if it were to be turned upside down.

AC Frequency

The speed or rate at which an AC voltage or current waveform repeats itself is called its *frequency*. Frequency is measured in units called *hertz*,

Figure 3-1
One cycle of a sinusoidal (sine) AC waveshape.

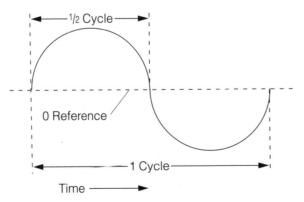

which is a method of relating frequency to time. One hertz is one cycle of AC occurring in a one-second time period.

If you have never been exposed to the concept of AC waveforms, this all might be rather abstract to you. To aid in understanding Fig. 3-1, assume it to be one cycle of common household AC power. Common household AC is usually labeled *120 volts AC, 60 hertz* (or "120 Vac, 60 Hz"). This means that the usable amplitude of the power is 120 volts, and it cycles at a rate of 60 times a second. If Fig. 3-1 represented this type of power, the time period from the beginning to the ending of one complete cycle would be about 16.6 milliseconds (ms). The time period of a half-cycle would be about one-half of the time period for a full cycle, or about 8.3 milliseconds.

There is an *inversely proportional* (opposite) relationship between frequency and the time period of one cycle. In other words, as frequency increases, the time period per cycle decreases, and vice versa. Equations (3-1) and (3-2) show how time is calculated from frequency, and frequency from time:

$$\text{Time period} = \frac{1}{\text{frequency}} \tag{3-1}$$

$$\text{Frequency} = \frac{1}{\text{time period}} \tag{3-2}$$

For example, to convert the common 60-Hz power frequency to its associated time period, you would use Eq. (3-1):

$$\text{Time period} = \frac{1}{\text{frequency}} = \frac{1}{60 \text{ Hz}} = 0.0166 \text{ second}$$

The time value 0.0166 second is commonly referred to as 16.6 milliseconds. If you knew the time period of one cycle, and wanted to calculate the frequency, you would use Eq. (3-2):

$$\text{Frequency} = \frac{1}{\text{time period}} = \frac{1}{0.0166} = 60 \text{ hertz}$$

NOTE: *Whenever a number is divided into 1, the result is called the reciprocal of that number. Therefore, it is proper to state the time period as being the reciprocal of frequency. Likewise, frequency is the reciprocal of the time period.*

AC Amplitude

As stated previously, one-half of a complete cycle is appropriately called a *half-cycle*. As shown in Fig. 3-1, for each half-cycle, the voltage or current actually passes through zero. In fact, it must pass through zero to change polarity or direction, respectively. If Fig. 3-1 is a voltage waveform, the voltage is shown to be positive during the first half-cycle, and negative during the second half-cycle.

The horizontal dotted line representing zero is called the *zero reference line*. As shown in Fig. 3-2, the maximum amplitude of either the positive, or negative, half-cycle (as measured from the zero reference line) is called the *peak voltage*. The total deviation from the negative peak to the positive peak is called the *peak-to-peak voltage*.

Defining the amplitude of pure DC voltages and currents is easy, because the level is constant and continuous. Defining AC voltage or current amplitudes is a little more complicated. As shown in Fig. 3-2, the peak voltage of this waveform is 10 volts. This represents an *instantaneous* voltage level. At other points in the waveform, the voltage level is less than 10 volts (sometimes even zero), and it can be of positive or negative polarity. For this reason, there are numerous ways to accurately define alternating voltage and current amplitudes, depending on the application.

One, somewhat controversial, method of defining an AC amplitude is by algebraically adding the negative half-cycle to the positive half-cycle. Because the positive half-cycle of a true sine wave is the "exact opposite and equal" of the negative half-cycle, the two cancel each other out, resulting in an average of zero. If this is difficult to understand, consider this analogy. A DC voltage or current is like driving a car in one direction. The passengers in this car will go somewhere, and end up in a different location than where they started. In contrast, an AC voltage or current is like driving a car back and forth in a driveway. The passengers in this car can ride all day long, and still end up where they started.

Figure 3-2
Relationship of peak, peak-to-peak, average, and rms voltages relative to a sinusoidal waveshape.

Peak = 10 Volts

0 Reference

Peak-to-peak = 20 Volts

Average = 0 volts
RMS voltage = 7.07

This is why a DC voltmeter (a test instrument for measuring DC voltage levels) will measure 0 volts if a pure 60-hertz AC voltage is applied to the test probes. The DC voltmeter cannot respond to the rapid AC polarity reversals. Therefore, it indicates the average level, or zero. In the beginning of this paragraph, I stated that this is a controversial definition.

In many electronics textbooks, the *average* AC amplitude for a sine wave is defined as *63.7% of the peak amplitude*. Technically speaking, this is incorrect; but it was generally adopted because many older type AC voltmeters would indicate this value when measuring a sine-wave voltage.

Regardless of definition, don't concern yourself with this situation. Average values of AC voltage are seldom used for any practical purpose. I included this explanation only to aid in understanding the operational physics of an AC sine wave.

If the AC voltage shown in Fig. 3-2 were applied to a load, as shown in Fig. 3-3, current would flow in a back-and-forth motion, proportionally following the voltage variations. Some quantity of power would be *dissipated* because current is flowing through the load. The power dissipated, while the current flows in one direction, does not negate the power dissipated when the current reverses (there is no such thing as negative power). Therefore, power is being dissipated continuously. Consider the previous analogy of the car driving back and forth in the driveway. Although it ended up where it began, energy (gasoline) was consumed, and work was performed (back and forth movement of the car).

A practical method of comparing AC voltage and current to a DC equivalent is the result of a complicated mathematical analysis called *root mean square* (rms). In other words, AC rms sources can be directly compared to DC sources in their ability to provide power, or to perform work. For example, the common 120 volts AC from household outlets is an rms rating. It can perform just as much work as 120 volts of DC. Values of rms are often called *effective* values.

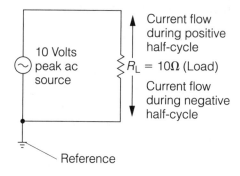

Figure 3-3
Simple AC circuit illustrating current flow reversal.

To calculate the rms value of a sine wave, simply *multiply the peak value by 0.707.* Referring to Fig. 3-2, the peak amplitude of the sine wave is 10 volts. To find the rms voltage, the 10 volts is multiplied by 0.707, providing the result of 7.07 volts. Figure 3-3 illustrates this same AC voltage being applied to a load. The power this load dissipates is the same power it would dissipate if a 7.07-volt DC source were applied to it.

AC Calculations

All of the following equations and calculation methods are applicable only to sinusoidal wave voltages and currents. Virtually all AC used for power sources will fall into this category, so you will be using this information frequently.

If you know the peak, or peak-to-peak, value of a sine-wave AC voltage or current, you can calculate the rms value using Eq. (3-3) or (3-4):

$$\text{rms} = (\text{peak value})0.707 \qquad (3\text{-}3)$$

$$\text{rms} = (\tfrac{1}{2}\,\text{peak-to-peak value})0.707 \qquad (3\text{-}4)$$

For example, using Eq. (3-4) to calculate the rms value of the peak-to-peak voltage shown in Fig. 3-2

$$\text{rms} = (\tfrac{1}{2}\,\text{peak-to-peak})0.707$$

$$\text{rms} = (\tfrac{1}{2}\,\text{of 20 volts})0.707$$

$$\text{rms} = (10\ \text{volts})0.707$$

$$\text{rms} = 7.07\ \text{volts}$$

The AC voltage shown in Fig. 3-2 can be accurately defined as being 7.07 volts$_{\text{rms}}$. This means it is equivalent to 7.07 volts DC. Rms current calculations are performed in the same manner.

If an rms voltage or current is known, and you wish to calculate the peak or peak-to-peak value, Eqs. (3-5) and (3-6) might be used:

$$\text{Peak} = (\text{rms value})1.414 \qquad (3\text{-}5)$$

$$\text{Peak-to-peak} = (\text{peak value})2 \qquad (3\text{-}6)$$

For example, common household outlet voltage is 120 volts AC rms. The peak and peak-to-peak values might be calculated as follows:

$$\text{Peak} = (\text{rms value})1.414$$

$$\text{Peak} = (120 \text{ volts})1.414 = 169.68 \text{ volts}$$

$$\text{Peak-to-peak} = (\text{peak value})2$$

$$\text{Peak-to-peak} = (169.68 \text{ volts})2 = 339.36 \text{ volts}$$

You can use any of the AC voltage or current terms in the familiar Ohm's law equations. The only stipulation is "they must be in common." In other words, you cannot mix up rms, peak, and peak-to-peak values, and still arrive at the correct answer. For example, if you wanted to use Ohm's law ($E =$ IR) to calculate a peak voltage, you must use the associated peak current value. Resistance is a constant value; therefore, it cannot be expressed in terms such as peak, peak-to-peak, or rms. For the following examples, refer to Fig. 3-3.

Example 1   Calculate the rms voltage applied to the load.

$$\text{Rms} = (\text{peak value})0.707 = (10)0.707 = 7.07 \text{ V}_{\text{rms}}$$

Example 2   Calculate the rms current flowing through the load:

$$I_{\text{rms}} = \frac{E_{\text{rms}}}{R} = \frac{7.07 \text{ volts}}{10 \text{ ohms}} = 0.707 \text{ amps}$$

Note that the rms voltage had to be used, in the previous equation, to calculate the rms current.

Example 3   Calculate the peak current flowing through the load:

$$I_{\text{peak}} = \frac{E_{\text{peak}}}{R} = \frac{10 \text{ volts}}{10 \text{ ohms}} = 1 \text{ amp peak}$$

Note that the peak voltage had to be used, in the previous equation, to calculate the peak current.

Example 4   Calculate the peak power being dissipated by the load:

$$P_{\text{peak}} = (I_{\text{peak}})(E_{\text{peak}}) = (1 \text{ amp})(10 \text{ volts}) = 10 \text{ watts peak}$$

The previous equation is simply the familiar $P = IE$ power formula, with the "peak" labels inserted.

Example 5 Calculate the rms (effective) power being dissipated by the load:

$$P_{rms} = (I_{rms})(E_{rms}) = (0.707 \text{ amp})(7.07 \text{ volts}) = 5 \text{ watts rms}$$

Inductance

At some point in your life (possibly in a school science class), you might have built a small electromagnet by twisting some insulated electrical wire around a nail, and then connecting the ends of the wire to a flashlight battery, as shown in Fig. 3-4. An electromagnet is an example of an inductor. An *inductor* is simply a coil of wire.

A typical inductor will consist of a coil (or multiple coils) wound on a metallic, ferrite, or phenolic core. The *metallic* and *ferromagnetic cores* concentrate the magnetic flux lines (lines of force), thus increasing the inductance value. Some coils used in high-frequency applications contain tunable ferrite slugs that allow adjustment of the inductance value. Other types of high-frequency, or special-purpose, inductors do not contain metallic cores. These are referred to as *air core* coils.

The electromagnet shown in Fig. 3-4 illustrates how a coil will develop an electromagnetic field around it. An *electromagnetic field* is made up of many lines of force called *flux lines*. The flux lines flow through the nail, causing it to become a temporary magnet. Electromagnetic flux lines are stationary only as long as the current flow through the coil is constant. If the current flow through the coil is increased, the electromagnetic field will expand, causing the flux lines to move outward. If the current flow is decreased, the electromagnetic field will collapse,

Figure 3-4
Simple
electromagnet.

Flux lines (stored energy) Nail (core)

South pole North pole

Insulated coil wire

causing the flux lines to move inward. The most important thing to understand is that the field will move (expand and contract) if the current flow changes. Remember this foundational rule of electricity:

> Whenever an electrical conductor (wire) cuts magnetic flux lines, an electrical potential (voltage) will be induced in the conductor.

It also stands to reason that if the conductor in the previous statement is part of a closed circuit, an induced current will flow whenever an induced voltage is generated. This is the basic principle behind the operation of any electrical generator. To generate electricity, magnetic flux lines must be cut by a conductor. This can be accomplished by moving either the conductor or the electromagnetic field.

As stated previously, if the current flow through a coil (inductor) is varied, the electromagnetic field will move in proportion to the variance. The movement of the electromagnetic field will cause the coil conductors to be cut by the changing flux lines. Consider these effects under the following three conditions:

1. A coil has a steady and continuous current flow through it; the applied voltage to the coil is held constant. Thus, the electromagnetic field surrounding the coil is also constant and stationary.

2. The applied voltage to the coil is reduced. The current flow through the coil tries to decrease, causing the electromagnetic field surrounding the coil to collapse by some undetermined amount. As the field collapses, the flux lines cut the coil wire generating a voltage that opposes the decrease in the applied voltage. The end result is that the coil tries to maintain the same current flow (for a period of time) by using the stored energy in the electromagnetic field to supplement the decreased current flow.

3. The applied voltage to the coil is increased. The current flow through the coil tries to increase, causing the electromagnetic field surrounding the coil to expand. As the field expansion occurs, the flux lines cutting the coil wire generate an opposing voltage. The end result is that the coil tries to maintain the same current flow (for a period of time) by storing energy into the electromagnetic field.

In essence, an inductor tries to maintain a constant current flow through itself, by either expanding or contracting its associated electromagnetic field. This expansion or contraction of the electromagnetic field generates a voltage that opposes any changes in the applied voltage. This opposing voltage is referred to as *counterelectromotive force* (cemf). *Electromotive force* is another name for voltage.

Inductors are storage devices; electrical energy is stored in the electromagnetic field surrounding an inductor. The quantity of energy that an inductor is capable of storing is called its *inductance* value. Inductance is measured in units called *henrys*. One *henry* is the inductance value if a current change of 1 ampere per second produces a *cemf* of 1 volt. The electrical symbol for inductance is *L*.

Although this might seem somewhat complicated at first, consider the circuit shown in Fig. 3-5. (Note the symbol for an inductor. The vertical lines to the right of the wire turns denote an iron core.) With the switch (S1) in the open position, obviously there cannot be any current flow through the inductor. The graph illustrated in Fig. 3-6 shows the response of the current flow immediately after the switch in Fig. 3-5 is closed. The inductor tries to maintain the same current flow that existed prior to the closing of the switch (which was zero in this case), by storing energy in its associated electromagnetic field. While the electromagnetic field is expanding, a reverse voltage (cemf) is being generated by the coil opposing the applied voltage. The end result, as shown in Fig. 3-6, is that the current takes time to change even though the voltage is applied instantaneously.

The time period required for the current to reach its maximum value is measured in seconds, and is defined by a unit called the *time*

Figure 3-5
Basic *LR* (inductive-resistive) circuit.

Figure 3-6
Circuit current response of Fig. 3-5.

constant, which is the time required (in seconds) for the current flowing through the inductor to reach approximately 63% of its maximum value. The time constant is found through dividing the inductance value by the circuit resistance. For example, the time constant for the circuit shown in Fig. 3-5 would be

$$Tc = \frac{L}{R} = \frac{10 \text{ henrys}}{10 \text{ ohms}} = 1 \text{ second}$$

Assume that the maximum obtainable current of the circuit shown in Fig. 3-5 is 1 amp. According to the previous time constant calculation, the circuit current will climb to a value of 0.63 amp (63% of 1 amp) after the switch has been closed for 1 second. During the next 1-second time interval, the current will increase by another 63% of the *difference between its present level and the maximum obtainable level*. In other words, if the current rose to 0.63 amp after the first time constant, that would leave a 0.37 amp difference between its present level and the maximum level of 1 amp (1 amp − 0.63 amp = 0.37 amp). Therefore, during the second time constant, there would be a current increase of 63% of 0.37 amp, or approximately 0.23 amp. This results in a circuit current of approximately 0.86 amp after two time constants. Likewise, during the third time constant, there would be another current increase of 63% of the difference between the present level and the maximum obtainable level, and so on.

In theory, the maximum steady-state current of an *LR* circuit can never be obtained; it would always be increasing by 63% of some negligible current value. In a practical sense, the maximum circuit current value is usually assumed to have been reached after five time constants. The current flow of the circuit shown in Fig. 3-5 would reach its maximum 1-amp level in approximately 5 seconds after S1 is closed.

The effect of the current taking time to catch up with the applied voltage is called the *current lag*. When the applied voltage is AC (changing constantly), the current will always lag behind the voltage.

Periodic AC voltages are divided up into *degrees*, like a circle. Just as a complete circle will contain 360 degrees, one complete cycle of AC is considered as having 360 degrees. Figure 3-7 illustrates how one complete cycle of sine-wave AC voltage is divided into 360 degrees. Note how the positive peak is at 90 degrees, the end of the first half-cycle occurs at 180 degrees, the negative peak is at 270 degrees, and the end of the cycle occurs at the 360-degree point. The waveform below the voltage waveform is the current waveform. Note how the current lags behind the

Figure 3-7
Voltage–current
phase relationship in
an inductive circuit.

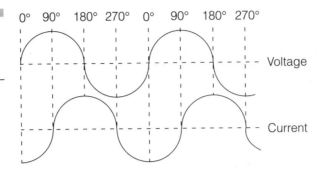

voltage, its first peak occurs at 180 degrees, the end of the first half-cycle is at 270 degrees, and so forth. In other words, the current is *out of phase* with the voltage by 90 degrees.

The waveforms shown in Fig. 3-7 are characteristic of any "purely inductive circuit" (a circuit consisting only of an AC source and an inductor, or multiple inductors). In a purely inductive circuit, the current will always lag the voltage by 90 degrees, regardless of the applied AC frequency. This voltage–current phase differential is often referred to as the *phase angle*, and it is also stated in terms of degrees.

An interesting consideration in purely inductive circuits is the *power consumption*—there is none! To understand why, refer back to Fig. 3-7. For discussion purposes, assume the peak voltage waveform levels to be 10 volts, and the peak current waveform levels to be 10 amps. At the 90-degree point, the voltage is at its peak level (10 volts) and the current is at zero. Using the power equation $P = IE$, the *instantaneous* power dissipation can be calculated:

$$P = IE = (0)(10 \text{ volts}) = 0$$

At 180 degrees, the voltage is at zero, and the current is at its peak of 10 amps. Using the same power equation

$$P = IE = (10 \text{ amps})(0) = 0$$

At 270 degrees, the voltage is at its negative peak of -10 volts, and the current is zero. Again, using the same power equation

$$P = IE = (0)(10 \text{ volts}) = 0$$

Because of the 90-degree phase differential between voltage and current, the power dissipation in any purely inductive circuit is considered to be virtually zero. The inductor continually stores energy and then regenerates this energy back into the source.

Assume that you were examining a purely inductive circuit possessing the voltage and current waveforms shown in Fig. 3-7. Again, assume the peak voltage amplitude to be 10 volts, and the peak current amplitude to be 10 amps. If you used a voltmeter to measure the applied voltage, it would read 7.07 volts; the rms value of 10 volts peak. Likewise, if you used an ammeter to measure the current flow, you might read 7.07 amps, the rms value of 10 amps peak. If you multiplied the measured voltage value by the measured current value, the answer should be the *rms (or effective) power* value ($P = IE$). The answer would be 50 watts rms (7.07 volts times 7.07 amps = 50 watts). This contradicts the previous statement regarding zero power dissipation in a purely inductive circuit. The reason for this discrepancy is that you did not take the voltage and current phase differential into consideration. In a purely inductive circuit, the power calculation based on the measured voltage and current values is called the *apparent power.* The actual power dissipated in an inductive circuit when the voltage and current phase differential are taken into consideration is called the *true power.*

If you wish to pursue higher mathematics, true power is calculated by finding the apparent power, and then multiplying it by the cosine of the differential phase angle. For any purely inductive circuit, the differential phase angle (as stated previously) is 90 degrees. The cosine of 90 degrees is zero. Therefore, zero times any apparent power calculation will always equal zero. (If you don't understand this, don't worry about it. Depending on your interests, you might never need to perform these calculations; but if you do, a good electronics math book will explain it in "easy" detail.)

Another term related to inductive circuits is called the *power factor.* The power factor of any circuit is simply the true power divided by the apparent power.

$$\text{Power factor} = \frac{\text{true power}}{\text{apparent power}}$$

DC Resistance

Earlier in this chapter, the time constant of the circuit in Fig. 3-5 was examined and the maximum obtainable current flow was found to be approximately 1 amp. The analysis and calculations of this circuit were performed by assuming the circuit to be "ideal"; that is, all components were considered to be perfect. With any circuit, and with all components, there are shortcomings that cause them to be something less than perfect in their operation. To become very technically accurate regarding the circuit in Fig. 3-5, you would have to consider such factors as the internal resistance of the battery (source resistance), the wiring resistance, the resistance of the switch contacts of S1, and the DC resistance of the inductor. In the majority of design situations, all of these factors are so low that they are considered negligible. But to fully understand the operation of inductors, the *DC resistance factor* should be discussed.

Referring back to Fig. 3-5, it was stated that after five time constants, the circuit current would reach its maximum amplitude. From this point on (assuming that S1 remains closed), both the voltage and current will remain stable. Therefore, the electromagnetic field around the coil will also remain stable. If the electromagnetic field does not move, the wire making up the coil cannot cut flux lines. This means there cannot be any generation of cemf. In effect, the only opposition to current flow posed by the inductor is the wire resistance of the coil. The coil's wire resistance (usually called the *DC resistance*) is normally very low and might be disregarded in most applications.

A term for defining the DC resistance of any coil relative to its inductance value is Q (abbreviation for quality). The Q of an inductor is the inductance value (in henrys) divided by the DC resistance (in ohms):

$$Q = \frac{L}{R}$$

For most inductors, this value will be 10 or higher.

Transformers

An inductor with two or more coils wound in close proximity to each other is called a *transformer.* (A single-coil inductor is often called a "choke.") If an AC voltage is applied to one coil of a transformer, its associated moving magnetic field will cause the magnetic flux lines to be cut by itself (causing *cemf*) and any other coil near it. As the other coils are cut by flux lines, an AC voltage is also induced in them. This is the basic principle behind *transformer action.*

If the multiple coils of a transformer are wound on a common iron core, the transfer of electrical energy through the moving electromagnetic field becomes very efficient. Iron-core transformers are designed for efficient transfer of power, and are consequently called *power transformers* or *filament transformers* (an older term from the vacuum-tube era). Properly designed, power transformers can obtain efficiencies as high as 99%.

Figure 3-8 illustrates some examples of transformer symbols. The coil on which the AC voltage is applied is called the *primary.* The transfer of power is *induced* into the *secondary.* Transformers might have multiple primaries and secondaries. The primaries and secondaries might also contain "taps" to provide multiple voltage outputs or to adapt the transformer to various input voltage amplitudes. The example shown on the left side of Fig. 3-8 is a transformer with one primary and one secondary. The illustration on the right side shows a transformer that has two primaries, one secondary, and a secondary tap. If a tap is placed in the exact center of a transformer coil, that coil is said to be *center-tapped* (ct).

The most important attribute of a power transformer is its ability to increase or decrease AC voltage amplitudes without a significant loss of energy. This is accomplished by means of the *turns ratio* designed into the transformer. The turns ratio is simply the ratio of the number of turns on the primary to the number of turns on the secondary. For example, if a transformer has a 1:1 turns ratio, it has the same number of turns on the primary, as it has on the secondary. A transformer with a 2:1 turns ratio has twice the number of turns on the primary, as on the secondary. A transformer with a 1:12 turns ratio has 12 times the number of turns on the secondary, as on the primary. In the ratio expression, the primary is always represented by the first number of the ratio, and the secondary is the second.

The turns ratio has a *directly proportional relationship* to the transformer's voltage amplitudes. If the secondary of a transformer has more turns than the primary, the secondary voltage will be proportionally higher in amplitude, or "stepped up." If the secondary contains less turns than the primary, the secondary voltage will be proportionally lower, or "stepped down." For example, if a transformer has a 2:1 turns

Figure 3-8
Transformer symbols.

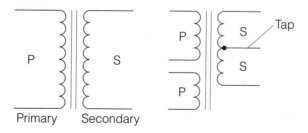

Primary   Secondary

ratio, the secondary voltage will be one-half the amplitude of the applied voltage to the primary. This is because there is only one-half the number of turns on the secondary as compared to the primary. The reverse is also true. If the transformer has a 1:2 turns ratio, the secondary voltage will be twice that of the primary.

A basic law of physics tells us that it is impossible to obtain more energy from anything than is originally put into it. For this reason, if the secondary voltage of a transformer is doubled, the secondary current rating will be one-half the value of the primary. Similarly, if the secondary voltage is only one-half the primary voltage, the secondary current rating will be twice the value of the primary. In this way, a transformer will always maintain an equilibrium of *power transfer ability* on both sides. The maximum power transfer ability of a transformer is called its *volt-amp* (VA) *rating*.

This principle is demonstrated in Fig. 3-9. Note that the transformer has a 10:1 turns ratio. This means that whatever voltage is applied to the primary (E_p) will be reduced by a factor of 10 on the secondary (E_s). The current flow is just the opposite. The current flowing through the load resistor (I_s) will be 10 times greater than the primary current flow (I_p). In this illustration, E_p is 100 volts rms. Because there is only one-tenth the number of turns on the secondary, E_s is 10 volts rms. According to Ohm's law, the current flow through the load resistor will be 10 volts rms divided by 10 ohms, or 1 amp rms. Therefore, the power transferred to the secondary load will be

$$P = IE = (1 \text{ amp rms})(10 \text{ volts rms}) = 10 \text{ watts rms}$$

Now consider the power being delivered to the primary. The current in the primary will be one-tenth that of the secondary, or 0.1 amp rms. Because E_p is 100 volts rms:

Figure 3-9
Transformer with secondary load demonstrating the relative primary and secondary currents.

$I_P = 0.1$ Amp $I_S = 1$ Amp

$E_P = 100$ Volts RMS $E_S = 10$ Volts RMS $R_L = 10\Omega$

10 : 1 Turns ratio

$$P = IE = (0.1 \text{ amp rms})(100 \text{ volts rms}) = 10 \text{ watts rms}$$

It is important to understand that the transformer is not dissipating 10 watts rms of power. It is transferring 10 watts rms of power from the primary to the secondary. The transformer itself is dissipating a negligible amount of power. If the secondary circuit were opened, the current flow in the secondary would consequently drop to zero. Similarly, the current flow in the primary would also drop. (A smaller current flow would still remain in the primary. The explanation and calculations for this condition will be covered in Chapter 15.)

Soldering

Enough of the theory stuff—if you please!! If you're thinking that about now, it just means you're human. If you have comprehended "most" of the theory covered in Chapter 2 and in this chapter, you are doing just fine. The fuzzy areas of understanding will clear up as you begin to physically work with components and circuits. The remainder of this chapter will deal with the beginning steps for constructing a laboratory quality power supply. This power supply will become the power source for many of the future projects in this book and, hopefully, many more of your own.

Before beginning any kind of electrical or electronics construction, it is essential to become proficient at soldering. By now, you should have accumulated the basic tools needed to solder (as covered in Chapter 1):

- A soldering iron with several size tips
- A soldering iron holder
- A damp sponge for tip cleaning
- A roll of good-grade, 60/40 rosin-core solder (if you have any "acid core" solder, throw it away or give it to a plumber!)
- A desoldering tool

In addition to these specific items, you will need a few small hand tools, a comfortable, well-lighted place to work, a little patience, and a worktable (please don't try this on an expensive piece of furniture!). A little steel wool and some isopropyl alcohol might come in handy for cleaning purposes.

Soldering Overview

Soldering is a process by which conductive materials are electrically and mechanically bonded together with a tin-lead alloy by the application of heat. For the bonding process to occur properly, the solder and the materials being soldered must be of sufficient temperature. The materials being soldered must also be clean and free of corrosion, oil, and dirt.

During the actual soldering process, the soldering iron is used to heat the material to be soldered (such as a component lead, bare wire, or the copper "artwork" on a printed circuit board) until it reaches a temperature above the melting point of the solder. Rosin-core solder is then placed on the material to be soldered and allowed to melt. As it melts, it will flow outward (called *wetting*) toward the soldering iron tip. When sufficient solder has flowed for a good electrical and mechanical bond, the soldering iron tip and the unused solder are removed simultaneously. The newly formed solder connection is given time to cool and solidify. After cooling, it should appear bright, shiny, and smooth. A rough or grainy-looking joint (a *cold joint*) is questionable, and should be redone.

Soldering for the First Time

If you have never tried soldering, I'm sure you probably think it is more difficult than it is. My experience in working with beginning students has taught me that if I just give them a few pointers and leave them alone, they will quickly get the hang of it.

Safety note: Wear safety glasses or goggles to protect your eyes when soldering. Also, work in a well-ventilated area, and don't inhale the smoke.

To begin, you will need something to practice on. I recommend a scrap of printed circuit board. *Printed circuit boards* (commonly called *PC boards*) are thin sheets of phenolic or fiberglass material with electronic components mounted on them. The electronic components interconnect on the board by means of copper foil "artwork," called *traces*, glued to the PC board. These are easy to come by. Virtually any kind of junk electronic equipment will contain at least one PC board. If it is difficult to obtain a scrap PC board, you can purchase a general-purpose "grid board" from your local electronics supply store, together with a bag of miscellaneous electronic components. Buy the cheapest you can find;

the board and the components will probably be scrap by the time you're comfortable with soldering.

When you've collected the tools, materials, and something to practice on, it's time to try your first soldering exercises. Plug in the soldering iron and allow it time to warm up. If you have a soldering iron with an adjustable tip temperature, set it to about 670 degrees. If your soldering iron is adjustable, but doesn't have a readout (in degrees), initially set it to a low temperature, and adjust it by trial and error. The tip should be hot enough to readily melt the solder, but not hot enough to "instantly" melt the solder while producing a lot of smoke.

If the tip of the soldering iron is not silver and shiny, it will need to be "tinned." Clean the tip with some steel wool and apply a little rosin core solder all around the tip and about a half-inch up on the tip. Make sure the solder flows evenly around the tip and then wipe the tip on the damp sponge to remove any excess solder. As the soldering iron is used, it will be necessary to periodically re-tin the tip whenever flux or other contaminates build up on the tip causing it to turn dark. If you remember to clean the tip frequently with the damp sponge, the need for re-tinning will be reduced. If the tip becomes dark quickly, even when not in use, the tip temperature is probably too high.

Soldering Procedure

If you have a scrap PC board to practice on, you will probably have to remove a few components before soldering them back. Skip this section and go on to the next section entitled *Desoldering Procedure*. Return to this section after you have finished desoldering.

Make up a few practice solder joints by placing some components randomly in the PC board. Before trying to solder, examine the joint to be soldered. It should be clean, free of corrosion, and mechanically stable (held in place). Place the soldering iron tip on *all* of the material to be soldered. If you are soldering on a PC board, this means the soldering-iron tip should be placed on the PC board foil material, and on the component lead at the same time. All of the material to be soldered must be heated, or the solder will not adhere properly. With the joint properly heated, apply the solder *to the joint* (not the soldering-iron tip). Allow the solder to flow evenly around the joint, and eventually flow to the tip. Remove the unused solder and the soldering iron from the joint, and allow it to cool.

Examine the newly soldered joint. It should be shiny and smooth, and the connection should be mechanically strong. If the solder balled up on the joint, it probably wasn't hot enough. A rough, gray-looking joint is also indicative of improper heating. If the soldering-iron tip is not properly tinned, it will not conduct heat adequately to the solderable material. A solder joint with poor electrical integrity is referred to as being "cold." A *cold solder joint* will not be mechanically solid, either. Wiggle the newly soldered component to verify it is tightly bonded together.

The real skill in soldering relates to the speed at which it can be performed. You should be able to solder a typical connection in well under 5 seconds. Heating a connection involving solid-state components for too long can destroy the components. Most electronic supply stores sell aluminum clamp-on *heatsinks* to conduct most of the heat away from the component when soldering. But the usefulness of these devices is limited. Most integrated circuits, and many miniature components, simply do not have an available lead length onto which to clamp.

Another point to consider, when soldering is *heat buildup.* Suppose that you are soldering a component with eight leads into a PC board. If you solder all eight connections, one immediately after another, the component might become very hot because of heat buildup. It becomes progressively warmer with each soldering because it did not thoroughly cool from the previous one. The solution is to allow sufficient time for the component to cool between solderings.

Desoldering Procedure

Desoldering, of course, is the opposite of soldering. Desoldering is required to change defective components, to correct construction errors, to redo wiring jobs, and to salvage parts.

One commonly used method of desoldering utilizes prefluxed copper braid to absorb molten solder by capillary action. The *desoldering braid,* or wicking, goes by a variety of brand names and is easy to use. The braid is simply placed in the molten solder, and the capillary action draws the excess solder up into the braid. The used piece of braid is then cut off and discarded. For smaller solder joints, the action is often improved by sandwiching the braid between the soldering iron tip and the joint to be desoldered.

Most people prefer to use a vacuum desoldering tool as described in Chapter 1. Desoldering is accomplished by "cocking" the tool: holding the tip close to the joint to be desoldered, melting the solder with a soldering iron, and pressing the "trigger" on the tool. The rapid suction action sucks the excess solder up into a holding chamber, from where it is removed later. This tool is often called a *solder sucker.*

If you plan on doing a lot of desoldering (for salvage purposes, for example), you might want to consider investing in a dedicated desoldering station. These units contain a dedicated vacuum pump and a specially designed soldering iron with a hollow tip. The molten solder is sucked up through the hollow tip of the iron and held in a filtered container for removal. The suction action is triggered by a foot or knee pedal. Unfortunately, dedicated desoldering stations are very expensive. Be sure you can justify the cost before investing in one.

Assembling and Testing the First Section of a Lab Power Supply

Now comes the fun part! In this section you will begin to build a lab-quality power supply that you can use as a power source for many other projects throughout this book. The building of this supply is also a learning process. You will perform checks and tests to drive home the theories and principles you have learned thus far. The experiments performed will also verify the accuracy of your understanding. The same procedure will be followed for virtually every project in this book, so upon completion, you not only have some "neat" projects to show off, but you also have the practical and theoretical knowledge to go with them (which is more important by far!).

Emphasis on Safety throughout This Book

Please don't think I sound like your mother, but if that's what it takes to save even one reader from having an unfortunate accident, the price is well worth it. You learn by doing, but you live to tell about it if you do it right! *Never compromise on safety.*

Before beginning construction, let me briefly explain some of the features that this power supply will have on completion. This supply will have two independently adjustable, voltage-regulated outputs; one positive, and one negative. The adjustment range for each will be from about 4 to 15 volts. It will be short-circuit protected (so you can make a "goof" without destroying part of the supply), and it will go into *current-limit mode* at about 1.5 amps. The current limit feature allows the power supply to double as a battery charger (with certain precautions). A handy feature will be two additional fuse protected "raw" DC outputs (+ and − 34 volts) for testing audio amplifiers, or powering higher voltage projects. In short, you will discover it to be a versatile and valuable piece of test equipment.

Materials Needed for Completion of This Section

Throughout the remainder of this book, I will always assume that you have a DVM (or equivalent), soldering tools, alligator clip leads, hook-up wire, mounting hardware, and the necessary hand tools. The materials list for each project will only list the materials and components actually used in the project itself.

Quantity	Item Description
1	$9 \times 9 \times 3$ ($w \times d \times h$)-inch (or larger) metal project box
1	Grounded (3-conductor) power cord
1	Power-cord strain relief
1	SPST 10-amp on-off switch
2	24-volt at 2-amp transformers (120-volt primary)
1	3AG size ($\frac{1}{4} \times 1\frac{1}{4}$-inch) fuse block or fuse holder
1	2-amp 250-volt slow-blow fuse
1	Locking terminal solder lug (see text)

The project box doesn't have to be fancy and expensive unless you're the type that wants to go first class all the way. The enclosure from a junked CD player or VCR should do nicely, provided you rework the front panel. (Front panels can be modified to accommodate almost any need by placing a sheet of chassis aluminum over the face of the original front panel. This hides the original front panel artwork, covers the holes, and provides a solid mounting for new hardware. Because the front panels on most consumer electronic equipment is made from plastic, the aluminum sheet can be easily mounted with self-tapping screws.) The project box can be much larger than specified; it is strictly a matter of personal preference. The bottom part of it must be metal, however, because it will be used for "heatsinking" purposes in a later construction section.

The power cord must have a ground wire; that is, it must have three prongs (ground, neutral, and hot). Also, be sure to use the proper size *strain relief* for the power cord. Do not use a rubber grommet in place of a strain relief; this could be dangerous in the long run.

Try to find an on-off switch that comes complete with a backplate to indicate its position, or some other means of indicating its status. It's easy to forget which way is *on* and which way is *off* with a bench full of parts and equipment.

This power supply design is a little unusual because it incorporates the use of two 24-volt transformers to form a single 48-volt center-tapped transformer. I decided to design it this way because it might be a little difficult to find a 48-volt ct transformer; however, 24-volt, 2-amp transformers are common.

The specified fuse is the common glass type, as used in a variety of electrical/electronics equipment, and in many automobiles ($1/4$ inch in diameter and $1^1/_4$ inch long). Be sure the fuse holder or fuse block is made for that type of fuse. I would personally recommend fuse blocks because they are easier to mount, but that will depend on your preference, and what you might already have in the junk box.

A *locking terminal solder lug* looks like a small internal tooth lock washer with an arm, or extension, sticking out of it. The extension will have one or two holes in it for wire connections. They can be bolted down to a chassis or PC board in the same manner as any other lock washer, but the extension provides a convenient solderable tab for connection purposes. One of these will be used in this project for connecting the ground lead of the power cord to the chassis.

For this particular project, the component placement, enclosure size, and various component construction parameters (such as physical dimension, mounting holes, color) are not critical to the functional operation of the finished project. This allows you the freedom to be creative and save some money, depending on what you might already have, or what can be obtained through salvaging.

I must assume, at this point in the book, that you are inexperienced at buying electronic parts (if I'm wrong, please forgive me). Chances are, you will purchase some (or all) of this material list from a local electronics supply store. This might be slightly traumatic for some people, so let me give you some encouragement and advice. Don't feel intimidated because you happen to be a novice! You are to be commended because you are learning and accomplishing something that many people are afraid to try. The personnel who work in an electronics supply store should be in total agreement with me on that point. Never be reserved about asking any questions you might have about purchasing the right materials and components for your projects. If any sales person ever makes you feel "low" because of your inexperience, I hope you bring this book into the store, and let that person read this paragraph. Then walk out, and take your hard-earned money to a different store that employs sales personnel with the right attitude.

Mounting the Hardware

Read this entire paragraph before physically mounting any components. Lay out the components in the enclosure for mounting. Figure 3-10 illustrates a top view of the approximate way the components should be mounted to the bottom of the enclosure; it is assumed that the top cover has been removed. The transformers should be placed side by side in either rear corner. Most transformers of this size are constructed with the primary connections on one side, and the secondary connections on the other. Be sure that the primary connection sides are facing the rear panel. The transformers should be placed far enough away from the rear panel to allow sufficient room for the fuse block, or the fuse holder, to be mounted. If you are using a fuse holder instead of a block, pay careful attention to how far it extends "into" the enclosure after mounting; this distance tends to be more than you would guess at first glance. If you are using a fuse block, it should be mounted close to the primary of each transformer. The power cord and strain relief assembly should come into a convenient location close to the fuse block or fuse holder.

The on-off switch is mounted to the front panel. Note how it should be on the same side (in the front) as are the transformers in the back. In regard to height, the power cord and strain-relief assembly, the fuse holder (if used), and the on-off switch should be placed about half way up the height of the enclosure. Double-check all of the mounting considerations in this paragraph; and if everything looks good, go ahead and mount the components in the enclosure, using whatever hardware is applicable (nuts, bolts, screws, washers, etc.).

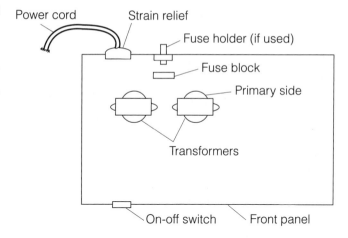

Figure 3-10
Approximate compo-
nent layout for first
section of power sup-
ply project.

Wiring and Testing Procedure

Do not attempt to apply power to the power supply until all of the tests are performed and understood. *Do not attempt to perform any tests using your DVM until you have read and understood the owner's manual for your particular DVM.* Even though a general procedure for using a DVM will be provided throughout this section, this is not a substitute for thoroughly understanding the particularities of your personal instrument.

Refer to Fig. 3-11 during the course of this wiring procedure. Look at the male outlet plug on the power cord. It should have three prongs; two will be flat "blade"—type prongs, and the third will be round. The round prong is the *earth ground* connection. The green wire on the other end of the power cord attaches to this round prong. Verify this using the DVM in the "ohms" position. Touch one DVM test probe to the round prong and the other to the green wire; you should read very close to zero ohms. Using a DVM in this way is called testing for *continuity.* In other words, you have proved that there is a continuous electrical path from the round prong, on the plug end of the power cord, to the green wire on the other end. If, by chance, you have a power cord without color-coded wires, test all three wires on the chassis end of the power cord to determine the one connected to the round prong. Solder the wire (connected to the round prong) to a locking terminal solder lug, run the cable through the strain-relief hole, and bolt the solder lug to the metal chassis bottom, underneath the power cord strain relief.

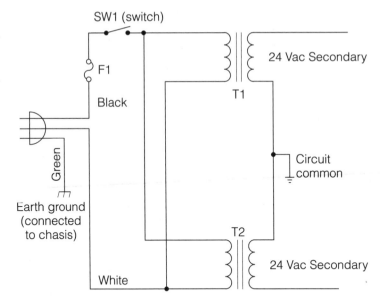

Figure 3-11
Schematic diagram of
first section of lab
power supply.

NOTES: 1. F1 = 2 Amp, 250
Volt "Slo-Blo" fuses
2. Do not connect circuit
common to chasis

Obtain some 18- to 20-gauge hookup wire with an insulation rating of at least 200 volts. Cut two lengths of this wire long enough to reach from the on-off switch to about 3 inches beyond the rear of the chassis, so they can be routed along the chassis bottom. Solder these two wires to the on-off switch connections. Connect one of the switch wires to one of the fuse block lugs. In a later step, the other fuse lug will be used to connect the power cord's hot lead.

Connect the other switch wire to one side of the T1 primary lugs, or wires. Some units will have metal lugs; others will have insulated and color-coded wire leads. Connect another piece of hookup wire from this same point (T1) to the same side of the T2 primary. Solder both of these connections. Now take a few moments to compare your wiring progress so far, with the schematic diagram illustrated in Fig. 3-11.

If you must connect two or more wires together to make these connections, be sure that the connection is well insulated with either insulated mechanical connectors, or soldered and covered with heat-shrink tubing. You might use an insulated "butt splice" connector, or an insulated "wire nut"—type connector. When soldering two wires together and using a length of heat-shrink tubing to insulate the connection, remember to install the tubing onto one of the wires before you twist and solder the ends. The tubing must be just large enough to fit over the twisted and soldered splice. Heat the tubing with a hair dryer, or a heat gun, until it shrinks tightly around the splice; tight enough to remain in position permanently. I do not recommend using electrical tape to insulate this type of connection.

Refer to Fig. 3-11 and note how one side of the F1 fuse block (or holder) is connected to SW1. Connect the other side of the F1 fuse block (or holder) to the black-wired hot side of the AC power cable. This black wire should show continuity to the smaller one of the two flat blades at the other end of the power cable.

You should have one white (neutral) power cord wire left. Connect it to the unused primary lug (or wire) of T2. Also, connect a piece of hookup wire from this connection to the unused side of T1. Double-check all wiring with the schematic diagram in Fig. 3-11, and solder all connections.

Note that the primaries of T1 and T2 are wired in parallel (Fig. 3-11). The bottom of T1's primary is connected to the bottom of T2's primary. And, the top of T1's primary is connected to the top of T2's primary. The secondaries of T1 and T2 are to be wired in series. The bottom of T1's secondary is connected to the top of T2's secondary. You must relate this to the way your transformers are constructed.

Let me attempt to walk you through this. Look at your transformer's secondary connections. If there are three connections for each secondary, it means they are *center-tapped* secondaries. Disregard the center-tap connections; they will not be used for this project. That leaves you with two connections for each secondary. Most transformers are constructed so that one connection will be on the left side of the transformer, and one on the right side. Simply connect the left side of one secondary to the right side of the other, and they should be in series, as shown in Fig. 3-11. Make this wiring connection to the best of your ability. If it's wrong, it won't hurt anything; and you can correct it in the following section.

If your transformers have leads, these wires will be coded. The two (usually) black wires are for the primaries. The secondary leads will normally exit the unit on the opposite side from the primary leads. These leads might be any solid color (red, green, yellow, etc.). If three leads are there, the multicolored (red/yellow; red wire with yellow stripes) lead is the center tap. Connect the "lefthand" red lead of one unit, to the "righthand" red lead of the other unit. This will *stack* them in series, and with the correct *phase*. Insulate each of the center-tap leads from each other, and from other circuit components, with wire nuts, or with shrink tubing.

Testing the First Section of the Lab Power Supply

There are two purposes for testing your work thus far. It is important, in regard to function and safety, to be sure that the construction you have completed to this point is correct. But it is just as important, in regard to theory and practicality, to understand the testing procedure. If you become confused, go back and review the area of your confusion. It will be time well spent.

Do not apply power to the power supply yet. If you have installed fuse F1 into the fuse block (or holder), remove it. Place the on/off switch in the "off" position. Adjust your DVM to read resistance ("ohms" position), and set it for the lowest resistance range available. Measure the resistance across the two blade prongs of the power cord. Take the same type of measurement from each blade to the round earth ground prong. All three of these readings should be "infinite" (meaning an open circuit or no continuity). Take the same measurement from the round earth ground prong to any point on the metal chassis. This reading should be very close to zero. This is because you connected the earth ground wire directly to the chassis through the locking terminal solder lug.

Now place SW1 in the "on" position. Again, take a resistance reading between the two blade prongs on the power cord. Again, this reading should be infinite because the fuse is not installed. Install F1 into its fuse block (or holder). Again, measure the resistance between the two blade prongs. This time, you should get a very low resistance reading (about 4 to 7 ohms). Referring back to Fig. 3-11, with SW1 closed and F1 installed, you are actually reading the parallel DC resistance of the T1/T2 primaries.

The resistance of each of these primaries is actually twice what this reading has indicated. To illustrate this fact, take two equal value resistors; 100 ohms each, for instance. Measure each resistor; they should each register approximately 100 ohms. Now, twist the ends of the resistors together, so that they are connected in parallel. The measured resistance should now be about 50 ohms. The paralleled resistors represent the paralleled primaries of the power transformers. Your blade-to-blade measurement across the two primaries, just as across the two resistors, shows a value of one-half of the actual primary resistance, and it demonstrates continuity. Your reading is the equivalent resistance (R_{equiv}) of the primaries.

Now, turning your attention to the transformer secondaries, measure the resistance between T1's secondary connections. This should be a very low reading, usually less than 1 ohm. Record this value. Similarly, measure T2's secondary resistance. Record this value. Now, measure the resistance from the unconnected side of the T1 secondary to the unconnected side of the T2 secondary. Because the two secondaries are in series, the last measurement should be the sum of the first two secondary measurements. As discussed in Chapter 2, to calculate the total series resistance, you simply add the individual resistance values.

As a final safety measure, take a resistance measurement between either secondary connection and the primary connection of T1/T2 at the power switch. This reading should be infinite. This proves that there is isolation (no direct wiring connection) between the primaries and the secondaries of T1 and T2. Do the same test between the secondary leads and the chassis ground, and verify that you achieve the same results. This verifies that the secondaries are not shorted (short-circuited) to ground. Consider the following safety tips before proceeding.

CAUTION   *Serious injuries, resulting from electrical shock, can occur in several ways.*

One form of injury is through the indirect effect of electrocution. On receipt of a high-voltage shock, the human body's automatic response is to involuntarily contract the muscles in and around the affected tissue. In other words, you jerk away from it really quick! If, for example, you receive a nasty shock to your arm, and your arm happens to be close to some sharp metal edges, you are likely to cut a major size gash in your arm as it jerks back against the sharp metal. This is an *indirect injury.* The arm wasn't actually cut by the electricity, but the effect of the electrical shock caused the injury. I learned a long time ago that you don't bend over and curiously watch someone working on a TV from behind. You could get your jaw broken!

Serious injuries resulting directly from electrical shock can be in the form of *burns.* This occurs more often in very high-voltage environments, such as those encountered by power company employees and industrial electricians.

But the most dangerous form of injury occurs when an accident or circumstance arises causing electrical current to pass through a person's vital organs; this is *electrocution*. This kills! (Contrary to what a lot of people think, voltage does not kill; current is what you need to fear.) I recommend the practice of the "*one hand*" *technique*. The principle is very simple. Electrical current needs a closed circuit to flow through. As long as a person is working close to a high-voltage source "with only one hand" (assuming that no other body extremity is grounded), it is impossible for a current to flow through that person's vital organs. A small capacitive current might "bite" your hand, but you'll live to talk about it. Therefore, when working with high-voltage sources (120 volts AC is definitely considered high voltage), get into a habit of putting one hand in your pocket, or laying it on your lap. Here are a few more safety tips to remember:

- Don't "snake" your hands and arms into tight-fitting places where electrocution could occur; there are tools available to perform those types of jobs.
- When working on equipment with power applied, always move slowly and deliberately; pay attention to what you are doing. Don't try to catch any falling objects.
- Never do any kind of electrical work while sitting on a metal seat of any kind, standing on metal grating or on a wet floor, or leaning or holding to any conductive or wet object.
- Don't stand too close to anyone working on high-voltage equipment.
- Read and follow the safety guidelines in the owner's manuals for all electronic test equipment you use.
- Perform "undervoltage" tests only when accompanied by an assistant who knows the location of the power shutoff switch, and who knows how to use it!

There are, of course, more safety considerations to follow than the previously listed ones, depending on the work environment. Check them out according to your personal situation. These considerations involve tool-related injuries, wire punctures, accidents that damage equipment, and so on. Just use common sense and forethought. Your brain and experience are two of the most important safety tools that you own. Use them well and often.

If you have understood everything thus far, all of the resistance measurements have been correct, and you feel lucky (only joking), it's time to make the final tests with the power applied. For safety reasons, please *don't jump ahead* in the following steps. And, if a fault is detected, *don't proceed until it is corrected.* Before applying power, be sure F1 is installed, and that SW1 is in the "off" position. If your transformers have leads, rather than lugs, place a piece of cardboard on the floor of the chassis, in such a manner that: It will *keep the free ends of the secondary leads from shorting to the chassis.* Tape the leads to the cardboard so that they cannot short to each other. As discussed in Chapter 1, you should have an outlet strip with a 15-amp circuit breaker and on/off switch attached to the work bench. Place the outlet strip switch in the "off" position. Plug the power supply circuit into the outlet strip.

Using only one hand, turn on the outlet strip. Using the same hand, place SW1 in the "on" position. Place SW1 back in the "off" position and observe the circuit. The fuse should not have blown, and there shouldn't have been any visual or audible indication of a fault. Place the outlet strip switch back in the "off" position.

Unplug the circuit from the outlet strip. Even though the outlet strip is turned off, unplugging a piece of electrical equipment, before working on it, is a good safety habit. Temporarily attach one end of an insulated alligator clip lead to the connection you made between the two secondaries of the transformers. This point is referred to as the circuit common in Fig. 3-11. Attach the other end of the clip lead to one test lead of your DVM. Adjust the DVM to read "AC volts," and set it to the 200-volt range, or higher. (If you have an autoranging DVM, it will automatically set its range on taking the voltage measurement.) Be sure that there is no possibility of an accidental short occurring by verifying the temporary connections are secure, and not extremely close to any other conductive points. Plug the circuit back into the outlet strip, and place the outlet strip switch to the ON position. Place SW1 in the "on" position.

Using the DVM test lead not attached to the circuit common, touch it to the "free" T1 secondary lead. The DVM should read about 24 volts AC (it is typical for this reading to be a little higher; possibly 26 to 29 volts AC). This is the T1 secondary voltage. Now, touch the DVM test lead to the T2 secondary lead. This is the T2 secondary voltage. T2's secondary voltage should be very close in value to T1's secondary voltage. Turn off the circuit and the outlet strip, and unplug the circuit.

Leave one end of the alligator clip lead attached to the DVM test probe, and attach the other end of the clip lead to the T1 secondary lead not connected to the circuit common. Plug the circuit back into the outlet strip; turn the outlet strip on; and place SW1 back in the "on" position. Using the DVM test lead that is not attached to the clip lead, touch it to the "free" T2 secondary lead that is not connected to the circuit common. This voltage reading should be the sum of the two secondary voltages, or about 50 volts.

If this last voltage measurement was close to zero, it means you have connected the two transformers *out of phase* with each other. In other words, either the primaries, or the secondaries, are connected in such a way that the AC voltage output of one secondary is upside down relative to the other. Another way of putting it is to say that the two secondaries are *bucking* each other, and that they are 180 degrees out of phase with each other. The easiest way to understand this phenomenon is to look at an instantaneous point of time in the transformer's operation.

Referring to Fig. 3-12, note the transformer secondaries shown in the left side of the illustration. Although you do not normally associate any *polarity* with AC voltages, assume this drawing to represent an instant of time when both secondaries are at a peak voltage output. The polarity of the secondary voltages shown would add to each other—in the same way flashlight batteries add to each other in a flashlight. Positive to negative, positive to negative, and so on.

However, if you accidentally wired the transformer secondaries as shown on the right side of Fig. 3-12, the two currents would cancel each other out, and no voltage would be regenerated. Note how the positive is connected to positive. The ability of a transformer to turn an AC voltage upside down (called *inversion*) is beneficial in many applications; but in our present application, it would render the power supply inoperative. If you measured approximately zero volts across both secondaries in your last reading, the correction is easy. Be certain that all applied power is off, and that the circuit is unplugged. Then, simply reverse the wiring connection on *only one* transformer secondary. Finally, repeat the last voltage check to verify that the 50 volts AC is present.

Figure 3-12
Possible phase relationship of two transformer secondaries.

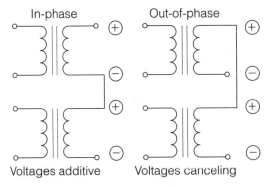

Turn off all applied power, and unplug the circuit. Going back to the resistance measurements performed earlier, the actual DC resistance of the transformer primaries was a very low value. For ease of calculation, assume it to have been about 12 ohms. When power was applied to the circuit, about 120 volts AC was applied to both primaries. If you used Ohm's law to calculate the primary current flow, the result would be

$$I_{primary} = \frac{E_{primary}}{R_{primary}} = 120 \text{ volts AC/12 ohms} = 10 \text{ amps??}$$

Obviously, this is not correct. F1 is only a 1-amp fuse, and it would have blown instantly, if the primary current flow had been that high.

Transformers are inductors. As discussed previously, inductors try to maintain a constant current flow, by storing and releasing energy from their associated electromagnetic field. Because the 120 volts applied to the primaries is an AC voltage, the storing and releasing of energy from their electromagnetic field are a constant and continuous process. This results in the generation of a cemf (counterelectromotive force) that opposes the applied voltage and reduces the primary current flow. The opposition an inductor poses to an AC current flow is called *inductive reactance*. Inductive reactance will be covered further in Chapter 15. The sum of this reactance and the DC resistance combine to limit the primary current to a much lower level.

Just as a point of interest, I measured the AC current flow through the primary of T1 in this circuit. It was about 37 milliamps. The AC current flow through the primary of T2 should be the same. The total current flowing through the fuse (F1) would then be about 74 milliamps, or 0.074 amps.

4

Rectification

Earlier in this book you learned the basics of electron flow. You discovered that electrical current is actually a flow of negative charge carriers called *electrons*. Electrons orbit around the nucleus of an atom, just as the earth orbits around the sun. Electrons are held in their orbital paths by their attraction to the positive nucleus, which contains the positive charge carriers called *protons*. Electrons are attracted to the positive nucleus because of a basic law of physics which states "unlike charges attract, and like charges repel." This same principle can be demonstrated by observing the attraction between two permanent magnets. The two north poles of the magnets will repel each other, but a north pole and south pole will attract. In the same way, the unlike charges of the negative electron and the positive proton attract one another.

Introduction to Solid-State Devices

Because one electron (negative-charge carrier) will equalize the effect of one proton (positive-charge carrier), a normal atom will be balanced in reference to the number of electrons and protons it contains. For example, if an atom has 11 electrons, it will also contain 11 protons. The end result is the negative charge of the electrons will be canceled out by the positive charge of the protons, and the atom will not present any external charge.

The orbital paths of the electrons around the nucleus follow a definite pattern of circular *shells* or *rings.* The maximum number of electrons in each shell is defined by the formula

$$2(n^2) \quad \text{where } n = \text{shell number}$$

For example, if you wanted to calculate the maximum number of electrons for the first shell, n would be 1:

$$2(1 \times 1) = 2$$

The maximum number of electrons in the first shell of any atom is 2, regardless of the total number of electrons in that particular atom. Similarly, the maximum number of electrons in the second shell is

$$2(2 \times 2) = 8$$

The maximum number of electrons in the second shell of any atom is 8. This goes on and on, with the maximum number of electrons in the third, fourth, and fifth shells as 18, 32, and 50, respectively.

In electronics, only the outermost shell of an atom is important because all electron flow occurs with electrons from this shell. The outermost shell of any atom is called the *valence shell.* If the atomic structure of any substance is made up of atoms containing the maximum number of electrons in their valence shells, the substance is said to be an *insulator* because the electrons are rigidly bonded together. If a substance is made up of atoms with valence shells far from being full, the electrons are easily loosened from their orbital bonds, and the substance is said to be a *conductor.* The principle can be compared to buying a new bottle of aspirin. In the beginning, when the aspirin bottle is full, you can shake the bottle, but produce very little movement of the pills because they are tightly packed together. But when you have only a few pills left in the bottle, even the slightest movement will produce a chorus of rattling and rolling pills. Think of this illustration the next time that the electrical-electronics field gives you a headache!

The atoms in some types of crystalline substances (such as silicon) fill their valence shells by overlapping the orbital paths of neighboring atoms. An isolated silicon atom contains four electrons in its valence shell. When silicon atoms combine to form a solid crystal, each atom positions itself between four other silicon atoms in such a way that the valence shells overlap from one atom to another. This causes each individual valence electron to be shared by two atoms. By sharing the electrons from four other atoms, each individual silicon atom appears to have eight electrons in its valence shell. This condition of sharing valence electrons is called *covalent bonding*.

In its pure state, silicon is an insulator because the covalent bonding rigidly holds all of the electrons leaving no (easily loosened) *free electrons* to conduct electrical current. If an impurity is injected into the pure silicon, having a valence shell containing five electrons, it cannot fit into the covalent bonding pattern of the silicon. The result is one free electron per impurity atom that can readily move and conduct current.

Similarly, if an impurity is injected containing only three electrons in its valence shell, the absence of the fourth electron (needed for proper covalent bonding) causes a free positive charge. (A free positive charge is another way of describing a *hole*, or the absence of an electron.)

In both cases, a *semiconductor* has been formed. The term semiconductor simply indicates the substance is neither a good insulator nor a good conductor; it is somewhere in between. (The term "semi-insulator" would be just as accurate as the term *semiconductor*.) The process of injecting an impurity into a substance to form a semiconductor is called *doping*.

In addition to becoming a semiconductor, the impure silicon will also possess a unique property, depending on whether it has been doped with a *pentavalent* impurity (an impurity with five electrons in its valence shell), or with a *trivalent* impurity (an impurity with three electrons in its valence shell). Silicon doped with a pentavalent impurity will become *N-type* material, and it will contain an excess of negative charge carriers (one free electron per impurity atom). Silicon doped with a trivalent impurity will become *P-type* material, and it will contain an excess of positive charge carriers (one "hole" per impurity atom).

Diode Principles

When N-type semiconductor material is sandwiched with P-type material, the resulting component is called a *diode*, which is a two-layer device that has an extremely low resistance to current flow in one direction, and an extremely high resistance to current flow in the other. Because it is a two-layer device, it can also be considered a *single-junction device* because there is only one junction between the P and N material, as shown in Fig. 4-1. A diode is often called a *rectifier.*

Ideally, you can consider a diode as being capable of passing current in only one direction. If the P-side voltage is positive, relative to the N side, by an amount greater than its *forward threshold voltage* (about 0.7 volt if silicon, and 0.3 volt if germanium), the diode will freely pass current almost like a closed switch. This diode is said to be *forward-biased.* If the P side is negative, relative to the N side, virtually no current will be allowed to flow, unless and until the device's breakdown voltage is reached. This condition is referred to as being *reverse-biased.* If the *reverse-breakdown voltage* is exceeded (the point at which reverse-biased current starts to flow) in most normal diodes, the diode may be destroyed.

The P side of a diode is called the *anode.* The N side is called the *cathode* (Fig. 4-2).

The principle behind diode operation is shown in Fig. 4-3. The diagram of a *forward-biased diode* demonstrates the operation of a diode in the *forward conduction mode* (freely passing current). With the polarity, or bias, of the voltages shown, the forward-biased diode will conduct current as if it were a closed switch. As stated previously, like charges repel each other. If a

Figure 4-1
Basic diode
construction.

Figure 4-2
Diode labels and
electrical symbol.

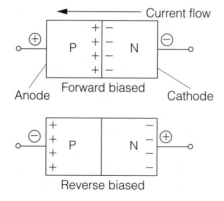

Figure 4-3
Diode operational
principles.

positive voltage is applied to the P material, the free positive-charge carriers will be repelled and move away from the positive potential toward the junction.

Similarly, the negative potential applied to the N material will cause the free negative-charge carriers to move away from the negative potential toward the junction. When the positive- and negative-charge carriers arrive at the junction, they will attract (unlike charges attract) and combine. As the positive- and negative-charge carriers combine at the junction, a new positive- and negative-charge carrier will be introduced to the semiconductor material from the source voltage providing the bias. As these new charge carriers enter the semiconductor material, they will move toward the junction and combine. Thus, current flow is established, and will continue for as long as the bias voltage remains above the forward-bias threshold.

The *forward-threshold voltage* must be exceeded before a forward-biased diode will conduct. The forward-threshold voltage must be high enough to loosen the charge carriers from their atomic orbit and push them through the junction barrier. With *silicon diodes,* this forward-threshold voltage is approximately 0.7 volt. With *germanium diodes,* the forward-threshold voltage is approximately 0.3 to 0.4 volt.

The *maximum forward current* rating of a diode is based on its physical size and construction. Diode manufacturers will typically specify this rating in two ways: the "maximum continuous (or average) forward current" and the "peak forward surge current." The *maximum continuous forward current* is precisely what the name implies: the maximum forward current the diode can conduct on a constant basis. The *peak forward surge current* is the maximum forward current a diode can conduct for 8.3 milliseconds. This last specification can be between 5 and 50 times higher than the continuous rating. 8.3 milliseconds is used as a standard reference for this rating, because it is the time period for one half-cycle of 60 hertz AC. Diodes used for power rectification experience a *surge current* on the initial application of power (covered further in Chapter 5).

Figure 4-3 also illustrates a diagram of a *reverse-biased diode*. As might be expected, the opposite effect occurs if the P material is negative-biased, relative to the N material. In this case, the negative potential applied to the P material attracts the positive charge carriers, drawing them away from the junction. Similarly, the positive potential applied to the N material draws the negative charge carriers toward it, and away from the junction. This leaves the junction area depleted; virtually no charge carriers exist there. Therefore, the junction area becomes an insulator, and current flow is inhibited.

The reverse-bias potential might be increased to the reverse-breakdown voltage for which the particular diode is rated. As in the case of the maximum forward current rating, the *reverse-breakdown voltage* is specified by the manufacturer. The reverse-breakdown voltage is much higher than the forward threshold voltage. A typical general-purpose diode might be specified as having a forward-threshold voltage of 0.7 volt, and a reverse-breakdown voltage of 400 volts. Exceeding the reverse-breakdown voltage is destructive to a general-purpose diode. [Some manufacturers refer to the reverse-breakdown voltage as the *peak inverse voltage* (PIV) or as the *peak reverse voltage* (PRV).]

Diodes are commonly used to convert alternating current (AC) to direct current (DC). This process is called *rectification*. A single diode used for rectification is called a *half-wave rectifier*. When four diodes are connected together, and are used to redirect both the positive and negative alternations of AC to DC, the four diode configuration is called a *diode bridge*, or a *bridge rectifier*. These configurations can be demonstrated in a few common types of circuits.

Figure 4-4 shows a simple *half-wave rectifier circuit*. Common household power (120 volts AC rms) is applied to the primary of a step-down transformer (T1). The secondary of T1 steps down the 120 volts AC rms to 12 volts AC rms. The diode (D1) will only allow current to flow in the direction shown (from cathode to anode). Diode D1 will be forward-biased during each positive half-cycle (relative to common). When the circuit current tries to flow in the opposite direction, the diode will be reverse-biased (positive on the cathode, negative on the anode), causing the diode to act like an open switch. As shown in Fig. 4-5, this results in a pulsating DC voltage applied across the load resistor (R_{load}). Because common household power cycles at a 60-hertz frequency, the pulses seen

Figure 4-4
Half-wave
rectifier circuit.

Figure 4-4
Half-wave
rectifier circuit.

Figure 4-5
Waveshapes across
D1 and R_L from
circuit illustrated
in Fig. 4-4.

across R_{load} will also be at 60 hertz. Figure 4-5 also shows the voltage waveform across D1.

During the positive half-cycle, D1 will drop the 0.7 volt forward-threshold voltage. R_{load} will drop the majority of the voltage, about 11.3 volts in this case. The term *drop* refers to the fact that a certain voltage difference appears across an electrical device, or component, as a current flows through it. In this example, 0.7 + 11.3 = 12 volts DC total. The entire negative half-cycle will be dropped across D1 while it is reverse-biased. The negative half-cycle is dropped across D1 because it looks like an open switch when reverse-biased. An open switch is an infinitely high resistance. Note that D1 is in series with R_{load}. As discussed previously, in a series circuit, the higher the resistance value of a component, the more of the source voltage it will drop. Because D1 looks like an infinitely high resistance when reverse-biased, it will drop the total source voltage (output of T1's secondary) during the negative half-cycle.

Consider the amplitude of the voltage developed across R_{load}. As shown in Fig. 4-4, the secondary of T1 is 12 volts AC rms (AC voltages are always assumed to be V rms values unless otherwise stated). Therefore, the peak voltage output from the T1 secondary is

$$Peak = (rms\ value)1.414 = (12\ volts\ AC)1.414 = 16.968\ volts\ peak$$

For discussion purposes, the 16.968 volts can be rounded off to 17 volts. The previous calculation tells us that for each full cycle, the T1 secondary will output one positive 17-volt peak half-cycle, and one negative 17 volt peak half-cycle. The negative half-cycles are blocked by D1, allowing R_{load} to receive only the positive half-cycles. The actual peak voltage across R_{load} will be the 17-volt positive peak, minus the 0.7-volt forward threshold voltage being dropped by D1. In other words, 16.3-volt positive peaks will be applied to R_{load}, as shown in Fig. 4-5.

The diode circuit illustrated in Fig. 4-4 is called a *half-wave rectifier*, because only one-half of the full AC cycle is actually applied to the load. However, it would be much more desirable to utilize the full AC cycle. The circuits shown in Fig. 4-6 are designed to accomplish this, and they are called *full-wave rectifiers*.

The *full-wave bridge rectifier* illustrated in Fig. 4-6 is the more common type of AC rectifier. It consists of four diodes. The operation of a full-wave bridge rectifier can be examined by referring to Fig. 4-7. When the transformer secondary outputs a half-cycle with the polarity shown in

Figure 4-6
Two types of full-wave rectifiers.

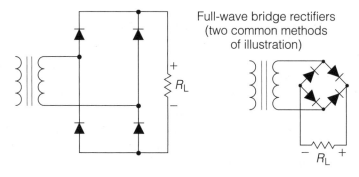

Full-wave bridge rectifiers
(two common methods
of illustration)

Full-wave rectifier
utilizing a center-tapped
transformer

■■ ■■ ■■ ■■
Figure 4-7
Current flow through
a full-wave bridge
rectifier.

*Darkened
diodes
are forward
biased

the top illustration, the current will follow the path indicated by the arrows. (Remember, forward current always flows through a diode from cathode to anode.) With this polarity applied to the bridge, diodes D1 and D4 are forward-biased, while D2 and D3 are reverse-biased.

The lower illustration of Fig. 4-7 shows the current path when the polarity on the transformer secondary reverses. With this polarity applied to the bridge, diodes D2 and D3 are forward-biased, while D1 and D4 are reverse-biased. In either case, the current always flows through the load resistor, R_{load}, in the same direction; meaning that the voltage polarity across R_{load} does not change. Essentially, the switching action of the diode bridge actually turns the negative half-cycle upside down.

Referring back to the circuit of Fig. 4-4, and to its associated waveforms shown in Fig. 4-5, note that R_{load} receives the positive half-cycles only of the AC being output by T1's secondary. These positive half-cycles occur at a 60-hertz rate because the full AC cycle is at a 60-hertz rate. Compare this to the circuit of Fig. 4-7, and to its associated waveshapes shown in

Fig. 4-9. Note how R_{load} will now receive the negative half-cycle turned upside down (converted to another positive half-cycle) as well as the regular positive pulse. In a sinusoidal waveform, the negative half-cycle is the exact inversion of the positive half-cycle; thus, the waveform across R_{load} will begin to repeat itself at the end of the first half-cycle. In simple terms, the frequency will double. The frequency of the transformer secondary is 60 hertz AC, but the frequency at the output of the diode bridge is 120-hertz *pulsating* DC.

You might be confused about the difference between AC and pulsating DC. As stated in the previous chapter, AC is characterized by a voltage polarity and current flow reversal. With pulsating DC, there is a large AC component, usually called *ripple*, but the current flow never changes direction through the load. Obviously, this means that the voltage polarity never changes either. In Fig. 4-9, note that the pulsating DC never crosses the *zero reference line* into the negative region.

The functional diagram of a *full-wave rectifier*, utilizing a transformer secondary with a center tap, is illustrated in Fig. 4-8. This functions just as well as the bridge rectifier discussed previously, and all of the wave-shapes shown in Fig. 4-9 are applicable to this circuit as well. The secondary center tap becomes the circuit common (or circuit reference). As shown in Fig. 4-10, the two outputs from each side of the secondary are exactly opposite to each other, in reference to the center tap (180 degrees out of phase). If one output is in the positive half-cycle, the other output must be in the negative half-cycle and vice versa. Thus, during each half-cycle (of the applied AC to the primary), one of the two secondary outputs will output a positive half-cycle.

The circuit of Fig. 4-8 is actually two half-wave rectifiers connected to the same load. The two outputs from the transformer secondary are *inverted* (opposite) from each other. Therefore, if D1 is forward-biased, D2 will be reverse-biased. When D1 becomes reverse-biased, D2 becomes forward-biased. The end result is that the inverted action of the two half-wave rectifiers are combined to form one full-wave rectifier. The effect is the same as that achieved with the circuit of Fig. 4-7. However, to obtain the same DC voltage amplitude to the load, the transformer's rated secondary voltage, in Fig. 4-8, must be twice the value of the transformer shown in Fig. 4-7. This is because the center-tap will divide the voltage in half when it is used as the common, or reference. For example, a 24-volt, center-tapped transformer secondary will measure 12 volts rms from the center tap to either side of the secondary.

A third common full-wave rectification circuit is illustrated in Fig. 4-11. This circuit is actually the same center-tapped full-wave rectifier

Figure 4-8
Current flow
through a
center-tapped
transformer
full-wave rectifier.

*Darkened diodes
are forward biased

Figure 4-9
Applied 60-hertz
AC waveshape
compared to the
120-hertz pulsating
DC output of a
full-wave rectifier.

circuit as shown in Fig. 4-8, but with two additional diodes incorporated. Referring to Fig. 4-8, the top diagram illustrates how the upper
half of the secondary output is applied to R_{load}, while the lower half
of the secondary output is blocked by D2. The reverse occurs in the
lower diagram. In both cases, half of the secondary output is
not used.

Figure 4-10
Two leads of a center-tapped transformer 180 degrees out of phase when ct is used as the reference.

180° Out of phase

Center-tapped reference

Figure 4-11
Dual-voltage rectification circuit.

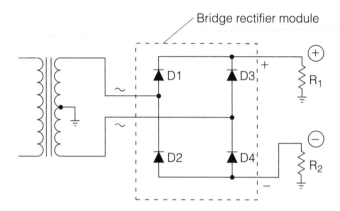

Bridge rectifier module

The circuit in Fig. 4-12 utilizes this unused portion of the secondary output to form another DC output of the opposite polarity. This can be more easily understood by referring back to Figs. 4-11 and 4-8. Figure 4-12 is exactly the same circuit (with the same component labeling) as illustrated in Fig. 4-11. Compare Fig. 4-12 with Fig. 4-8. Notice how the positive full-wave rectification section of Fig. 4-12 is identical to the circuit illustrated in Fig. 4-8. The circuit of Fig. 4-12 incorporates two additional diodes (D4 and D2) to form another full-wave output from the unused half-cycles. Because of the orientation of the diodes, it will be negative in respect to the circuit common. *Dual-polarity power supplies* are very common because of their extensive use in operational amplifier and audio circuits; both of which will be discussed later in this book. As in the case of the center-tapped full-wave rectifier circuit of Fig. 4-8, the voltages across R1 and R2 will be only half of the amplitude of the secondary rating.

Referencing

By now, you might be beginning to appreciate the importance of wave-shapes in the electrical and electronics fields. All of the waveshapes

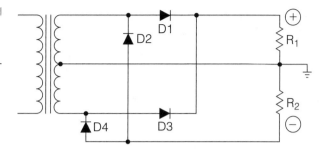

Figure 4-12
Redrawn schematic
of Fig. 4-11.

shown in the previous chapters are shown just as they would appear if they were viewed with an oscilloscope. When discussing voltage wave-shapes, they are always said to exist with respect to some point of "reference." As pointed out in Chapter 2, voltage readings are always taken in respect to some *common point of reference.*

To help in understanding this, consider the following analogy. If an airplane is said to be flying at an altitude of 10,000 feet, it is always assumed that the point of reference is sea level. If the point of reference was changed to a 5000-foot mountaintop, the plane would then be said to be flying at a 5000-foot altitude. If the plane were to be flying close to the ground, it could be said that it was flying at nearly a "negative" 5000-foot altitude.

Referring back to Fig. 4-10, the center tap essentially splits the transformer secondary into two halves when it is used as the common reference. In the same way that you could make the airplane fly at a negative altitude by moving the reference point up, you can provide an inverted output by moving the circuit reference up to the halfway point of the secondary output.

Unless otherwise stipulated, voltage amplitudes and waveshapes will always be in reference to the circuit common (usually the chassis "ground") on all schematics and electrical drawings.

Assembly and Testing of Second Section of a Lab Power Supply

The following materials are needed:

Quantity	Description
1	6-amp, 200-volt PIV bridge rectifier module
2	10-kohm, ½-watt resistors

The type of bridge rectifier module specified is commonly available. Although the case styles vary somewhat with different manufacturers, it should be a square or rectangular block, about an inch on each side, with a mounting hole in the middle. (There are some rectifiers with these ratings, meant for PC board installation; they do not have a hole in the middle, and they will be harder to mount.) The two 10-kohm resistors will be used for testing purposes only. Their tolerance rating, and other parameters, are not important.

Testing Bridge Rectifier Modules

Before mounting the bridge rectifier module, it is good practice to test its functional operation. This will help you understand it and provide you with more experience at using your DVM.

A DVM measures resistance by applying a low voltage to the unknown resistance value, measuring the current through the unknown resistance, and converting the current reading to a resistance reading. The older forms of DVMs, usually called *vacuum-tube voltmeters* (VTVMs), would apply about 1.5 volts to the unknown resistance value for measurement purposes. When checking resistance values on solid-state equipment, it is often undesirable to use 1.5 volts as a measurement standard, because this is above the forward threshold voltage of junction devices (diodes, transistors, etc.), causing them to "turn on," and to interact with the component being measured. Interaction of this sort will cause errors in resistance measurements. For this reason, most modern DVMs use only a few tenths of a volt as the measurement standard. When the need arises to check a semiconductor junction, the DVM must be set to the "diode test" position. In this position, the DVM will apply about 1.5 volts to the semiconductor junction under test. The black DVM test lead is of negative polarity, and the red lead is positive. (Be sure you have the DVM test leads plugged into the correct holes on the instrument; black to common.)

When testing a diode, touch the black DVM test lead to the cathode (the "banded" side of the diode body), and the red lead to the anode. This forward-biases the diode, and the resistance reading should be low [the actual resistance reading is irrelevant; when testing semiconductor junctions, you are concerned only with "high (or infinite)" versus "low" readings]. By reversing the DVM leads (red on cathode; black on anode), the diode is reverse-biased, and the resistance reading should be infinitely high.

Be aware that some *volt-ohm-milliammeters* (VOMs) are reverse-polarized for their resistance checks; the positive DC potential being on the black probe. In this instance, your readings will always be reversed. Test about a dozen common diodes (1N4001 or 1N4148/1N914). If they all test "good" only when the probes are reversed, then you have positive volts on your black probe.

The bridge rectifier module you obtained for this project actually contains four diodes connected in a bridge configuration, as shown in Fig. 4-11. Four individual diodes would function just as well. The four internal-module diodes can be tested with a DVM just as though they were four standard diodes.

The bridge rectifier module should have four connection terminals, or leads, extending from it. Each terminal, or lead, should have an identification symbol associated with it, on the case of the module. Two of the terminals will be marked with an "AC" or the symbol for a sine wave; these are the AC input terminals from the transformer. The other two terminals will be specified with polarity symbols ("−" or "+"); these are the DC output terminals. Figure 4-11 illustrates the actual external and internal connections associated with the bridge.

Referring to Fig. 4-11, if you wanted to test the internal diodes labeled D1 and D3, you would place the black (negative) DVM lead on the + terminal of the diode module, and the red (positive) DVM lead to either one of the AC terminals of the module. A low resistance reading would indicate either D1 or D3, respectively, was functioning correctly. You actually wouldn't know which one of the diodes you were testing because the module will not specify any difference in the AC terminals. The issue is totally unimportant; if either one of the diodes is defective, the whole bridge module must be replaced.

Rather than detailing each individual diode measurement, Table 4-1 will make it easy for you to check any bridge rectifier module. Use this table, in conjunction with Fig. 4-11, to understand how the measurements will check each internal diode.

Assuming that there were no problems with the bridge rectifier module, mount it close to the T1 and T2 secondaries, with the appropriate hardware. Wire and solder the connections from the T1 and T2 secondaries to the AC input terminals on the bridge module as illustrated in Fig. 4-13. Using an alligator clip lead, temporarily connect one side of test resistor R_{T1} to the positive terminal of the bridge rectifier. Connect the other side of R_{T1} to the circuit common connection between the two transformer secondaries with another clip lead. Using two more clip leads, connect one side of test resistor R_{T2} to the negative terminal of the

TABLE 4-1

Testing Bridge
Diodes

Bridge Terminals (or leads)				
+	–	AC	AC	**Results**
Black		Red		Low resistance
Black			Red	Low resistance
Red		Black		Infinite resistance
Red			Black	Infinite resistance
	Red	Black		Low resistance
	Red		Black	Low resistance
	Black	Red		Infinite resistance
	Black		Red	Infinite resistance

Figure 4-13
Schematic diagram of
the first and second
sections of the lab
power supply.

bridge, and the other side to the same circuit common point as R_{T1}. Verify that your wiring is the same as is shown in Fig. 4-13. *Be sure that all four clip leads are secure, and not touching anything except the desired connection point.*

Turn off your lab outlet strip, and set SW1 to the "off" position. Plug the circuit into the outlet strip, and (using only one hand) turn on the outlet strip. Again, using only one hand, set SW1 to the "on" position. If F1 blows, turn off all power immediately; unplug the circuit; and double-check all wiring. If the wiring is good, the fault must be in the bridge rectifier (assuming that you followed the test procedures for the first section of the power supply in the previous chapter).

Hopefully, F1 didn't blow, and you can leave the power applied to the circuit and continue with the test. Set your DVM to indicate "DC volts" on at least the 100-volt range. Use one more clip lead (and only one hand!) to connect the black DVM test lead to circuit common. Touch the red DVM test lead to the side of R_{T1} that is connected to the positive terminal of the bridge rectifier. You should read about 24 volts DC. Touch the red lead to the side of R_{T2} that is connected to the negative terminal of the bridge rectifier. You should read about -24 volts DC. Leave the black DVM test lead connected to the circuit common, and set the DVM to indicate "AC volts" on the same range. Again, measure the voltage across R_{T1}, and then R_{T2}. Both readings should be about 12 volts AC. Turn SW1 off, turn the outlet strip off, and unplug the circuit.

Referring back to Fig. 4-9, the DC voltage across R_{T1} and R_{T2} was pulsating DC as shown in the bottom waveform illustration. The DVM did not measure the peaks of the DC pulses; it read the effective DC level. You might think of it as "cutting off the peaks, and using that energy to fill in the valleys." The AC voltage across the resistors is called the *AC component*, or the *ripple*. The DVM measured the rms value of this AC component.

If you own an oscilloscope, read the owner's manual, and be sure you understand how to operate it correctly and safely. Then, use it, in conjunction with this power supply circuit, to observe these waveforms. They should appear just as shown in the various illustrations. *Do not try to view the waveforms of the transformer primaries without using an isolation transformer.*

Resistors R_{T1} and R_{T2} were used for testing purposes only. They can now be removed from the circuit, together with the alligator clip leads used to temporarily connect them.

Capacitance

A capacitor is one of the most common components used in electronics, but is probably one of the least understood. As in the case of inductors, a capacitor is a storage device. An inductor stores electrical energy in the form of an electromagnetic field, which collapses or expands to try to maintain a constant current flow through the coil. In comparison, a *capacitor* stores an electrostatic charge, which increases or decreases to try to maintain a constant voltage across the capacitor.

Capacitor Types and Construction

A capacitor consists of two conductive plates with an insulator placed between them called the *dielectric*. The size of the plates, the thickness of the dielectric, and its dielectric constant all combine to determine the *capacity*, or energy storage capability, of the capacitor. The capacity can be increased through the use of larger conductive plates, thinner dielectric material, or a dielectric material with a higher dielectric constant. Thinner dielectric yields higher capacity, but it also lowers the maximum voltage rating. Because it is impractical to manufacture (or to try to use) extremely large metal plates, capacitors are usually manufactured by rolling the foil plates (with the plastic dielectric sheets interleaved between them) into a round tubular form. Alternatively, layers of foil plates can be stacked and sandwiched, as when shuffling a deck of cards, with the dielectric sheets (usually mica, ceramic, or plastic) interleaved between them.

Basically, all capacitors can be divided into two categories: polarized and nonpolarized. If a capacitor is polarized (a step in the manufacturing process), the correct voltage polarity must be maintained when using it. A *polarized capacitor* will indicate which lead is to be connected to a positive polarity, and/or which lead is to be connected to a negative polarity, by means of labels or symbols imprinted on its body. Accidental reversal of the indicated polarity will destroy a polarized capacitor, and *it could cause it to explode. Nonpolarized capacitors* do not require any observance of voltage polarity.

The vast majority of capacitors use conductive metal foil as the plate material. The only exceptions are adjustable capacitors and trimmers. The big difference in capacitor types is based on the dielectric material used. The two important properties of dielectric materials are called *dielectric strength* and *dielectric constant*. The *dielectric strength* defines the insulating quality of the material, and is a key factor in determining the capacitor's voltage rating. *Dielectric constant* will be explained a little later in this chapter.

Paper and mica were the standard dielectric materials for many years. Mica is used in special applications, and paper is still used quite often for general-purpose use. The paper is impregnated with a wax, or a special oil, to reduce air pockets and moisture absorption.

Plastic films of polycarbonate, polystyrene, polypropylene, and polysulfone are used in many of the newer *large-capacity, small-size capacitors*. Each film has its own special characteristic, and is chosen to be used for various applications according to its unique property.

Ceramic is the most versatile of all the dielectric materials because many variations of capacity can be created by altering it. Special capacitors (that increase in value, stay the same, or decrease value with temperature changes) can be made using ceramics. If a ceramic disk capacitor is marked with a letter "P" (positive change), such as "P100," then the value of the capacitor will increase 100 parts per million per degree centigrade increase in temperature. If the capacitor is marked "NPO" (neg/pos/zero) or "COG" (change zero), then the value of capacity will remain relatively constant with an increase or decrease in temperature. If it is marked with an "N" (negative), such as "N1500," it will decrease in capacity as the temperature increases.

The term defining the manner in which a component is affected by changes in temperature is called the *temperature coefficient*. If a component has a *negative temperature coefficient*, its value decreases as the temperature increases, and vice versa. If it has a *positive temperature coefficient*, its value increases as the temperature increases, and vice versa. A capacitor's temperature coefficient is critical for circuits in which minor changes of capacitance can adversely affect the circuit operation. One reason why ceramic capacitors are the most commonly used type is the versatility of their different temperature coefficients. The other main reason for their widespread use is cost; they are very inexpensive to manufacture.

A ceramic capacitor marked "GMV" means that the marked value on the capacitor is the "guaranteed minimum value" of capacitance at room temperature. The actual value of the capacitor can be much higher. This type of capacitor is used for applications in which the actual value of capacitance is not critical.

Aluminum electrolytic capacitors are very popular because they provide a large value of capacitance in a small space. Electrolytic capacitors are polarized and the correct polarity must always be observed when using or replacing these devices. For special-purpose applications (such as crossover networks in audio speaker systems and electric motors), *nonpolarized electrolytics* are available, which will operate in an AC environment.

The aluminum electrolytic capacitor is constructed with pure aluminum foil wound with a paper soaked in a liquid electrolyte. When a voltage is applied during the manufacturing stage, a thin layer of aluminum oxide film forms on the pure aluminum. This oxide film becomes the dielectric because it is a good insulator. As long as the electrolyte remains liquid, the capacitor is good, or it can be "re-formed" by applying a DC voltage to it for a period of time (while observing the correct polarity). If the electrolyte dries out, the leakage increases and the capacitor loses capacity. This undesirable condition is called *dielectric absorption*.

Dielectric absorption can happen to aluminum electrolytics even in storage. Sometimes, if an electrolytic capacitor has been sitting on the shelf (in storage) for a long period of time, it may need to be reformed to build up the oxidation layer. This can be easily accomplished by connecting it to a DC power supply for approximately an hour. Remember to observe the correct polarity, and do not exceed its voltage rating, or the capacitor can explode. (Reversing the polarity on an electrolytic capacitor causes it to look "resistive" and build up internal heat. The heat causes the electrolyte to boil, and the steam builds up pressure. The pressure can cause the capacitor to explode if it is not equipped with a "safety plug.") Few electrolytic capacitor manufacturers guarantee a shelf-life of more than 5 years.

Another type of polarized electrolytic capacitor is called the *tantalum capacitor*. Because they have the physical shape of a water drop, they are commonly referred to as "teardrop" capacitors. Tantalum capacitors have several advantages over aluminum electrolytics; lower leakage, tighter tolerances, smaller size for an equivalent capacity, and a much higher immunity to electrical "noise" (undesirable "stray" electrical interference from a variety of sources). Unfortunately, their maximum capacity values are limited and they are more expensive than comparable aluminum electrolytics. For applications requiring large capacitance values, aluminum electrolytics are still preferred.

Basic Capacitor Principles

When a DC voltage is placed across the two plates of a capacitor, a certain number of electrons are drawn from one plate, and flow into the positive terminal of the source (the positive potential attracts the negative-charge carriers). At the same time, the same number of electrons flow out of the negative terminal of the source and are pushed into the other plate of the capacitor (the negative potential repels the negative charge carriers). This process continues until the capacitor is charged to the same potential as the source. When the full charge is completed, all current flow ceases, because the dielectric (an insulator) will not allow current to flow from one capacitor plate to the other. In this state, the capacitor has produced an *electrostatic field*, that is, an excess of electrons on one plate and an absence of electrons on the other.

Figure 5-1 illustrates a capacitor in series with a resistor and an applied source voltage of 10 volts DC. Assuming that C1 is not initially charged, when the switch (SW1) is first closed, a circuit current will begin to flow. The rate of the current flow (which determines how rapidly the capacitor will charge) will be limited by the series resistance in the circuit, and the difference in potential between the capacitor and the source. As the capacitor begins to charge (build up electrical pressure), the rate of current flow from the source to the capacitor begins to decrease. The graph of Fig. 5-2 illustrates the voltage across the capacitor (C1) relative to time. The capacitor eventually charges to the full source potential of 10 volts. When this happens, all circuit current will cease because there can be no current flow through the dielectric. If SW1 is opened at this time, the capacitor will hold this static charge until given a discharge path of lesser potential. (The word *static* means "stationary"; hence, the term *electrostatic field* means an electrically stationary field.)

As stated previously, with SW1 closed, the capacitor will eventually charge to the source potential. The *capacity,* or the quantity of charge it can hold at this potential, is determined by the physical characteristics of the capacitor.

Here is an analogy to help clarify the previous functional aspects of capacitor theory—electrical references will be to the circuit shown in Fig. 5-1. Imagine that you have a very large tank of water filled to a 10-foot level (this is analogous to the battery at a 10-volt potential). From this large tank, you wish to fill a small tank to the same 10-foot level (the small tank is analogous to the capacitor). You connect a water pipe and valve from the bottom of the large tank to the bottom of the small "empty" tank. When you first open the water valve to allow water to flow from tank to tank, the flow rate will be at its highest because the level differential will be at its greatest (the water flow is analogous to electric current flow). As the water begins to fill up the small tank, it also begins to exert a downward pressure opposing the flow of incoming water. Consequently, the flow rate begins to decrease. As the water level in the small tank gets higher, the flow rate continues to decrease until the small tank is at the same 10-foot water level as the large tank.

Figure 5-1
Basic *RC*
(resistive-capacitive)
circuit.

As soon as the two levels are equal, all water flow from tank to tank will cease. As the old proverb states, "water seeks its own level."

If you were to monitor the level increase in the small tank with respect to time, you would notice that the level does not rise in a linear fashion; that is, it rises at a slower rate as it approaches the 10-foot top. If you charted the rise in "level versus time" in the form of a graph, the curve would be identical to the curve shown in Fig. 5-2.

The small water tank will have a holding capacity associated with it. For instance, it can be specified as capable of holding 40 gallons. Capacitors are also rated with respect to the quantity of charge they can hold.

The *capacity* (quantity of charge) of a capacitor is measured in units called *farads*. A *farad* is the amount of capacitance required to store one coulomb of electrical energy at a 1 volt potential. A *coulomb* is a volume measurement unit of electrical energy (charge). It is analogous to other volume measurement units such as quart, pint, or gallon. A gallon represents 4 quarts. A coulomb represents 6.28×10^{18} electrons (or 6,280,000,000,000,000,000 electrons). The basic unit of current flow, the *ampere*, can be defined in terms of coulombs. If one coulomb passes through a conductor in one second, this is defined as one ampere of current flow. Thus, 1 amp = 1 coulomb/second.

A farad is generally too large of a quantity of energy to be stored by only one capacitor. In the 1930s, a Gernsback magazine calculated the size requirements for building a 1-farad paper-foil capacitor. Completely fill the Empire State Building, from bottom to the top, with a stack of paper and foil layers!

Therefore, the capacity of most capacitors is defined in terms of *microfarads* (1 μF = 0.000,001 farad) or *picofarads* (1 pF = 0.000,000,000,001 farad). As stated previously, capacitors also have an associated voltage rating. If this *voltage rating* is exceeded, the capacitor could develop an internal short (the term *short* means an undesired path of current flow.

Figure 5-2
Capacitor voltage
response of Fig. 5-1.

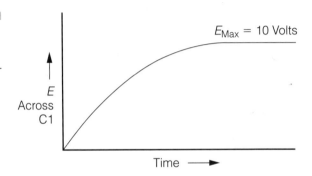

In reference to capacitors, the short would occur through the dielectric, destroying the capacitor in the process).

As the graph in Fig. 5-2 illustrates, the capacitor in the circuit of Fig. 5-1 charges *exponentially.* This means that it charges in a nonlinear fashion. An *exponential curve* (like the charge curve in Fig. 5-2) is one that can be expressed mathematically as a number repeatedly multiplied by itself. The exponential curve of the voltage across a charging capacitor is identical to the exponential curve of the current increase in an *LR* circuit as shown in Chapter 3 (Figs. 3-5 and 3-6).

As discussed in Chapter 3, the current change in an *LR* (inductive-resistive) circuit, relative to time, is defined by the *time constant* and expressed in seconds. Similarly, in an *RC* (resistive-capacitive) circuit, the voltage change across the capacitor is defined by the time constant, and it is also expressed in seconds. An *RC time constant* is defined as the amount of time required for the voltage across the capacitor to reach a value of approximately 63% of the applied source voltage. The *RC* time constant is calculated by multiplying the capacitance value (in farads) times the resistance value (in ohms). For example, the time constant of the circuit shown in Fig. 5-1 would be

$$Tc = RC = (1{,}000{,}000 \text{ ohms})(0.000{,}001 \text{ farad}) = 1 \text{ second}$$

The principle of the time constant, that applies to inductance, also applies to capacitance. During the first time constant, the capacitor charges to approximately 63% of the applied voltage. The capacitor in Fig. 5-1 would charge to approximately 6.3 volts in one second after SW1 is closed. This would leave a remaining voltage differential between the battery and capacitor of 3.7 volts (10 volt − 6.3 volts = 3.7 volts). During the next time constant, the capacitor voltage would increase by an additional 63% of the 3.7-volt differential; 63% of 3.7 volts is approximately 2.3 volts. Therefore, at the end of two time constants, the voltage across the capacitor would be 8.6 volts (6.3 volts + 2.3 volts = 8.6 volts). Five time constants are required for the voltage across the capacitor to reach the value usually considered to be the same as the source voltage.

In the circuit shown in Fig. 5-1, the approximate source voltage (10 volts) would be reached across the capacitor in 5 seconds. If SW1 is opened, after C1 is fully charged, it would hold the stored energy (1 microfarad) at a 10-volt potential for a long period of time. A "perfect" capacitor would hold the charge indefinitely, but in the real world, perfection is hard to come by. All capacitors have internal and external leakage characteristics, which are undesirable. It would be reasonable, however, to expect a well-made capacitor to hold a charge for several weeks, or even months.

Consider the current and voltage relationship in Fig. 5-1 when SW1 is first closed (assuming C1 is discharged). Immediately after SW1 is closed, the capacitor offers virtually no opposition to current flow (just like the small empty water tank in the earlier analogy). This maximizes the current flow, and minimizes the voltage across the capacitor. As the capacitor begins to charge, the current flow begins to decrease. At the same time, the voltage across the capacitor begins to increase. As stated earlier, this process continues until the voltage across the capacitor is equal to the source voltage, and all current flow stops.

The important point to understand is that the peak current occurs before the peak voltage is developed across the capacitor. Simply stated, the *voltage lags behind the current in a capacitive circuit.* As a matter of convenience, professionals in the electrical and electronics fields tend to describe this phenomenon as "the current leading the voltage." This is very much akin to deciding how to describe a glass containing water at the 50% level; is it half full or half empty?

As discussed in Chapter 3, the current lags the voltage by 90 degrees in a purely inductive circuit. In comparison, the current leads the voltage by 90 degrees in a purely capacitive circuit. As in the case of purely inductive circuits, purely capacitive circuits do not dissipate any "true" power because of the 90-degree phase differential between voltage and current.

Filter Capacitors

Capacitors used in filter applications remove an AC component from a DC voltage. Some filter capacitors are implemented to remove only a frequency-dependent part of an AC signal, but these applications will be discussed in a later chapter. In this section, you will examine how filter capacitors are used in DC power supply applications.

Figure 5-3 illustrates a simple half-wave rectifier circuit with R1 and C1 acting as a filter network. For a moment, review what you already know about this circuit. T1's secondary is rated at 12 volts AC. This is an *rms* voltage, because no other specification is given. The peak voltage output from this secondary will be about 17 volts (12 volts × 1.414 = 16.968 volts). The negative half-cycles (in reference to circuit common) will be blocked by D1, but it will act like a closed switch to the positive half-cycles, and apply them to the filter network (R1 and C1). The amplitude of the applied positive half-cycles will not be the full 17 volts,

Figure 5-3
A DC filter circuit.

because the 0.7-volt forward threshold voltage must be dropped across D1. This means that the positive half-cycles applied to the filter network will be about 16.3 volts in peak amplitude.

These positive half-cycles will occur every 16.6 milliseconds (the reciprocal of the 60-hertz power-line frequency). The positive *half-cycle duration* will be about 8.3 milliseconds, because it only represents one-half of the full AC cycle. [Refer back to Chapter 4 (Fig. 4-5) and the related text if this is confusing.] Going back to Fig. 5-1, the calculated time constant for the circuit was 1 second. Because the same component values are used in the filter network of Fig. 5-3, the time constant of this filter is also 1 second. In power supply circuits, this time constant is called the *source time constant*.

Assuming that C1 is fully discharged, examine the circuit operation of Fig. 5-3 from the moment that power is first applied to the primary of T1. When D1 is forward-biased by the first positive half-cycle from the T1 secondary, the 16.3-volt peak half-cycle is applied to the filter network of R1 and C1. C1 will begin to charge to the full 16.3-volt peak amplitude, but it will not have time to do so. Because the filter's *RC* (resistor-capacitor) time constant is 1 second, and the positive half-cycle only lasts for 8.3 milliseconds, C1 will only be able to charge to a very small percentage of the full peak level.

During the negative half-cycle, while D1 is reverse-biased, C1 cannot discharge back through D1; because C1's charged polarity reverse-biases D1. Therefore, C1 holds its small charge until the next positive half-cycle is applied to the filter network. During the second positive half-cycle, C1 charges a little more. This process will continue, with C1 charging to a little higher amplitude during the application of each positive half-cycle, until C1 charges to the full 16.3-volt peak potential. When C1 is fully charged, all current flow within the circuit will cease; C1 cannot charge any higher than the positive peak, and D1 blocks all current flow during the negative half-cycles. At this point, the voltage across C1 is pure DC. The AC component (ripple) has been removed, because C1 remains charged to the peak amplitude during the time periods when the applied voltage is less. Even if all circuit power is removed from the T1 primary, C1 will still remain charged because it doesn't have a discharge path. Notice that it could not discharge back through D1, because its charged polarity is holding D1 in a reverse-biased state.

The operation of the circuit illustrated in Fig. 5-3 can be compared to pumping up an automobile tire with a hand pump. With each downward stroke of the hand pump, the pressure in the tire increases a little bit. Think of the tire as being C1. Air is inhibited from flowing back out of the tire by a small one-way air valve in the base of the pump. The one-way air valve allows air to flow in only one direction: into the tire. D1 allows current to flow in only one direction, causing C1 to charge. When the tire is pumped up to the desired pressure, it will hold this pressure even if the pump is removed. Once C1 is charged up to the peak electrical pressure (voltage), it will retain this charge, even if the circuit power is removed.

Although the circuit of Fig. 5-3 is good for demonstration purposes, it is not very practical as a power supply. To understand why, examine the circuit illustrated in Fig. 5-4. You should recognize T1 and the four diode network as being a *full-wave bridge rectifier.* As discussed in Chapter 4, for each half-cycle output of T1's secondary, two diodes within the bridge will be forward-biased, and thus conduct "charging" current in only one direction through C1. With SW1 opened, C1 will charge to the peak voltage, minus 1.4 volts (since two bridge diodes must drop their forward threshold voltages at the same time; 0.7 volt + 0.7 volt = 1.4 volts). The peak voltage of a 12-volt AC secondary is about 17 volts, so C1 will charge to 17 volts − 1.4 volts, or about 15.6 volts. When C1 is fully charged, the voltage across it will be pure DC.

Note the symbol used for C1 in Fig. 5-4. The positive symbol close to one plate indicates that it is an electrolytic capacitor, meaning that it is polarized. The value also indicates this capacitor type because a capacity of 1000 μF is too large for a conventional, nonpolarized capacitor. In this circuit, the negative side is connected to circuit common, which is the most negative point in the circuit. C1 also has an associated voltage rating of 25 *working volts DC* (WVDC), which means that this is the highest direct voltage that can be safely applied to the capacitor. Because the peak voltage applied to it will be only 15.6 volts, you are well within the safe operating parameters in this circuit.

Referring back to Fig. 5-3, you calculated the source time constant by multiplying the resistance value of R1 by the capacitance value of C1.

Figure 5-4
Full-wave filtered DC
power supply.

This *RC* time constant was 1 second. Going back to Fig. 5-4, it might appear that C1 would charge "instantly" because there doesn't seem to be any resistance in series with it. In an "ideal" (perfect) circuit, this would be true. In reality, the DC resistance of the T1 secondary, the wiring resistance, and a small, nonlinear "current-dependent" resistance presented by the diode bridge will be in series with C1. Of these three real-world factors, the only one that is really practical to consider is the DC resistance of the T1 secondary.

For illustration, assume that the DC resistance of T1's secondary winding is 1 ohm. To calculate the source time constant, the 1-ohm resistance would be multiplied by the capacity of C1:

$$Tc_{source} = (1\ ohm)(0.001\ farad) = 0.001\ second\ (or)\ 1\ millisecond$$

Because the time duration of each half-cycle is approximately 8.3 milliseconds, for all practical purposes, you can say that C1 will be fully charged by the end of the first positive half-cycle. During the rapid charging of C1, a very high *surge current* will flow through the diode pair that happens to be in the forward conduction mode at the time. This is because C1 will initially look like a direct short until it is charged to a sufficient level to begin opposing a substantial portion of the current flow.

As you may recall, one of the forward current ratings for diodes (discussed briefly in Chapter 4) was called the *peak forward surge current* and was based on an 8.3-millisecond time period (for 60-hertz service). The purpose for such a rating should now become apparent. Diodes, used as power supply rectifiers, will be subjected to high surge currents every time the circuit power is initially applied. 8.3 milliseconds is used as a basis for the peak forward surge current rating, because that is the time period of one half-cycle of 60-hertz AC. A rule-of-thumb method for calculating the peak forward surge current is to estimate the maximum *short-circuit secondary current* of T1 based on its peak output voltage and DC resistance. The peak output voltage of a 12-volt secondary is about 17 volts, and you have assumed the DC resistance to be 1 ohm. Therefore, using Ohm's law:

$$I_{\text{peak surge}} = \frac{E_{\text{peak}}}{R} = \frac{17 \text{ volts}}{1 \text{ ohm}} = 17 \text{ amps peak}$$

In reality, a transformer with a 12-volt secondary, and a secondary DC resistance of 1 ohm, would not be capable of producing a 17-amp short-circuit current (remember about inductive reactance being additive to resistance), so this gives us a good margin of safety.

After the first half-cycle, C1 is assumed to be fully charged. As long as SW1 is open (turned off), the only current flow in the circuit will be a very small *leakage current* which is considered negligible. The voltage across C1 is pure DC at about 15.6 volts.

A power supply would be of no practical value unless it powered something. The "something" that a power supply powers is called the *load*. The load could be virtually any kind of electrical or electronic circuit imaginable; but in order to operate, it must draw some power from the power supply. In Fig. 5-4, a load is simulated by the resistor R_{load}. By closing SW1, the load is placed in the circuit, and the circuit operation will be somewhat changed.

In power supply design, it is important to consider the *source time constant* (calculated previously) as well as the *load time constant*. When C1 is charged, it cannot discharge back through the bridge rectifier and the T1 secondary, because its charged polarity reverse-biases all of the diodes. By closing SW1, a discharge path is provided through R_{load}. The load time constant (sometimes called the *discharge time constant*) can be calculated by multiplying the capacitance value of C1, by the resistance value of the load (R_{load}):

$$TC_{\text{load}} = (0.001 \text{ farad})(1000 \text{ ohms}) = 1 \text{ second} \quad (1000 \text{ } \mu\text{F} = 0.001 \text{ farad})$$

The importance of the load time constant becomes apparent by examining the "exaggerated" illustration of Fig. 5-5. The waveshape shown in dotted lines is the 120-hertz, full-wave rectified waveshape that would appear across R_{load} if C1 were not in the circuit. The solid line represents the DC voltage levels across C1 with the charge/discharge amplitudes exaggerated for the sake of illustration. The charge periods represent the source time constant showing C1 receiving a charge (electrical energy) from the bridge rectifier circuit. As calculated earlier, this time constant is very short (only 1 millisecond). So, for illustration purposes, this charge is shown to be concurrent with the applied 120-hertz half-cycles.

The discharge periods represent the load time constant. During these periods, C1 is supplying its stored energy to power the load, causing its voltage level to drop by some percentage until the next charge period replenishes the drained energy. The variations in voltage level between the charge and discharge periods is called *ripple*. In a well-designed power supply, ripple is a very small, undesirable AC component "riding" on a DC level. Ripple can be specified as a peak-to-peak value, an rms value, or a percentage relationship, as compared to the DC level.

As stated earlier, the illustration in Fig. 5-5 is exaggerated; the ripple variations would not be nearly as pronounced, because the load time constant (1 second) is much longer than the peak charging intervals, which occur every 8.3 milliseconds. As a general rule of thumb, the *load time constant should be at least 10 times as long as the charge interval*. In the case of a full-wave rectifier circuit, as in Fig. 5-4, the charge interval is 8.3 milliseconds, so the load time constant should be at least 83 milliseconds. With a half-wave rectifier circuit, the load time constant would have to be twice as long for the same quality of performance.

The power supply illustrated in Fig. 5-4 is referred to as a *raw DC power supply*. This simply means that it is not voltage- or current-regulated. Regulated power supplies will be discussed in Chapter 6.

Designing Raw DC Power Supplies

Every type of electrical or electronic apparatus needs a source of electrical energy to function. The source of electrical energy is called the

Figure 5-5
Charge/discharge
cycle of C1
in Fig. 5-4.

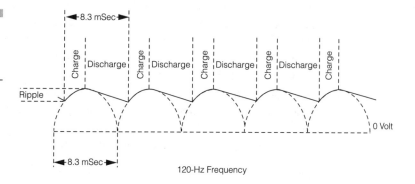

power supply. The two main classifications of power supplies are *line-operated power supplies* (operated from a standard 120-volt AC wall outlet) and *battery supplies* (electrical energy is provided through a chemical reaction). It is relatively safe to say that any device capable of functioning properly from a battery power source, can function equally well from a properly designed "raw" DC power supply, receiving its energy from a wall outlet. This is important because you will probably run into many situations where you will want to test or operate a battery-powered device from standard household power. As an exercise to test all you have learned thus far, here is a hypothetical exercise in designing a raw DC power supply for a practical application.

Assume that you own an automobile CB (citizen's band) radio that you would occasionally like to bring into your home and operate as a "base station." In addition to installing an external stationary antenna (which is irrelevant to our present topic of discussion), you would have to provide a substitute for the automobile battery as a power source. The CB radio is specified as needing "12 to 14 volts DC at 1.5 amps" for proper operation.

The CB radio power supply will have three primary parts: the transformer, a rectifier network, and a filter. You should choose a transformer with a secondary "peak" (not rms) voltage rating close to the maximum desired DC output of the power supply. In this case, a 10-volt AC secondary would do nicely, and they are commonly available. The peak voltage output of a 10-volt secondary would be

$$Peak = 1.414(\text{rms}) = 1.414(10 \text{ volts}) = 14.14 \text{ volts}$$

The rectifier network will drop about 1 volt, so that would leave about 13 volts (peak) to apply to the filter capacitor. The current rating of the transformer secondary could be as low as 1.5 amps, but a 2-amp secondary current rating is more common, and the transformer would operate at a lower temperature. A transformer with a 10-volt AC at 2-amp secondary rating can also be specified as a 10-volt 20-volt-amp transformer. The *volt-amp* (VA) rating is simply the current rating multiplied by the voltage rating (10 volts × 2 amps = 20 VA).

A full-wave bridge rectifier can be constructed using four separate diodes, or it can be purchased in a module form. Bridge rectifier modules are often less expensive, and are easier to mount. The average forward current rating should be at least 2 amps, to match the transformer's secondary rating. The peak reverse-voltage rating, or PIV, would have to be at least 15 volts (the peak output voltage of the transformer is 14.14 volts), but it is usually prudent to double the minimum PIV as a safety margin. However, a 30-volt PIV rating is uncommon, so a good choice would be diodes (or a rectifier module) with at least a 2-amp, 50-volt PIV rating.

If you purchase these rectifiers from your local electronic parts store, don't be surprised if they don't have an associated "peak forward surge current" rating. Most modern semiconductor diodes will easily handle the surge current if the average forward current rating has been properly observed. This is especially true of smaller DC power supplies, such as the hypothetical one presently being discussed. If it is desirable to estimate the peak forward surge current, measure the DC resistance of the transformer secondary, and follow the procedure given earlier in this chapter. (The secondary DC resistance can be difficult to measure with some DVMs because of its very low value.)

In order to choose a proper value of filter capacitance, you can equate the CB radio to a resistor. Its power requirement is 12 to 14 volts at 1.5 amps. Using Ohm's law, you can calculate its *apparent resistance:*

$$R = \frac{E}{I} = \frac{12 \text{ volts}}{1.5 \text{ amps}} = 8 \text{ ohms (worst case)}$$

As far as the power supply is concerned, the CB radio will look like an 8-ohm load. Note that the 8-ohm calculation is also the worst-case condition. If the upper voltage limit (14 volts) had been used in the calculation, the answer would have been a little over 9.3 ohms. Eight ohms is a greater current load to a power supply than 9.3 ohms (as the load resistance decreases, the current flow from the power supply must increase).

You now know two variables in the load time constant equation: the *apparent load resistance* (R_{load}) and the *desired time constant* (83 milliseconds with a full-wave rectifier). To solve for the *capacitance value*, the time constant equation must be rearranged. Divide both sides by R:

$$\frac{Tc}{R} = \frac{R(C)}{R}$$

The R values on the right side of the equation cancel each other, leaving

$$\frac{Tc}{R} = C \quad \text{or} \quad C = \frac{Tc}{R}$$

By plugging our known variables into the equation, it becomes

$$C = \frac{83 \text{ milliseconds}}{8 \text{ ohms}} = 0.010375 \text{ farad} \quad or \quad 10,375 \text{ } \mu F$$

According to the previous calculation, the filter capacitor needed for the CB radio power supply should be about 10,000 μF. A capacitor of this size will always be electrolytic, so polarity must be observed. The voltage rating should be about 20 to 25 WVDC (this provides a little safety margin over the actual DC output voltage), depending on availability. If this calculation had been based on a half-wave rectifier circuit, the required capacitance value for the same performance would have been about 20,000 μF.

Assembly and Testing of Third Section of a Lab Power Supply

Materials needed for the completion of this section are (1) three phenolic-type, 2-lug solder strips (see text) and (2) two 4400-μF, 50-WVDC electrolytic capacitors. The phenolic solder strips are specified only because they are inexpensive and effective. If your local electronics parts store doesn't have these in stock, there are many good alternatives. Any method of providing three chassis mountable, *insulated tie points* that will hold the filter capacitors firmly in place, and allow easy connections to the capacitor leads, will function equally well.

Figure 5-6
Schematic diagram of the first, second, and third sections of the lab power supply.

Figure 5-7
Approximate physical
layout of the major
components for
the lab power
supply project.

Referring to Figs. 5-6 and 5-7, mount the solder strips to the chassis, being careful to space them far enough apart to allow room for the capacitors (C1 and C2). Connect the capacitors to the insulated lugs and crimp the leads to hold them in place, until the remaining wiring is completed. Connect a piece of hook-up wire, from the joint connection of C1 and C2, to the circuit common point between the secondaries of T1 and T2. Use another piece of hook-up wire to connect the positive side of C1 to the positive terminal of BR1. Connect another piece of hook-up wire from the negative side of C2 to the negative terminal of BR1. *Do not connect circuit common to chassis ground.* At this stage of the project, the only connection to circuit common should be the junction of C1 and C2. Double-check all wiring connections, and be sure all of the voltage polarities are correct. Solder all of the connections.

The power supply you have constructed thus far is called a *dual-polarity, 34-volt raw DC power supply.* It is the same type of power supply as is illustrated in Fig. 5-4 (minus R_{load} and SW1). Power supplies similar to this one are commonly used in audio power amplifiers. This particular power supply could provide the electrical energy that a power amplifier would need to drive an 8-ohm speaker at about a 50-watt rms level. Audio amplifiers will be covered further in Chapter 8. Chapter 6 will discuss how to add an adjustable regulator section, to this design, for improved lab performance.

Testing the Power Supply

To limit redundancy, it is assumed at this point that you are practicing all of the safety procedures that have been discussed previously. In the successive chapters, I will mention only the special safety considerations which can apply to unique situations. *Review all of the safety recommendations presented thus far, and put them to use all of the time.*

Set your DVM to measure "DC volts" on the 100-volt range (or higher). Plug the power supply into the outlet strip. Set SW1 to the "off" position and turn on the outlet strip. Briefly, turn SW1 to the "on" position, and then back to the "off" position. Turn the outlet strip off. Measure the DC voltage across C1 and C2, paying close attention to the polarity (C1 should be positive, and C2 should be negative, in reference to circuit common). The actual amplitude of the voltage is not important at this point in the test.

You have simply "pulsed" the power supply on and off, to verify that the capacitors are charging and in the correct polarity. As you measured the DC voltages, they should have been decreasing in amplitude as the charge was draining off. The draining of the charge is caused by the internal leakage inherent in all electrolytic capacitors (new capacitors can be very leaky until they have the chance to re-form during circuit operation). Also, the capacitors will discharge, to some degree, through the internal input impedance of the DVM while you are measuring the voltage.

If you measured some voltage level across C1 and C2, with the correct voltage polarities, reapply power to the circuit and measure the DC voltages across C1 and C2. The calculated voltage across each capacitor should be about 33 volts (the peak value of the 24-volt secondaries is about 34 volts, minus an estimated 1-volt drop across the rectifier). In reality, you will probably measure about 36 to 38 volts across each capacitor. There are several reasons for this higher level. Transformer manufacturers typically rate transformers based on minimum worst-case conditions, so it is common for the secondaries to measure a little high under normal conditions. Also, you are measuring the voltage levels under a no-load condition (often abbreviated N.L. in data books). If you took these same measurements while the power supply was operating under a full load (abbreviated F.L.), they would be considerably lower.

Leaving power applied to the power supply circuit, set your DVM to measure "AC volts" beginning on the 100-volt (or higher) range. Measure the AC voltage across C1 and C2. If you get a zero indication on the 100-volt range, set the range one setting lower and try again. Continue this procedure until you find the correct range for the AC voltage being measured. (When measuring an unknown voltage or current, always begin with a range setting higher than what you could possibly measure and work your way down. Obviously, if you are using an "autoranging" DVM, you won't have to worry about setting the range.) If the circuit is functioning properly, you should measure an AC component (ripple) of about 5 to 20 mV. Turn off the circuit.

Many of the more expensive DVMs are specified as measuring true rms. If you are using this type of DVM to measure the ripple content, the indication you'll obtain will be the true rms value. The majority of DVMs, however, will give an accurate rms voltage measurement of sine-wave AC only. Referring back to Fig. 5-5, note that the ripple waveshape is not a sine wave; it is more like a "sawtooth" (you'll learn more about differing waveshapes in succeeding chapters). The point here is that there can be some error in the ripple measurement you just performed. High accuracy is not important in this case, but you can experience circumstances in the future, where you must consider the type of AC waveshape that you are measuring with a DVM, and compensate accordingly.

As an additional test, I used a 100-ohm, 25-watt resistor to apply a load to the circuit. If you have a comparable resistor, you might want to try this also, but be careful with the resistor; it gets *hot*.

The resistor is connected across each capacitor, and the subsequent AC and DC voltage measurements are taken. The loading effect was practically identical between the positive and negative supplies, which is to be expected. The DC voltage dropped by about 3.1 volts, and the ripple voltage increased to about 135 mV. These effects are typical.

Food for Thought

Throughout this chapter, I have followed a more traditional, and commonly accepted, method of teaching and analyzing capacitor theory. I suggest that you continue to comprehend capacitor operation from this perspective. However, in the interest of accuracy, you will find the following story to be of interest.

Michael Faraday, the great English chemist and physicist, had a theory that more closely approaches the way a capacitor really works. His theory stated that the charge is actually contained in the dielectric material—not the capacitor's plates. Inside the dielectric material are tiny *molecular dipoles* arranged in a random fashion. Applying a voltage to the plates of a capacitor stresses these dipoles causing them to line up in rows, storing the energy by their alignment. In many ways, this is similar to the physical change occurring in iron, when it becomes a temporary magnet by being exposed to magnetic flux lines. When a capacitor is discharged, the dipoles flex back like a spring, and their energy is released.

The fact that the stored energy within a capacitor is actually contained within the dielectric explains the reason why different dielectric materials have such a profound effect on the capacity value. Dielectric materials are given a *dielectric constant* rating (usually based on the quality of air as a dielectric; air = 1.0) relative to their overall effect on capacity. A dielectric material with a rating of 5, for example, would increase the *capacitance value* of a capacitor to a value 5 times higher than air, when all other variables remained the same.

6

Transistors

Before getting into transistor theory, there are a few definitions that are better discussed in advance.

Preliminary Definitions

Gain is a term used to describe a ratio of increase. The most common types of gain are *current gain, voltage gain,* and *power gain.* Gain is simply the ratio of the input (voltage, current, or power) to the output (voltage, current, or power). For example, if a 1-volt signal is applied to the input of a circuit and, on the output, the signal amplitude has been increased to 10 volts, you say that this circuit has a voltage gain of 10 (10 divided by 1 = 10). It is possible to have a gain of less than 1. For example, if the output of a circuit is only one-half of the value of the input, it can be said this circuit has a gain of 0.5. However, it is preferable to say that the output is being *attenuated* (reduced) by a factor of 2.

The symbol for gain as used in equations and formulas is *A.* Typically, the uppercase *A* (symbolizing gain) will be followed by a lowercase suffix letter designating the gain type. For example, A_e symbolizes "voltage gain."

Power gain is a specific type of gain indicating that more *energy* is being delivered at the output (of a device or circuit) than is fed into the input. A transformer, for instance, is capable of providing voltage gain, if it is a step-up transformer; but the secondary current is reduced by the same factor (turns ratio) as the voltage is increased. Because power is equal to *voltage times current,* the equation seems to balance—equal power in and power out—but this is still not a gain. Additionally, all components have losses. The transformer's efficiency loss is called its *efficiency ratio.* A good unit will have about a 90% ratio, and the primary to secondary power transfer ratio will always be less than 1.

Therefore, a transformer is not capable of producing power gain. Electronic components capable of providing power gain are called *active devices.* These include transistors, vacuum tubes, some integrated circuits, and many other devices. Electronic components that cannot produce power gain are called *passive devices.* Some examples of passive components are resistors, capacitors, transformers, and diodes.

Introduction to Transistors

The development of the transistor was the foundational basis for all modern solid-state electronics. William Shockley, John Bardeen, and Walter Brattian discovered *transistor action* while working at the Bell

Telephone Laboratory in 1947. The term *transistor* started out as a combination of the phrase "transferring current across a resistor." The important developmental aspect of the transistor is that it became the first "active" solid-state device and opened a new perspective of design ideology. It's hard to imagine what our lives would be like today without it!

A *transistor* is a solid-state, three-layer semiconductor device. Figure 6-1 shows the basic construction of a transistor, and compares transistor construction to diode construction. Note that a *diode* contains only one junction, whereas a transistor contains two junctions. Because a transistor contains two junctions, it is often referred to as a (dual-junction) *bipolar device*. Figure 6-1 also illustrates how bipolar transistors can be constructed in either of two configurations: NPN or PNP.

Bipolar transistors have three connection points, or leads. These are called the *emitter*, the *base*, and the *collector*. The symbols for bipolar transistors are shown in Fig. 6-1. The only difference between the NPN and PNP symbols is the direction of the arrow in the emitter lead.

Figure 6-1
Transistor construction and symbols.

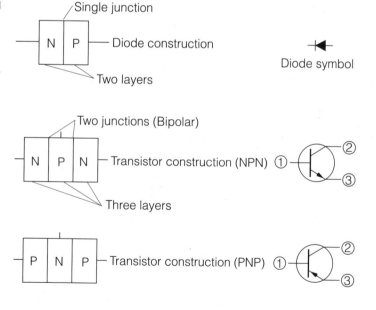

Transistor Principles

Figure 6-2 shows an NPN transistor connected in a simple circuit to illustrate basic transistor operation. A PNP transistor would operate in exactly the same manner, only the voltage polarities would have to be changed. Note that the emitter lead is connected to circuit common (the most negative potential in the circuit) through the *emitter resistor* (R_e), the base lead is connected to a potentiometer (P1), and the collector lead is connected to 30 volts, through the *collector resistor* (R_c).

A transistor actually consists of two diode junctions: the *base-to-emitter junction* and the *base-to-collector junction*. Assume that P1 is adjusted to provide 1.7 volts to the base (note how P1 could provide any voltage to the base from 0 to 30 volts). The 1.7-volt potential applied to the P-material base, in reference to the N-material emitter at 0 volts (circuit common), creates a *forward-biased* diode and causes current to flow from emitter to base. However, because of a phenomenon known as *transistor action*, an additional current will also flow from emitter to collector.

To understand transistor action, we have to consider several conditions occurring simultaneously within the transistor. First, notice that the base-to-collector junction is reverse-biased. The collector is at a much

Figure 6-2
Simple circuit to
illustrate basic
transistor operation.

higher positive potential than the base, causing the base to be negative in respect to the collector. Therefore, current will not flow from base to collector. The high positive potential on the collector has the tendency to attract all of the negative-charge carriers (electrons) away from the base-collector junction area. This creates a *depletion area* of negative charge carriers close to the base layer. The depletion area seems very positive for two reasons: (1) all of the negative-charge carriers have been drawn up close to the collector terminal and (2) because it is part of the collector layer, it is connected to the highest positive potential. Keeping this condition in mind, turn your attention back to the base-emitter junction.

Referring again to Fig. 6-2, as stated previously, the base-emitter junction is forward biased and current is flowing. However, note how the base layer is much thinner than the emitter layer. The emitter has many more *negative-charge carriers* (electrons) than the thin base material has *holes* (absence of electrons) to combine with. This causes an "overcrowded" condition of electrons in the base layer. These crowded electrons have two directions in which to flow (or *combine*, which gives the appearance of flowing); some will continue to flow out to the 1.7-volt base terminal, but the majority will flow toward the very positive-looking depletion area created in the collector area close to the base layer.

The end result is a much higher current flow through the collector than is flowing through the base. The parameter (component specification) that defines the ratio of the base current to the collector current is called *beta* (abbreviated B or H_{FE}). In essence, beta is the maximum possible current gain that can be produced in a given transistor. Typical beta values for small-signal transistors are in the range of 100 to 200. In contrast, power transistors can have beta (β) values of 20 to 70. The equation for calculating beta is

$$\beta = \frac{I_c}{I_b}$$

This equation states that beta is equal to the collector current divided by the base current. The important point to recognize about transistor current gain is that the higher collector current is controlled by the much smaller base current.

Going back to Fig. 6-2, assume that the transistor illustrated has a beta of 100. With 1.7 volts applied to the base, about 0.7 volt will be dropped across the base-to-emitter junction (like any other forward-biased silicon diode). The remaining 1 volt (1.7 volts − 0.7 volt = 1 volt) will be dropped across the emitter resistor (RE). Because we know the resistance value of

RE and the voltage across it, you can use Ohm's law to calculate the current flow through it:

$$I = \frac{E_{RE}}{R} = \frac{1 \text{ volt}}{1000 \text{ ohms}} = 1 \text{ milliamp}$$

The 1 milliamp of current flow through RE is the "sum" of the *base current* and the *collector current*. You can think of the emitter as the layer that "emits" the total current flow. The majority of this current is "collected" by the collector, and the overall current flow is controlled by the base. Our assumed beta value tells us that the collector current will be 100 times larger than the base current. Therefore, the base current flow will be about 9.9 microamps, and the collector current will be about 990 microamps (990 microamps + 9.9 microamps = 999.9 microamps, or about 1 milliamp). Because the collector current must flow through RC, the voltage drop across the collector resistor (RC) can be calculated using Ohm's law:

$$E = IR = (990 \text{ microamps})(10,000 \text{ ohms}) = 9.9 \text{ volts}$$

There are three individual voltage drops in Fig. 6-2 that must be analyzed to understand the action of the transistor. Two of these have already been calculated: the voltage across the emitter resistor (RE), and the voltage dropped by the collector resistor (RC).

The third important voltage drop occurs across the transistor itself. All three of these voltage drops are in series; the emitter resistor is in series with the transistor, which is in series with the collector resistor. Going back to our discussion of simple series circuits, you know that the sum of these three voltage drops must equal the source voltage of 30 volts. Because you already know the value of two of the voltage drops, you can simply add these two values, subtract the sum from the source voltage, and the difference must be the voltage drop across the transistor:

$$1 \text{ volt (RE)} + 9.9 \text{ volts (RC)} = 10.9 \text{ volts}$$

$$30 \text{ volts source} - 10.9 \text{ volts} = 19.1 \text{ volts}$$

Although 19.1 volts are being dropped across the transistor, this is not the collector voltage. Unless otherwise noted, all voltage measurements are always made in reference to circuit common (or ground, whichever is applicable). Therefore, to calculate the collector voltage, the voltage drop across the transistor is added to the voltage drop across the emitter resistor (RE). This must be done because, from a circuit common point of reference, the emitter resistor voltage drop is in series with the transistor voltage drop. If this is confusing, look at it this way. If you made the source voltage your point of reference and measured the collector voltage, you would actually be measuring the voltage drop across RC. If you made the transistor's emitter lead your point of reference, and measured the collector voltage, you would be measuring the voltage across the transistor. By making the circuit common your point of reference, you are actually measuring the voltage across the transistor and the voltage drop across RE. Therefore, the collector voltage would be

$$1 \text{ volt (RE)} + 19.1 \text{ volts (transistor voltage drop)} = +20.1 \text{ volts}$$

Notice that this is the same value you could have calculated by simply subtracting the voltage drop across the collector resistor (RC) from the source voltage.

Now, observe transistor action by changing the base potential.

NOTE   *To eliminate redundancy, many of the previous calculations and discussions will not be repeated.*

Assume that P1 is adjusted to increase the base bias potential to 2.7 volts. The base-emitter junction will still drop about 0.7 volts, so the voltage across RE will increase to 2 volts. The emitter current increases to 2 milliamps. With a beta of 100, about 19.8 microamps of the emitter current will flow through the base lead. The remaining 1.98 milliamps of collector current will flow through the collector lead, dropping about 19.8 volts across RC. Subtracting the voltage drop across RC from the 30-volt source produces a collector voltage of 10.2 volts.

When the base voltage was originally set to 1.7 volts, the collector voltage was 20.1 volts. By increasing the base voltage by 1 volt (up to 2.7 volts), the collector voltage decreased to 10.2 volts. In other words, the 1-volt change in the base voltage resulted in a 9.9-volt change in the collector voltage (20.1 volts − 10.2 volts = 9.9 volts). This is an example of voltage gain. In this circuit, we have a voltage gain (abbreviated A_c) of 9.9 (9.9-volt change at the collector divided by the 1-volt change at the base = 9.9). There was also an "internal" current gain equal to beta. Obviously, this results in power gain, because both the voltage and the current increased.

If you applied a 1-volt peak-to-peak signal to the base lead of this circuit (Fig. 6-2), it would be increased to a 9.9-volt peak-to-peak signal at the collector lead. However, the amplified output would be inverted, or 180 degrees out of phase, with the input signal. As you might have noticed, as the base voltage increased (from 1.7 to 2.7 volts), the collector voltage dropped (from 20.1 to 10.2 volts).

There are three "general" transistor configuration methods: the common base, the common emitter, and the common collector. In the previous discussion, involving Fig. 6-2, you looked at the operation of a common emitter configuration. In this configuration, the output is taken off of the transistor collector, and the signal is always inverted.

If you haven't had any prior experience with transistors, your head is probably "buzzing" with all of the voltages and currents relating to Fig. 6-2. Fear not. As you progress and gain experience, the haze will clear, and these principles will seem simple. This section is necessary to provide you with a good working knowledge of basic transistor operational principles.

Now, for practical purposes, you can simplify things according to the transistor configuration. However, before proceeding, there is a very important principle to establish. A bipolar transistor is a "current" device. Although there will always be voltages present in an operating circuit, the "effect" produced by a transistor is *current gain,* and the controlling "factor" of a transistor is the input current.

Consider a transistor as being similar to a water valve. A water valve controls the flow of water. Obviously, water pressure must exist to push the water through any type of system, but we never think of a water valve in terms of controlling water pressure (even though pressure changes will occur with differing valve adjustments). A water valve is always used as a device to control the flow of water. Similarly, always think of a bipolar transistor as a device used to control electric current flow.

Common Transistor Configurations

Transistors are "active" devices because they are capable of producing power gain. In actuality, bipolar transistors are current amplifiers, but an increase of current while maintaining the same voltage is an increase of power ($P = IE$). Depending on the circuit configuration a transistor is designed into, the output can produce current gain, voltage gain, or both. The following transistor circuit configurations will differ in their ability to provide voltage, current, and power gain, but to avoid confusion, a parallel analysis will be given at the end of this section.

Adding Some New Concepts

Before getting into more circuit analysis, it is helpful to understand a few new terms and concepts that will make the whole process simpler.

The remainder of this text contains a lot of discussion involving impedance. While examining some of the basic operational fundamentals of inductors and capacitors, you learned they have a certain *frequency-dependent* nature (this will be discussed further in Chapter 15). We call such components "reactive." In a *reactive component*, the opposition to AC current flow is frequency dependent, and these components exhibit a special form of AC resistance called *reactance*. For example, with inductors, as the frequency of an applied voltage rises, the inductor's opposition to that voltage increases. Capacitive action is just the opposite. In other words, a DC voltage applied to a reactive circuit will promote a current flow dependent on the resistance in the circuit, but an applied AC voltage of the same amplitude might cause a totally different current flow, depending on its frequency. For this reason, you need a term to describe the "total" opposition to AC current flow; taking both the resistive and reactive components into consideration. This term is *impedance*. Impedance is defined in ohms, just as resistance; it is composed of DC resistance, plus the inductive reactance and/or the capacitive reactance in a circuit; and its symbol, as used in equations and formulas, is Z.

In practical transistor circuits, capacitors are used extensively for a variety of purposes. The following transistor circuits use capacitors for "coupling" (sometimes referred to as "blocking") and "by-pass" functions. Remember back to our discussions on basic capacitor operation; you know that current cannot pass through the dielectric (insulating) material under normal operation. However, the "effect" of a changing electrostatic field on one plate can be transferred to the other plate, even though no actual current is passing through the dielectric. Technically speaking, the subatomic distortion occurring on one plate, and in the dielectric, must cause a subsequent distortion in the other plate.

In effect, a capacitor can appear to pass an AC current with virtually no opposition, while totally "blocking" any DC current. A capacitor used to block DC while passing an AC signal current is called a *coupling capacitor*. There are some situations where we want just the opposite to occur: the DC voltage present with no AC component in it. Capacitors used for this function are called *bypass capacitors*. (The previous chapter covered filter capacitors. Filter capacitors are actually a type of bypass capacitor; they maintain the DC voltage, while reducing the AC component, or ripple.)

In the following transistor configuration circuits, it would be advantageous for you to start thinking of them as "building blocks" to more complex circuits. The most sophisticated electronic equipment can be broken down into simpler subassemblies, and, in turn, these subassemblies can be broken down into basic circuit blocks. In performing electronic design, we start with basic blocks and put them together in such a way that they will collectively produce a desired result. Electronic equipment manufacturers will often include "block diagrams" as part of the overall documentation package for their products. This provides an efficient method of becoming thoroughly familiar with the technical operation of the products, without having to analyze the operation down to a component level. As you gain more experience in the electrical and electronics fields, you'll appreciate the usefulness of a block approach in design, analysis, or troubleshooting ventures.

The Common-Emitter Configuration

As stated previously, you analyzed the circuit in Fig. 6-2 as though it were a common-emitter amplifier for the purpose of demonstrating voltage gain. In actuality, Fig. 6-2 can be either a common-emitter amplifier or a common-collector amplifier depending on whether you use the output from the collector or the emitter, respectively. However, a practical common-emitter transistor amplifier would probably require some improvements, as illustrated in Fig. 6-3.

Bipolar transistors have a negative temperature coefficient; that is, as a transistor's temperature increases, its internal resistances decrease. Also, temperature increases cause an increase in undesirable "leakage" currents that can further add to temperature buildup. In high-power transistor circuits, this chain-reaction temperature effect can lead to a condition called *thermal runaway*, which renders the circuit inoperative. In small-signal transistor circuits, as we are presently examining, varying temperatures will cause shifts in operating points.

In addition to temperature considerations, typical bipolar transistors do not have precise beta values. Manufacturers specify beta values within minimum and maximum ranges for each transistor type. Consequently, two transistors with the same exact part number can have dramatically differing beta values.

A third problematic variable to consider is the *source voltage*. Transistor circuits intended to be powered from a battery power supply must be capable of operating with a relatively broad range of supply variance.

Figure 6-3
Practical example of
a common-emitter
transistor amplifier.

Even 120-volt AC line-powered transistor circuits might experience voltage variations if the power supply is not well regulated. A "perfect" transistor amplifier circuit would be self-correcting with temperature
variations, unaffected by the transistor's beta value, and totally immune
to source voltage variations. Although you can't achieve perfection, you
can come close to it with the circuit in Fig. 6-3.

Notice that P1 (Fig. 6-2) has been replaced with two fixed resistors to
set the *base bias.* C1 is a coupling (or blocking) capacitor to keep the DC
base bias voltage from being applied to the input source. Similarly, C3
serves the same function of keeping the DC collector voltage from
being applied to the output. C2 is a bypass capacitor which effectively
"shorts" the AC emitter (input signal) voltage to circuit common, while
leaving the DC emitter voltage unaffected.

The *bias voltage divider,* consisting of R1 and R2, keeps the correct *percentage* of source voltage applied as a base bias, regardless of the *actual*
value of the source voltage. In other words, the actual resistor values in
Fig. 6-3 have been chosen so that about 9% of the source voltage will be
dropped across R2 (without considering the parallel base impedance).
Regardless of the actual value of the source voltage, about 9% of it will
be applied as a base bias. This has the effect of keeping the bias voltage

optimized, even with wide variations in source voltage. Temperature stability is also improved by this method. However, R2 is not absolutely necessary, and it has the undesirable effect (for most applications) of lowering the input impedance. For these reasons, R2 is not incorporated into all common-emitter designs.

Although R4 is commonly called the *emitter resistor,* from an operational perspective, R4 is also a negative-feedback resistor. *Negative feedback* is the term given for applying a percentage of an amplifier's output back into the input. This improves amplifier stability, but it also reduces gain. In Fig. 6-3, the negative feedback provided by the voltage developed across R4 makes the circuit less dependent on the individual transistor current gain (beta), and aids in increasing temperature stability.

As stated earlier, C2 "couples" (short-circuits) the AC signal to circuit common, but it has little effect on the DC emitter voltage. In reality, this causes two individual gain responses within the circuit. From a DC perspective, C2 doesn't exist. The negative feedback produced by R4 causes the DC voltage gain to be a ratio of the value of the collector resistor (R3), divided by the value of the emitter resistor (R4). With the circuit illustrated, the DC voltage gain is approximately 10. From an AC perspective, R4 doesn't exist, because C2 looks like a short tying the transistor emitter directly to circuit common. Consequently, the AC voltage gain becomes the ratio of the collector resistor value divided by the internal base-emitter junction resistance. The junction resistance of a forward-biased semiconductor junction is very low, so the AC voltage gain is reasonably high, typically in the range of 200, using the 2N2222 transistor as illustrated. This high gain is advantageous, but is highly dependent on individual transistor characteristics. If C2 were removed from the circuit, the AC voltage gain would become the same as the DC voltage gain, or about 10.

The DC input impedance of this circuit (Fig. 6-3) can be considered infinite, because C1 blocks any DC current flow. The AC input impedance consists of three parallel resistive elements: R1, R2, and the forward-biased base-emitter junction resistance multiplied by beta. The forward-biased junction resistance is a low value, typically only a few ohms. Even after multiplying this value by a high beta value, the product would still be much lower than a typical R2 value. Therefore, for practical analysis purposes, you can usually eliminate R1 and R2 from consideration and estimate the input impedance to be the base-emitter junction resistance times the beta value. If C2 is removed from the circuit, the AC input impedance then becomes the parallel resistance value of R1, R2, and the value of R4 multiplied by the beta value. Because R4 times beta is typically a high-resistance value, and R1 is usually a high-resistance value, a reasonably close estimate of the input impedance with C2 removed is the value of R2. The output impedance of this circuit can be considered to be equal to the value of the collector resistor (R3).

Figure 6-3 is a good, stable design with reasonably high voltage gain and wide-range immunity from beta, temperature, and source voltage variations. You might use this circuit as a good building block for most voltage amplification applications, or you can tailor the component values according to specific needs.

The Common-Collector Configuration

Figure 6-4 is a practical example of a common-collector transistor amplifier. Note that the output is taken off of the emitter instead of the collector (as in the common-emitter configuration). A common-collector amplifier is not capable of voltage gain. In fact, there is a very slight loss of voltage amplitude between input and output. However, for all practical purposes, we can consider the voltage gain at unity. Common-collector amplifiers are *noninverting*, meaning the output signal is in phase with the input signal. Essentially, the output signal is an exact duplicate of the input signal. For this reason, common-collector amplifiers are often called *emitter-follower amplifiers*, because the emitter voltage follows the base voltage.

Common-collector amplifiers are current amplifiers. The current gain for the circuit illustrated in Fig. 6-4 is the parallel resistance value of R1 and R2, divided by the resistance value of R3. R1 and R2 are both 20 Kohms in value, so their parallel resistance value is 10 Kohms. This 10 Kohms divided by 1 Kohm (the value of R3) gives us a current gain of 10 for this circuit. Because the voltage gain is considered to be unity (1), the *power gain* for a common-collector amplifier is considered equal to the current gain (10, in this particular case).

The input impedance of common-collector amplifiers is typically higher than the other transistor configurations. It is the parallel resistive effect of R1, R2, and the product of the value of R3 times the beta value. Because beta times the R3 value is usually much higher than that of R1 or R2, you can closely estimate the input impedance by simply considering it to be the parallel resistance of R1 and R2. In this case, the input impedance would be about 10 Kohms. The traditional method of calculating the output impedance of common-collector amplifiers is to divide the value of R3 by the transistor's beta value. Although this method is

Figure 6-4
Practical example of
a common-collector
(emitter-follower)
transistor amplifier.

still appropriate, a closer estimate can probably be obtained by consider-ing the output impedance of most transistors to be about 80 ohms. This 80-ohm output impedance should be viewed as being in parallel with R3, giving us a calculated output impedance of about 74 ohms (80 ohms in parallel with 1000 ohms).

Resistors R1 and R2 have the same function within a common-collector amplifier as previously discussed with common-emitter amplifiers. The high negative feedback produced by R3 provides excellent temperature sta-bility and immunity from transistor variables. As in the case of Fig. 6-3, the circuit illustrated in Fig. 6-4 can be a valuable building block toward future projects.

The Common-Base Configuration

I have included the common-base transistor amplifier configuration in this text for the sake of completeness, but the applications for it are few. They are often used as the first RF amplifier stage, amplifying signals from radio antennas, but are seldom seen otherwise.

Figure 6-5 illustrates a practical example of a *common-base amplifier.* Common-base amplifiers have the unique characteristic of a variable input impedance dependent on the emitter current flow. The equation for calculating the input impedance is

$$Z_{in} = \frac{26}{I_e}$$

where I_e is the emitter current in milliamps.

As can be seen from the previous equation, the input impedance is low. Furthermore, common-base amplifiers have a *high output impedance,* and a *power gain* slightly higher than common-emitter amplifiers.

Transistor Amplifier Comparisons

The common-emitter configuration is used in applications requiring reasonably high voltage and power gains. The output is inverted. Common-emitter amplifiers have low input impedances, and high output impedances.

The common-collector configuration is used for impedance-matching applications. Common-collector amplifiers have high input impedances with low output impedances. Voltage gain is considered at unity, and the output is noninverted.

Figure 6-5

Practical example of a common-base transistor amplifier.

The common-base configuration is seldom used because of its very low input impedance and high output impedance. Common-base amplifiers also have a very unstable nature at high gain values.

Impedance Matching

Figure 6-6 illustrates the importance of correctly matching input and output impedances between circuit stages. The AC signal source illustrated in the left half of Fig. 6-6 could be the output of a common-emitter amplifier, a laboratory signal generator output, the "line" output of a FM radio receiver, or thousands of other possible sources. The important point is that all signal sources will have an internal impedance; shown as Z_s in the illustration. Internal source impedances can range from fractions of an ohm to millions of ohms in value. In this example, assume the internal impedance to be 1000 ohms. The signal level you want to apply to R_L is 10 volts peak to peak. Notice that Z_s is in series with R_L. You know that, in a series circuit, the voltage drop will be proportional to the resistance values. Therefore, in this circuit, about 9.1 volts P-P will be dropped across the internal source impedance, and only about 0.9 volt P-P will actually be applied to R_L. This is undesirable because over 90% of the signal has been lost.

The circuit illustrated in the right half of Fig. 6-6 shows the same signal source connected to the input of a common-collector amplifier stage. (Note the triangular symbol for the common-collector amplifier. This is the symbol used for amplifiers in most block diagrams.) Because the amplifier has an input impedance of 10,000 ohms, only 0.9 volt P-P will be dropped across the internal source impedance, and 9.1 volts P-P will be applied to the amplifier. This is much better because the signal loss is only about 9%.

Here are a few general rules to remember regarding impedance matching. For the *maximum transfer of voltage* (as discussed in the previous example), the output impedance should be as low as possible and the input impedance should be as high as possible. For the *maximum transfer of power*, the output impedance should be the same value as the input impedance. For the *maximum transfer of current*, the output impedance and input impedance should be as low as possible.

Figure 6-6
Demonstration
of impedance
matching.

ac Signal source

Transistor Workshop

The previous sections on transistor fundamentals can be rather abstract during your first read-through, so if you are somewhat confused regarding several of the discussions of transistor fundamentals at this point, don't feel bad. The field of electronics is somewhat analogous to finding your way around in a big city for the first time. It's inevitable that you'll get lost occasionally, but with persistence, you soon begin to feel quite at home.

This section is dedicated to illustrating the previous transistor fundamentals in practical circuitry. Although the basic action and physics involved with transistor operation remains the same, the way in which we can utilize these actions is quite varied. In addition, many of the devices and circuits detailed in previous chapters will be brought into the context of this transistor workshop. By spending some time studying and understanding the circuit examples in this section, you will be applying the fundamentals of transistor operation in practical ways, and familiarizing yourself with some basic electronic building blocks at the same time!

Referring to Fig. 6-7a, note that a transistor can be thought of, in an equivalent sense, as two back-to-back diodes.

NOTE: *You cannot construct a functional transistor from two diodes. The equivalent circuit illustrated is for "visualization" purposes only.*

If you have a DVM with a "diode test" function, you can perform a functional test of a typical NPN transistor in the same manner as if it were two common diodes connected together at their anodes. By placing the red (positive) lead of your DVM on the base and the black (negative) lead on the emitter, you should see a low resistance, just like a typical forward-biased diode would provide. Likewise, leaving the red lead connected to the base and placing the black lead on the collector, you should see another low resistance, indicative of another forward-biased semiconductor junction. By reversing the lead orientation (i.e., placing the black lead on the base and the red lead on the emitter or collector), you should read an infinite resistance in both cases. And finally, connecting the DVM leads from emitter to collector, in either orientation, should provide a reading of infinite resistance.

Figure 6-7b illustrates the equivalent circuit of a PNP transistor. The same exact principles and physics apply; the only difference is in the polarity of voltages and the corresponding opposite direction of current flow. In the case of PNP transistors, placing the black lead of your DVM (in diode test mode) on the base with the red lead on either the emitter or collector will provide a low resistance, typical of any forward-biased semiconductor junction. Infinite resistance will result from placing the red lead on the base with the black lead connecting to the emitter or collector. And finally, as in the case of "any" bipolar transistor, a resistance measurement from emitter to collector should provide an infinite resistance in either test lead orientation.

It is a good experience-oriented learning task to purchase a grab bag of assorted transistors from any electronics supply house and use your DVM to determine the general type (i.e., either NPN or PNP) and the operational condition (i.e., either "good" or "defective"). Since transistor lead designation is seldom marked on the case, you'll have to make numerous trial-and-error guesses until you can find the base lead and determine the general transistor type. With a little practice, this routine becomes rather second-nature to most electronics hobbyists. It's an inexpensive exercise, and you can always use a variety of transistor types for future experimentation and project building.

Figure 6-7a
NPN transistor equivalent circuit and schematic symbol.

COLLECTOR

BASE

SCHEMATIC
SYMBOL

EMITTER

Figure 6-7b
PNP transistor
equivalent circuit
and schematic
symbol.

COLLECTOR

BASE

SCHEMATIC
SYMBOL

Figure 6-7c
Example circuit
to illustrate DC
conditions.

V1
30 V

RC
10 kΩ

COLLECTOR VOLTAGE

Q1

VARIABLE DC SUPPLY
0 VOLT TO 4 VOLTS

V2

EMITTER VOLTAGE

RE
1 kΩ

Referring to Fig. 6-7c, note that Q1 is connected to two power supply sources (illustrated as batteries). The V1 source is set at 30 volts DC and is fixed (i.e., it doesn't vary). The V2 source is connected to the base of Q1 and is variable from 0 volts DC to 4 volts DC. We will examine this circuit in several conditions to understand the concept of the *current amplification factor,* or "beta," of Q1.

Continuing to refer to Fig. 6-7c, assume the beta parameter for this transistor (Q1) to be 99. As stated previously, *beta* simply defines the ratio of the base current to the collector current. A beta parameter of 99 means that the collector current will be 99 times higher than the current flowing in the base circuit.

NOTE: *Typical transistor beta values can vary from 20 to over 300. I chose 99 for these example problems to make the calculations easier.*

If V2 is adjusted for a base voltage of 1 volt, approximately 0.7 volt will be dropped across the forward-biased base-emitter junction, leaving about 0.3 volt to be dropped across the emitter resistor (RE). Since the value of RE is known (1 Kohm) and the voltage across it is known (0.3 volt), the current flow through RE can be easily calculated using Ohm's law:

$$I = \frac{E}{R} = \frac{0.3 \text{ volt}}{1000 \text{ ohms}} = 0.0003 \text{ amps} \quad or \quad 300 \text{ microamps}$$

The current flow through RE is the same variable as the "emitter current flow" of Q1. Therefore, since the total current flow for both base and collector comes from the emitter, you know that the sum of the base current and collector current will equal 300 microamps. In addition, since the beta for Q1 is assumed to be 99, you know that the base current will be 99 times smaller than the collector current. Therefore, the base current will be approximately 3 microamps and the collector current will be approximately 297 microamps (297 microamps divided by 3 microamps = 99), with both current flows adding up to the total emitter current flow of 300 microamps.

Under the previous conditions, the emitter voltage of Q1 has already been determined to be 0.3 volt (i.e., the voltage across RE). The collector voltage of Q1 will be the V1 source voltage "minus" whatever voltage is dropped by the collector resistor (RC). The voltage drop across RC can be calculated using Ohm's law by multiplying the collector current (297 microamps) by the resistance value of RC (10,000 ohms):

$$E = IR = (0.000297 \text{ amps})(10,000 \text{ ohms}) = 2.97 \text{ volts}$$

Therefore, the collector voltage will be

$$30 \text{ volts} - 2.97 \text{ volts} = 27.03 \text{ volts}$$

Therefore, with V2 adjusted for 1 volt and an assumed beta of 99 for Q1, the collector voltage will be "about" 27 volts and the emitter voltage will be "about" 0.3 volt. Remember, all specified voltages in a typical schematic will be in reference to circuit common, or ground potential, unless the voltage is accompanied with a specific qualifying statement. Therefore, when measuring the collector voltage of Q1, you are actually measuring the voltage across RE as well as the voltage from emitter to collector of Q1.

Now, assume that all other circuit conditions for Fig. 6-7c remain the same, with the exception that V2 is adjusted to 2 volts. Subtracting the typical 0.7-volt base-emitter junction drop, the voltage across RE (which is the emitter voltage) will be about 1.3 volts. Using Ohm's law to calculate the emitter current:

$$I = \frac{E}{R} = \frac{1.3 \text{ volts}}{1000 \text{ ohms}} = 0.0013 \text{ amp} \quad or \quad 1.3 \text{ milliamps}$$

Again, assuming the beta of Q1 to be 99, the base current flow will be about 0.013 milliamps and the collector current flow will be about 1.287 milliamps (1.287 milliamps divided by 0.013 milliamp = 99). The voltage dropped by RC under these conditions will be

$$E = IR = (1.287 \text{ milliamps})(10{,}000 \text{ ohms}) = 12.87 \text{ volts}$$

Therefore, the collector voltage will be

$$30 \text{ volts} - 12.87 \text{ volts} = 17.13 \text{ volts}$$

Therefore, with V2 adjusted for 2 volts and an assumed beta of 99, the collector voltage will be "about" 17 volts and the emitter voltage will be "about" 1.3 volts. As you recall, when V2 was adjusted for 1 volt, the collector voltage was approximately 27 volts and the emitter voltage was approximately 0.3 volt. In other words, an *increase* of 1 volt to the base of Q1 resulted in a *decrease* of about 10 volts at the collector (27 volts − 17 volts = 10 volts) and an *increase* of about 1 volt on the emitter. Imagine that the V2 voltage cycled up and down, like a seesaw, from 1 volt to 2 volts, back to 1 volt, back to 2 volts, and so on. In other words, it is continually changing by 1-volt levels. Under these conditions, the collector voltage of Q1 would be changing by 10-volt levels, *but it would be always be going in the opposite direction of the base voltage.* If the base voltage increased, the collector voltage would decrease, and vice versa. Therefore, it is said that the collector voltage is 180 degrees out of phase, or *inverted*, compared to the base voltage. In gist, the collector voltage shows a voltage *gain* of 10, but it is inverted. In contrast, the emitter voltage is in phase, or *noninverted*, compared to the collector, but it exhibits no voltage gain. However, it does provide *current gain* (i.e., the emitter current is approximately equal to beta times the base current).

You have just examined the fundamental operation of a common-emitter amplifier and a common-collector amplifier as discussed in the previous section. Although some variations will occur in these basic circuit configurations to accommodate cost, efficiency, and performance factors, the fundamental difference between a common-collector amplifier and a common-emitter amplifier is simply the decision to use the signal output from either the emitter or collector, respectively. In other words, if you use the signal from the collector, the amplifier is a common-emitter configuration. If you use the signal from the emitter, the amplifier is a common-collector configuration.

Again referring to Fig. 6-7c, you should be familiar with two more common fundamentals of transistor operation. Suppose that V2 were adjusted to only 0.1 volt. This low base voltage is not sufficient to forward-bias the base-emitter junction of Q1. Therefore, there is no significant current flow in either the base or emitter. Likewise, since there is no emitter current, there cannot be any consequential collector current. Without any current flow through RC, the voltage drop across RC is zero. Because the collector voltage of Q1 will equal the V1 voltage (30 volts) minus the voltage drop across RC (zero, in this case), the collector voltage will become the entire V1 voltage, or +30 volts. Likewise, since there isn't any emitter current flow, the voltage drop across RE is zero, and the emitter voltage is therefore zero. A transistor in this condition, where the base current is zero due to insufficient "drive" voltage to promote current flow, is said to be *cut off*, or in a condition of *cutoff.*

In contrast to cutoff, suppose that V2 in Fig. 6-7c is increased to 4 volts. Subtracting the normal base-emitter drop of 0.7 volt, this leaves 3.3 volts across RE. Therefore, the emitter current is approximately 3.3 milliamps. Neglecting the small base current and assuming the collector current to be "close" to the emitter current, this would mean that RC would "try" to drop about 33 volts (3.3 milliamps × 10 Kohm = 33 volts). However, this is quite impossible since the V1 power supply is only 30 volts. A transistor circuit in this condition, wherein an increase of the base current (or voltage) causes no further change in collector voltage (or current) is said to be in *saturation*. As should be readily apparent, cutoff and saturation are the two opposite operational extremes of any functional transistor circuit.

As an aid to visualizing the concepts of cutoff and saturation, assume you adjusted V2 in Fig. 6-7c down to 0.1 volt and slowly began increasing it. Q1 would remain in cutoff until you reached a V2 voltage of about 0.7 volt. As you continued to slowly increase V2 above 0.7 volt, the transistor circuit would be in its *active* operational region, because the associated voltages and currents would be changing with the changing base voltage in a linear, or *proportional,* manner. However, as you continued to increase V2 above 3 volts, you would soon reach a point wherein the transistor circuit would saturate, causing any further V2 increases to have little or no effect on the collector voltage or current.

The transistor circuit of Fig. 6-7c is useful for demonstration purposes, but practical amplifiers are called on to amplify AC voltages that periodically change polarities. As illustrated in Fig. 6-7c, Q1 would fall into its cutoff region whenever an AC signal (applied to Q1's base) started to go into the negative region. For this reason, AC voltages intended to be amplified must be "lifted up" to some DC level that falls within the *active operational region* of a transistor amplifier circuit to facilitate amplification of the *entire AC waveform*. Figure 6.7d illustrates the basic principle of how this is accomplished. Note that Fig. 6.7d is essentially the same circuit as Fig. 6-7c with the addition of an AC signal source inserted in series with the V2 DC source of the base lead. The amplitude of this AC signal source is 2 volts peak to peak, which means that it normally rises to 1-volt peaks above the positive level and falls to 1-volt peaks in the negative direction. However, in this case, the AC signal source is in series with a constant DC *bias* of 1.7 volts. Therefore, as the base "sees" the composite signal, it rises to 2.7-volt peaks in the positive direction (1.7-volt bias + 1 volt positive peak = 2.7 volts) and falls to minimum levels of 0.7 volts when the AC voltage reaches its maximum negative-going peak (1.7 volt bias − 1 volt negative peak = 0.7 volt). In other words, by causing the AC voltage to "ride" on a steady-state DC voltage, or *bias*, the base of Q1 never goes below its required 0.7-volt cutoff extreme, and the entire AC signal is amplified by a factor of 10 at the collector of Q1. Therefore, the AC collector voltage signal will be amplified to 20 volts peak to peak, but it will be inverted in respect to the original signal on the base. The emitter voltage will be about 2 volts peak to peak (P-P) (minus a negligible loss of amplitude), but it will be noninverted and the current gain will be

Figure 6-7d
Example circuit to illustrating AC operation.

approximately equal to beta. (The importance of this current gain factor will become clear in a few circuit examples presented later.)

The principle of establishing various steady-state DC levels to place transistorized circuits into their most optimized active operational regions is called *biasing*. The steady-state DC levels used in establishing a correct bias are typically called the *quiescent* settings. The term *quiescent* refers to the "steady-state conditions" of any electronic circuitry, in contrast to the term *dynamic*, which applies to the "changing conditions" of electronic circuitry.

Although Fig. 6-7*d* illustrates the principle of establishing a functional base bias for Q1, Fig. 6-7*e* illustrates a more practical method of accomplishing the same thing. Note that a simple voltage divider, consisting of R1 and R2, has replaced the V2 bias source of Fig. 6-7*d*. With the resistance values chosen, the positive quiescent voltage applied to the base of Q1 will be about 2.2 volts (2.2 volts is a little more "optimum" than 1.7 volts for providing the largest possible collector voltage swings). Since the voltage divider receives its operational power from the 30-volt V1 supply, only one power supply is required for operation. Rather than placing the AC source in series with the bias voltage, coupling capacitor C1 is used to *sum* the AC voltage onto the quiescent DC base voltage. Remember, a capacitor will appear to easily pass AC voltages while blocking DC voltages. Therefore, the DC voltage on the positive plate of C1 is the 2.2-volt base bias, but the DC voltage on the negative plate will be zero. Thus, the AC signal voltage is superimposed on (i.e., added to) the DC quiescent base bias; the end result is essentially the same as if the AC source were in series with the base bias voltage.

Figure 6-7e

Example circuit to illustrating a practical method of base bias.

Again referring to Fig. 6-7e, capacitors C2 and C3 provide the same type of coupling action as C1. The DC quiescent voltages on the collector and emitter of Q1 will be blocked by capacitors C2 and C3, respectively, providing only the "amplified" pure AC signal at the outputs. Note that the AC signal source of Fig. 6-7e is set to 0.707 volt RMS at a frequency of 1 kHz. Also, 0.707 volt rms is the same level of AC sinewave voltage as 2 volts P-P (0.707 volt rms = 1 volt peak = 2 volts P-P. If this is confusing, you should review the basic concepts of AC waveshapes provided in the beginning of Chapter 3 of this textbook.). The 1-kHz frequency denotes that 1000 complete cycles of AC voltage occur in every 1-second time period.

If you were to construct the Fig. 6-7e circuit exactly as illustrated, applying an AC signal input at the amplitude and frequency shown, you would obtain an *inverted* replication of the base signal on the "collector AC output" connection, with the amplitude increased to a level of 7.07 volts rms (a gain factor of 10). In addition, you would obtain an "exact" copy of the base signal at the "emitter AC output" connection (i.e., noninverted), and the amplitude would be almost identical to the base signal (there will be a slight, usually negligible, voltage amplitude "loss"). The amplifier circuit of Fig. 6-7e is a practical and well-performing *single-stage* (i.e., only one transistor stage is used) transistor amplifier that can be used in a variety of real-world applications. It will function well using a wide variety of transistor types and power supply voltages. Keep in mind that transistor amplifiers are seldom designed for the purpose of using the outputs from both emitter and collector simultaneously. If voltage gain is desired, the output will be taken exclusively from the collector, which classifies the amplifier circuit as a *common-emitter* configuration. In contrast, if only current gain is desired, the output will typically be taken from the emitter, which classifies the amplifier circuit as a *common collector* (or *emitter-follower*) design. Output coupling capacitors (such as C2 or C3 in Fig. 6-7e) may or may not be incorporated, depending on whether it is important to remove the "DC" component from the output signal.

As discussed briefly in the previous section, the voltage gain of the Fig. 6-7e amplifier circuit is established by the ratio of the collector resistor and emitter resistor. The 10-Kohm collector resistor (RC) divided by the 1-Kohm emitter resistor (RE) equals 10; therefore, the voltage gain is 10. If RE were reduced in resistance value, the voltage gain would rise proportionally, and this process would continue until the voltage gain approached the value of beta (H_{fE}). Since voltage gain is actually a product of the current gain (beta) converted to a voltage drop across the collector resistor, the voltage gain at the collector of Fig. 6-7e could never go above the beta value of Q1. However, there are practical limits to the gain obtainable in a single-stage amplifier that relate to both *repeatability and stability.*

Examining the concepts of stability first, suppose you needed a voltage amplifier with a voltage gain of 200, so you reduced the value of RE in Fig. 6-7e to 50 ohms to cause the RC/RE ratio to equal 200. Provided the beta value of Q1 were well above 200 and the bias voltage to the base were reduced, you could achieve your desired gain, but the circuit would become very *unstable.* This is because Q1 would be amplifying all changes to the base voltage, both AC and DC, by a factor of 200. As stated previously in this chapter, transistors, as well as most solid-state components, are sensitive to temperature changes. Typical ambient-temperature changes around Q1 would create small DC bias voltage changes on the base, which would be amplified by a factor of 200, and result in large DC shifts in the quiescent operating points. In other words, the circuit might work at 70 degrees, but become totally inoperative at 80 degrees. Therefore, it is said to be *unstable,* because its proper operation is easily hampered by normal, anticipated environmental changes.

The unstable attributes of single-stage transistor amplifiers operating at high voltage gain levels can be corrected by the modifications illustrated in Fig. 6-7f. Note that RE has been *bypassed* with a capacitor (C3) and the output signal is taken from Q1's collector. This modification essentially separates the AC and DC gain factors. Since a capacitor blocks any DC current flow, C3 looks like (resembles) an open circuit to the DC quiescent emitter voltage, so the DC gain of the Fig. 6-7f circuit is simply the RC/RE ratio, which comes out to 10. In contrast, C3 looks like a short-circuit to AC voltages, so the AC voltage gain factor increases to the maximum level allowed by Q1's beta. Depending on Q1's beta value, the actual AC voltage gain of this circuit will probably be about 200. This circuit

Figure 6-7f
Example circuit illustrating the effect of adding a bypass capacitor.

will be very stable as illustrated, because the voltage gain associated with the DC quiescent operational levels is held at 10, but it will not be *very repeatable.*

Repeatability applies to the capability of being able to reproduce the identical set of operating characteristics within multiple identical circuits. For example, suppose you constructed the circuit of Fig. 6-7*f*, compared the input signal to the output signal, and discovered the actual AC voltage gain to be 180. If you then removed Q1 from the circuit and replaced it with another identical 2N3904 transistor, would the AC voltage gain still be 180? In all probability, it would not. Transistors are not manufactured with exact beta values; rather, they are specified as meeting a required *range* or *minimum* beta value. Two identical transistors with the same part number could vary by more than 100% in their actual beta values. Since the AC voltage gain of Fig. 6-7*f* is largely dependent on the beta value of Q1, you could construct a dozen identical circuit copies of Fig. 6-7*f* and achieve variable AC voltage gains ranging from 120 to 250. Therefore, the *repeatability* of such a circuit is said to be poor.

In its simplest form, you can think of any single-stage transistor amplifier circuit as a "block" with an input and an output. Like all electronic circuits having inputs or outputs, there will always be some finite input impedance and output impedance. So far in this section, you have examined the internal functions of transistor amplifiers, but it is also important to understand the effect such amplifiers have on *external* devices or circuits.

Referring back to Fig. 6-7*e*, the AC signal source in this illustration could represent a wide variety of devices. It could be a lab signal generator, a previous transistor amplifier stage, a signal output from a radio, the electrical "pickup" from an electric guitar, and the list goes on and on. It represents any conceivable AC voltage that you want to amplify for any conceivable reason. Regardless of what this AC signal source represents, it will have an *output impedance.* In order for transistor amplifiers to function well in a practical manner, the output impedance of the intended signal source must be compatible with the *input impedance* of the transistor amplifier. Otherwise, your intended amplifier may turn out to be an attenuator (i.e., a signal reducer).

Before going into a detailed description of input and output impedances, it is important to understand that all power supplies look like a low impedance path to circuit common (or ground potential) to AC signals. In other words, the *internal impedance* of all high-quality power supplies must be very low. As a means of understanding this principle, you can try a little experiment with a common 9-volt transistor battery (the small rectangular type used in low-power consumer products) and a 100-ohm resistor. First, measure the DC voltage at the battery terminals with your DVM. If the battery is new, you should read about 9.4 volts, or a little higher. Now, using a couple of clip leads, connect a 100-ohm, $1/_2$-watt resistor across the battery terminals and quickly measure the voltage across the resistor (the resistor will become hot if you leave it connected to the battery for more than a few seconds). You should see a noticeable "drop" in the battery output voltage. This "drop" in output voltage represents the voltage dropped across the "internal impedance" of the battery. When I tried this experiment with a new battery, I measured 9.46 volts "unloaded" (i.e., without any load placed on the battery) and 9.07 volts "loaded" (i.e., with the 100-ohm resistor connected across the battery's terminals). You will probably obtain similar results. By performing a simple Ohm's law calculation, you can use your unloaded voltage and loaded voltage measurements to determine the internal impedance of the battery. For example, in my case, the difference between 9.46 volts (unloaded) and 9.07 volts (loaded) is 0.39 volt. This tells me that while 9.07 volts was being dropped across the external 100-ohm resistor, 0.39 volt was being dropped across the internal impedance of the battery. The current flow through the 100-ohm resistor while it was connected to the battery was:

$$I = \frac{E}{R} = \frac{9.07 \text{ volts}}{100 \text{ ohms}} = 0.0907 \text{ amp} \quad or \quad 90.7 \text{ milliamps}$$

Since the battery and 100-ohm resistor were in series with each other while connected, the previous calculation tells me that the internal impedance of the battery was a resistance value that caused 0.39 volt to be dropped when the current flow was 90.7 milliamps. Using these two known variables, I can now calculate the actual internal impedance of the battery:

$$R = \frac{E}{I} = \frac{0.39 \text{ volt}}{0.0907 \text{ amp}} = 4.299 \text{ ohms}$$

If you happen to try the same experiment with a used battery, you will discover that the difference between its "unloaded" and "loaded" voltages is much more extreme. In other words, as batteries become drained of their energy-producing capabilities, their internal impedance rises. An ideal power supply would exhibit zero ohms of internal impedance.

Taking the principle of the low impedance characteristic of power supplies one step further, refer back to Fig. 5-4 of the previous chapter.

Imagine you were going to use this simple power supply circuit to provide operational power to a transistor amplifier stage, similar to that shown in Figs. 6-7e or f. The transistor amplifier stage would take the place of R_{load} as illustrated in Fig. 5-4, so capacitor C1 would be in parallel with the entire amplifier stage. As stated previously, capacitors look like a short to AC voltages (typically), so from the perspective of any AC signal voltage, the positive side of C1 would appear to be connected directly to circuit common. Therefore, AC line-operated DC power supplies also appear to have extremely low internal impedances from the perspective of any AC signal voltages.

Keeping the aforementioned principles in mind, refer to Fig. 6-7e once again. Note that the top end of R1 connects directly to the 30-volt power supply. However, from the perspective of the AC signal source, the top end of R1 connects directly to circuit common, because the internal impedance of the power supply is assumed to be very low.

Figure 6-7g is an equivalent circuit used to illustrate the AC input impedance of the transistor amplifier circuit of Fig. 6-7e, where Z_s symbolizes the *output impedance* of the AC signal source. Although C1 is shown, note that it is shorted from the negative to the positive plate, because it looks like a short to AC signals. R1 is shown in parallel with R2 because, from the perspective of the AC signal source, the top end of R1 is connected to circuit common, which places it in parallel with R2. And finally, the base-emitter input impedance of Q1 will look like the value of RE (1 Kohm) multiplied by the beta value of the transistor. If Q1's beta happens to be around 200, then this impedance value will be approximately 200 Kohm.

The total *equivalent impedance* of the three parallel impedances of Fig. 6-7g (i.e., R1, R2, and beta × RE) can be calculated using Eq. (2-5) of Chapter 2. (All of the fundamental laws and equations applicable to resistance

Figure 6-7g
Equivalent circuit of the AC signal input impedance of Fig. 6-7e.

values apply equally to impedance values.) In other words, the conductance value of each impedance is calculated, the three conductance values are summed, and the reciprocal of the sum is calculated. If you perform this calculation on the three parallel impedances of Fig. 6-7g, you should come up with approximately 7.3 Kohm. Therefore, from the perspective of the AC signal source, the entire circuit of Fig. 6-7e simplifies down to a simple series circuit, consisting of the AC signal source in series with its own internal impedance, which is in series with the 7.3-Kohm input impedance of the amplifier stage.

To understand the importance of considering the input impedance of typical transistor amplifiers, consider the following hypothetical circumstances. Suppose that the AC signal source of Fig. 6-7g has an internal impedance (Z_s) of 100 ohms, and the amplitude of the signal is 1 volt rms. Under these conditions you have a 1-volt AC source in series with a 100-ohm impedance (Z_s) and a 7.3-Kohm impedance (the equivalent input impedance of the amplifier stage). Using the simple ratio method of calculating voltage drops (as detailed in Chapter 2), you will discover that only 13.5 millivolts would be dropped across the internal AC source impedance (Z_s), with the remaining 0.9865 volt applied to the amplifier stage. This example situation represents a good, practical impedance match, because over 98% of the AC signal source is applied to the amplifier stage.

In contrast, many AC signal sources, such as *piezoelectric transducers* or certain types of *ceramic transducers*, have very high internal impedances. Looking at another hypothetical situation, suppose Z_s in Fig. 6-7g were 100 Kohm, with the AC signal amplitude remaining at 1 volt rms. Again using the ratio method of calculating the AC voltage drops, you'll discover that about 0.932 volt will be dropped across the internal impedance of the AC signal source, and only 68 millivolts will be applied to the input of the amplifier stage. Since the amplifier of Fig. 6-7e provides a voltage gain of 10 (if the output is taken from the collector), the end result would be an "amplified" output voltage of only about 680 millivolts (i.e., 0.68 volt). In other words, your amplified signal is "lower" in amplitude than your original applied AC signal voltage! This condition results because the impedance of the AC signal source is not very compatible with the input impedance of the amplifier stage.

The procedure of insuring a practical and efficient impedance compatibility from one stage (or source) to another stage (or source) is commonly called *impedance matching*, or simply *matching*. The concept of impedance matching does not imply that two impedances are "equal." Rather, it denotes the fact that two impedances are properly chosen for the desired results. For example, the ideal situation for the maximum transfer of a voltage signal, as illustrated in Fig. 6-7e, is for the internal impedance of the AC signal source (Z_s) to be "zero," with the input impedance of the amplifier stage as high as possible. In this way, virtually "all" of the voltage is applied to the amplifier stage and none is lost across the internal impedance of the source.

Figure 6-7h is an equivalent circuit illustrating the AC input impedance of the amplifier stage illustrated in Fig. 6-7f. Note that the parallel effect of R1 and R2 remained unchanged, but the incorporation of a bypass capacitor in Fig. 6-7f changed the base-emitter impedance as seen by the AC signal source. Since a capacitor typically looks like a short to an AC signal, incorporating C3 in Fig. 6-7f caused the emitter of Q1 to look as though it were shorted to circuit common. In other words, the AC signal no longer sees the effect of resistor RE. Therefore, in this case, the base-emitter impedance becomes the internal base-emitter junction impedance (called r'_e) multiplied by the beta parameter of the transistor; r'_e is typically approximated with the equation $r'_e = 25/I_c$, where I_c = quiescent collector current in milliamps.

The quiescent DC base voltage established by R1 and R2 in Fig. 6-7f is approximately 2.2 volts. Subtracting the typical 0.7-volt drop across the base-emitter junction leaves about 1.5 volts across RE (remember, capacitor C3 looks like an open circuit to DC voltages, so the quiescent operating voltages are not affected by C3); 1.5 volts across emitter resistor RE indicates that about 1.5 milliamps of current is flowing through the emitter. Considering the base current to be negligible, this means that approximately 1.5 milliamps will be flowing through the collector as well. Using the previous equation to calculate r'_e, we obtain

$$r'_e = \frac{25}{I_c} = \frac{25}{1.5} = 16.67 \text{ ohms}$$

(Remember, I_c must be in terms of "milliamps" for this calculation.)

Figure 6-7h
Equivalent circuit of the AC signal input impedance of Fig. 6-7f.

If you look up a 2N3904 transistor in a data book, you'll discover that its typical beta (H_{fE}) parameter is about 200. Therefore, the base-emitter impedance of Fig. 6-7f will be r'_e (16.67 ohms) multiplied by the beta (200), which equals about 3.33 Kohm. Now that all three parallel impedance values of Fig. 6-7h are known, you can calculate the equivalent impedance in the same manner as detailed for Fig. 6-7g. In this case, it comes out to about 2315 ohms, or about 2.3 Kohms.

If you recall, the AC input impedance for the transistor amplifier illustrated in Fig. 6-7e was 7.3 Kohm. By incorporating bypass capacitor C3, as illustrated in Fig. 6-7f, the AC input impedance was reduced to 2.3 Kohms. However, by incorporating the bypass capacitor (C3), the amplifier's AC voltage gain was greatly increased.

Thus far in this "transistor workshop," you have examined transistor fundamentals as they applied to signal amplification circuits. Transistors are also commonly used in various types of *regulator* circuits. *Regulation* is a general term applied to the ability to maintain, or "hold constant," some circuit variable, such as voltage or current. An older term synonymous with regulation is *stabilization,* which is still used commonly in some European countries (especially the United Kingdom).

Figure 6-8a illustrates a *constant-current source.* As the name implies, constant-current sources provide a *regulated* current flow. In other words, the current flow is held at some constant level even though the resistance of the load varies greatly. Constant-current sources are of vital importance to a great variety of circuits, but their uses will be covered in later chapters of this textbook. For now, only their theory of operation is discussed.

Continuing to refer to Fig. 6-8a, note that diodes D1 and D2 are kept in a condition of continuous "forward bias" through resistor R1. As you

Figure 6-8a

An example of a simple constant current source.

may recall from Chapter 4, a silicon forward-biased diode will produce a reasonably constant voltage drop of about 0.7 volt, regardless of how much forward current it is passing. Since diodes D1 and D2 are held in a constant forward bias from the V_{cc} power supply (through R1), the base voltage of Q1 will be the sum of the two 0.7-volt drops, or approximately 1.4 volts DC. The base-emitter junction of Q1 will drop about 0.7 volts of this 1.4 volt base bias, leaving approximately 0.7 volts across RE. Therefore, using Ohm's law, the emitter current of Q1 is

$$I = \frac{E}{R} = \frac{0.7 \text{ volt}}{700 \text{ ohms}} = 0.001 \text{ amp} \quad or \quad 1 \text{ milliamp}$$

Of course, about $\frac{1}{200}$th of this 1-milliamp emitter current will flow through the base (assuming that Q1 has a beta of 200), but if you consider this small base current to be negligible (which is appropriate in many design situations), you can say that about 1 milliamp of current must also be flowing through the collector. Note that the variable controlling the collector current flow is the constant voltage dropped across RE, which is held constant by the stable voltage drops across the two diodes. In other words, the collector *resistance* has nothing to do with controlling the collector *current*. You should be able to adjust the 10-Kohm potentiometer (i.e., the collector resistance) from one extreme to the other with almost no change in the approximate 1-milliamp collector current flow.

You may want to construct the circuit of Fig. 6-8a for an educational experiment. If so, construct RE a 620-ohm resistor (700 ohms is not a standard resistor value—I chose this value in the illustration for easy calculation). If you don't have the 1N4148 diodes, almost any general-purpose diodes will function well. When I constructed this circuit, my actual collector current flow came out to 0.9983 milliamps at the minimum setting of the potentiometer (I was lucky—actual results seldom turn out that close on the first try). By adjusting the potentiometer to its maximum resistance value, the collector current decreased to 0.9934 milliamps. This comes out to a regulation factor of 99.5% (i.e., the regulated current varied by only 0.5% from a condition of minimum load to maximum load), which is considered very good.

Many types of electronic circuits require a very accurate and steady amplitude level of DC operational voltage. The problem with a simple "raw" (i.e., unregulated) power supply, such as the types discussed in Chapter 5, is that the output voltage(s) will vary by about 10–30% as load demands change. Consequently, *voltage regulator* circuits are needed to hold voltage levels constant regardless of changes in the loading conditions. Figure 6-8*b* illustrates a simple method of maintaining a constant voltage across a load. A raw DC power source is applied to the collector of Q1, with a *reference voltage source* of 5.7 volts applied to the base. Allowing for the typical 0.7-volt drop across Q1's base-emitter junction, about 5 volts should be dropped across the emitter load (think of the emitter load as being an emitter resistor, such as RE in Fig. 6-8*a*). For illustration purposes, a potentiometer is shown as a variable load in Fig. 6-8*b*. Note that the setting of the potentiometer does not control the voltage drop across the variable load. Rather, it is maintained at about 5 volts as a function of the constant 5.7 volts applied to the base of Q1 and the effect of Q1's beta, which tries to keep the emitter voltage equal to the base voltage (minus the 0.7-volt base-emitter drop). By adjusting the potentiometer, both the emitter current and the collector current will change radically in proportion to differences in the load resistance, but the voltage across the variable load will remain relatively constant. Because Q1 is connected in series between the raw DC power supply and the load, it is often referred to as a *series-pass transistor.*

At this point, you may be wondering why it is advantageous to incorporate Q1 in the first place. Why not simply provide operational power to the variable load directly from the stable reference voltage source? The answer to this very reasonable question is the simple fact that almost all high-quality sources of reference voltages have very limited output current capabilities. In other words, as soon as you begin to draw higher operational currents from the reference voltage supply, the voltage output will drop and it will cease to be a "reference voltage." Therefore, you need to make the reference voltage into a "controlling factor," while drawing the actual operational "power" from a different source. The regulator circuit of Fig. 6-8*b* utilizes the current amplification factor (beta) to control the "large" emitter-collector current flow with only a small base current flow provided by the reference voltage. In this way, the reference voltage and the voltage applied to the load remain stable, while the raw DC power supply provides almost all of the operational power to the load (i.e., almost all of the load current is provided by the raw power supply). For example, if you assume the beta parameter of Q1 to be 200, then the current requirements placed on the reference voltage will only be $1/200$th of the actual operational current supplied to the load.

Figure 6-8c illustrates how two transistors and a few passive components can be configured into a "voltage-regulated, adjustable DC" regulator. This type of regulator circuit allows you to adjust the level (or amplitude) of the DC output voltage with potentiometer P1. Once the output voltage is adjusted to the desired level, the regulator circuit will maintain this voltage level regardless of major variations in the load current (i.e., the voltage will be "regulated" at whatever level it is set to by P1).

The fundamental principle of operation for Fig. 6-8c is essentially the same as for Fig. 6-8b. That is, a reference voltage is applied to the base of the series-pass transistor (Q1) that is about 0.7 volt higher (in amplitude) than the desired voltage applied to the load. The raw DC power supply is connected to the collector of Q1, which will supply the majority of operational power delivered to the load. Transistor Q2, diodes D1 through D3, R2, and P1 form an "adjustable reference voltage" that is applied to the base of Q1, which, in turn, provides an "adjustable output voltage" at the emitter of Q1.

Figure 6-8c

More sophisticated adjustable voltage regulator circuit.

As a means of understanding the operation of Fig. 6-8c, imagine that Q2 and P1 are removed from the circuit, with the base of Q1 unconnected to anything except R1. Under these conditions, when the unregulated DC voltage is applied to the collector of Q1, the base-emitter current flow of Q1 will be forward-biased through resistor R1. The emitter voltage will rise to the same voltage as the unregulated DC voltage, minus a small drop across Q1. This is because the base of Q1 is at about the same voltage as the collector (the collector voltage is applied to the base by R1). The positive emitter voltage will forward-bias D1, D2, and D3 through R2. The voltage at the anode of D1 will be approximately 2.1 volts (0.7 volt + 0.7 volt + 0.7 volt = 2.1 volts; the three forward threshold voltages of the three diodes summed). As you recall, the forward threshold voltage of a forward-biased diode is relatively "constant" regardless of the forward current flow. So the 2.1 volts produced by the forward-biased diodes serves the function of a "voltage reference," since the voltage drop across R2 nor the current flow through the diodes will alter this voltage by any great degree.

Now imagine adding P1 to the circuit of Fig. 6-8c. Since P1 is connected from the emitter of Q1 to circuit common, the voltage at the "tap" of the potentiometer can be at any level, from circuit common to the full emitter voltage, depending on how the potentiometer is adjusted. Finally, imagine that Q2 is now incorporated into the complete circuit as illustrated, with P1 adjusted to "tap off" about 50% of Q1's emitter voltage. Under these conditions, when the unregulated DC voltage is applied to the collector of Q1, R1 will forward-bias the base-emitter junction of Q1, causing Q1's emitter voltage to begin to rise. As Q1's emitter voltage rises slightly above 2.1 volts, diodes D1, D2, and D3 go into forward conduction (through R2), applying about 2.1 volts to the emitter of Q2. As Q1's emitter voltage reaches a level of about 5.6 volts, the "tap voltage" of P1 is applying 2.8 volts to the base of Q2 (i.e., 50% of 5.6 volts is 2.8 volts). Since the emitter of Q2 is biased at 2.1 volts from the diode reference voltage source, and 2.8 volts is applied to its base by P1, the base of Q2 is now 0.7 volt "more positive" than the emitter, causing Q2 to begin to conduct. A type of balance occurs at this point. As the emitter voltage of Q1 tries to continue to rise, the voltage to the base of Q2 also rises, causing the collector-emitter current flow of Q2 to increase, which steals current away from the base of Q1 and restricts its emitter voltage from rising any higher than 5.6 volts. In other words, Q1's emitter voltage (which is the regulator's output voltage) is *stabilized*, or *regulated*, at this voltage level.

Now that the fundamentals of the circuit operation of Fig. 6-8c are understood, consider the effects of regulation when a hypothetical load is connected to its output. First, recognize that the conditions described in the previous paragraph applied to the no-load condition of the Fig. 6-8c circuit. In other words, no circuit or device of any type is drawing operational power from the output of the regulator. Now, assuming P1 to be left at its previous 50% setting (causing a regulated output voltage of about 5.6 volts), imagine connecting a 5.6-ohm resistor from the output terminal to circuit common. According to Ohm's law, if 5.6 volts is placed across a 5.6-ohm resistor, 1 amp of current will flow through that resistor. As soon as the load resistor is connected, the voltage at the output of the regulator will try to drop. However, as soon as it begins to drop, the voltage to the base of Q2 starts to drop also. This action decreases the collector-emitter current flow of Q2, causing an increase of base current to Q1, which, in turn, raises Q1's emitter voltage back to the balanced state of 5.6 volts. In other words, even with the extreme contrast of a no-load condition to a 1-amp load, the voltage output of the regulator circuit remained relatively constant. However, it is likely that the unregulated DC voltage applied to the collector of Q1 dropped by 1 or 2 volts.

A few final principles of the Fig. 6-8c circuit should be understood. First, the regulated output voltage can be adjusted to any voltage level between the extremes of the reference diode voltage (i.e., 2.1 volts) and the level of the unregulated DC input voltage, by adjusting the setting of P1. However, regulation will become very poor when the output voltage level gets close to the unregulated input voltage level. Also, in the previous functional descriptions, I described voltages "rising, falling, trying to fall," and so forth. It should be understood that these changes take place in a few microseconds, so don't expect to see such actions without the assistance of a high-quality oscilloscope. And finally, there isn't any such thing as a "perfect" regulator circuit. When I constructed this circuit and adjusted P1 for a 5-volt output, there was a 0.3-volt drop in my output voltage from a no-load to a 1.5-amp loaded condition. By dividing the *loaded voltage* by the *unloaded voltage*, and then multiplying by 100, my *percent regulation* for this circuit came out to 94% regulation (4.7 volts divided by 5 volts $\times$ 100 = 94%). This regulation factor can be improved by utilizing transistors with higher H_{fE} parameters.

If you would like to construct the regulator circuit of Fig. 6-8c, the raw DC power supply illustrated in Fig. 5-4 of the previous chapter will function very well as the unregulated DC voltage input source. SW1 and R_{load} in Fig. 5-4 would not be used for this application. Remember to properly "fuse-protect" the primary of T1. There will be one disadvantage to the completed power supply, however. If the load resistance goes too low, or if you accidentally short-circuit the output to circuit common, you will probably destroy the series-pass transistor, Q1. This is because there isn't any form of *current-limit protection* incorporated into the regulator circuit.

All common transistors have a *maximum collector current* parameter that cannot be exceeded without destroying the transistor. While common types of fuses can adequately protect many electrical and electronic devices (such as motors, transformers, and resistors), they cannot provide "overcurrent" protection to transistors, because the transistor will be destroyed before the fuse has time to react. Also, in many situations a certain electronic application may require *current limiting* to a special type of load or circuit. For such applications, a rapid solid-state method is needed to freely allow current flow up to a limit, but restrict it at the limit if it tries to rise any higher. Such circuits are appropriately called *current-limit circuits* or *current-limit protection*.

Figure 6-8d illustrates how a transistor (Q1) can provide *current-limit protection* for a transistor (Q2) or a load (RC). Assume, for example, you wanted to set the maximum collector current flow for Q2 at 10 milliamps. If you assume the base current to be negligible, you can view the emitter current as being approximately equal to the collector current (this is a reasonable approximation with high-beta transistors). At emitter current levels below 10 milliamps, the emitter-resistor (RE) voltage will drop to less than 0.7 volt, so the forward base-emitter voltage of Q1 cannot be obtained, and Q1 has no effect on the circuit. However, if Q2 reaches 10 milliamps of emitter current flow, the voltage drop across RE reaches the typical forward base-emitter voltage threshold of Q1 (i.e., 0.7 volt), initiating base-emitter current flow through Q1. As

Figure 6-8d
A simple method
of current limit
protection.

soon as some amount of base-emitter current flow begins, Q1 begins to promote a collector current equal to its beta parameter multiplied by its base current. Since Q1's collector current has to be provided by the *base input* to Q2, a certain portion of Q2's *base drive current* will be diverted to circuit common through transistor Q1, pulling the base input down to a level that will not cause more than 10 milliamps of emitter current flow through Q2. In other words, a *current limit* will occur when the voltage across RE reaches about 0.7 volt (corresponding to about 10 milliamps of emitter current flow, which also corresponds to about 10 milliamps of collector current flow). As the base drive current to Q2 tries to increase beyond this point, Q1 will divert more of the drive current to circuit common to maintain the 0.7-volt drop across RE. Within practical limits, the emitter current of Q2 can never exceed 10 milliamps. It should also be noted that the current flow through RC cannot exceed 10 milliamps, even if RC is reduced to zero ohms. Therefore, Q1 provides current-limit protection for both Q1 and the collector load of RC.

You may want to construct the circuit illustrated in Fig. 6-8d to experiment with. If so, RE can be 68 ohms (70 ohms is not a standard value) and you can place a 1-Kohm potentiometer (connected as a rheostat, as illustrated) in the collector circuit to act as a variable load. You can supply the base input to Q2 with a variable DC power supply or battery, but be careful not to allow the "base drive current" to exceed the maximum collector-current parameter of Q1. Otherwise, Q1 could be destroyed.

Although transistors with very high H_{fE} (beta) parameters are available, in certain circumstances you may desire extremely high current gain factors. In these situations, a *Darlington pair* can be used (sometimes referred to as *beta-enhanced pairs*). Figure 6-8e illustrates two common methods of producing a Darlington pair. In its simplest form, a Darlington pair is created by simply connecting the emitter of one transistor to the base of a second, and connecting the two collector leads together. This is illustrated in the lefthand illustration of Fig. 6-8e. Darlington pairs can be regarded and used, as though they were a "single" transistor with three leads, as illustrated. However, the beta parameter of the first transistor is multiplied by the beta of the second. In other words, if you made a Darlington pair out of two transistors, each having a beta value of 100, the Darlington pair would act like a single transistor with a beta of 10,000 (i.e., $100 \times 100 = 10,000$).

Unfortunately, there are a few disadvantages with Darlington configurations. Since the beta value is so extraordinarily high, complications with temperature instability may arise. A more temperature-stable Darlington configuration is shown in the righthand illustration of Fig. 6-8e.

Figure 6-8e

Several examples of Darlington (i.e. beta-enhanced) transistor pairs. The circuit on the left is the more standard configuration, while the circuit on the right provides improved DC and temperature stability characteristics.

However, by adding the temperature-stabilizing resistor, the base impedance of the Darlington pair is lowered considerably. Since the base input of a Darlington pair literally consists of "two" base-emitter junctions, the required forward threshold voltage is doubled (i.e., the base lead would have to be biased with about 1.4 volts for base-emitter current flow, instead of the normal 0.7-volt minimum for a single transistor). Depending on the type of circuit in which a Darlington pair is being used, there can be other problems as well. In high-frequency applications, the speed limitations of one transistor tends to add to the speed limitations of the second, making Darlington pairs relatively slow in comparison to single transistors. Also, with some critical circuitry, the apparent difference in the internal capacitances can adversely affect operation.

Final Comments   Even though the uses of transistors are quite varied, the basic fundamentals of transistor operation are quite simple. Within every circuit you have examined in this section, the actual transistor operation consisted of the following basics:

1. When the transistor was forward-biased, there was a 0.7-volt drop across the emitter-base junction.

2. The base current was multiplied by the beta (H_{fE}) parameter promoting a collector current flow equal to the product of the two.

3. The emitter voltage was the same as the base voltage, minus the 0.7-volt base-emitter voltage.

4. The "total" current flow through the transistor originated in the emitter, with a minor portion flowing through the base circuit, while the vast majority flowed through the collector circuit.

As you have seen, these basic operational characteristics of transistors can be manipulated into providing DC voltage amplification, AC voltage amplification, current regulation, voltage regulation, current limiting, and beta enhancement (i.e., Darlington pairs). These basic fundamental building blocks of transistor applications can be combined in various ways to provide a large variety of more complex functions and circuitry.

All of the circuits in this transistor workshop section have incorporated NPN transistors. PNP transistors can be used in the identical manner if all of the circuit voltage polarities are reversed.

Throughout this book, I specify the "typical" forward threshold voltage of any forward-biased semiconductor junction as 0.7 volt. You may want to keep in mind that it is a little more technically accurate to consider this voltage to be 0.68 volt (i.e., 0.68 volt will be a little more accurate in the majority of cases). I elected to approximate this voltage at 0.7 volt for ease of calculation.

If this is your first reading of this section, I recommend that you now go back to the beginning of this chapter and study the entire contents up to this point again. The information contained in this chapter is concentrated and quite abstract to most individuals on their first exposure to it. Don't become disheartened if you are confused on some points. Most of these fundamentals will become clearer during your second study, and most of your questions will probably be answered in later chapters.

Assembly and Testing of Last Section of a Lab Power Supply

The following materials are needed to complete this section:

Quantity	Schematic Reference	Item Description
6	D1 through D6	1N4001 (NTE 116) general-purpose silicon diode
2	R1 and R2	1000-ohm, 2-watt resistor
2	R3 and R4	2.2-Kohm, 2-watt resistor
2	R7 and R8	470-ohm, $1/2$-watt resistor
2	R5 and R6	0.4-ohm, 5-watt resistor
2	P1 and P2	1000-ohm, linear taper 2-watt potentiometer
2	C3 and C4	0.1-μF ceramic disk capacitor
2	Q3 and Q5	TIP 31C (NTE 291) transistor
2	Q4 and Q6	TIP 32C (NTE 292) transistor
1	Q1	2SC3281 (NTE 2328) transistor

Quantity	Schematic Reference	Item Description
1	Q2	2SA1302 (NTE 2329) transistor
2	(Not illustrated)	3AG size ($\frac{1}{4} \times 1\frac{1}{4}$-inch) fuse blocks (optional)
2	(Not illustrated)	2-amp slow-blow fuses (optional)
3	See Fig. 6-10	Insulated banana jack binding posts (Radio Shack 274-661A or 274-662A or equivalent)
1	See Fig. 6-9	Universal grid board (Radio Shack 276-168A or 276-150A or 276-159A or equivalent)

The fuse blocks and fuses are listed in case you want to fuse-protect the positive and negative outputs. This is a good idea, but it is not mandatory. The fuse blocks can be mounted in any convenient location prior to the binding posts. The fuse/fuse block assemblies are simply placed in series with the output wires going to the positive and negative binding posts (see Fig. 6.12).

Figure 6-9
Examples of universal-type circuit boards that can be used to permanently construct many projects within this book.

Figure 6-10
Examples of the most popular styles of insulated binding posts.

It might be a little difficult to obtain the *series-pass transistors* (Q1 and Q2) on a local basis. However, they are readily available through catalog suppliers.

Assembling the Circuit Board

The specified grid boards are universal types, designed for building one-of-a-kind circuits or prototypes. They consist of three-point solder pads and long strips, or rails, called "buses." Circuits are assembled using the three-point solder pads as connection points, and the bus lines are used for common connections to power supply voltages or circuit common.

To begin laying out the circuit board, refer to Fig. 6-11. The first step is to recognize that all of the components shown in the schematic will not be placed on the circuit board. C1 and C2 are already installed on solder strips connected to the chassis (this was performed in the last chapter). Q1 and Q2 will be externally mounted to the chassis bottom for heatsinking. P1 and P2 will be mounted to the front panel, as your voltage adjustment controls.

As the second step, notice how the *positive regulator* is almost a mirror image of the *negative regulator* (the circuitry above the circuit common

Figure 6-11
Regulator section of the lab power supply project.

line is the positive regulator; the negative regulator is below the circuit common line). Divide the grid board in half and concentrate on the positive regulator section first. On completion, the negative regulator is simply a duplication on the first half of the grid board.

By the time you have completed the prior steps, you are left with four resistors, two transistors, three diodes, and one capacitor (the circuit board mountable components for the positive regulator). It is not difficult to arrange these few components for proper connections (Fig. 6-11). There is not a critical nature to these parts placements, so you might install them in any way that is easy and organized.

When you finish with the positive regulator section, copy the parts layout for the negative regulator. It is usually easier to wait until the entire circuit board is "populated" and double-checked, before permanently soldering any components into the board. Simply insert the components and bend the leads back to hold them in place temporarily. Don't forget to leave some empty holes in the pads for connecting the external components (Q1, Q2, P1, and P2) and the external wiring (raw DC inputs, regulated DC outputs, and the circuit common).

Figure 6-12 shows a suggested method of physically mounting all of the components together with the circuit board. You might have to move things around a little to accommodate your enclosure. If you choose to fuse the outputs, the fuse blocks should be wired in series between the regulated DC outputs and the binding posts.

Regarding the external wiring to the circuit board, it should be stranded, 18- to 22-gauge insulated hook-up wire. An itemized list of the external wiring is as follows:

- Three wires from the raw DC power supply (raw +DC, raw −DC, and circuit common)
- Three wires from Q1 (emitter, base, collector)
- Three wires from Q2 (emitter, base, collector)
- Six wires from P1 and P2 (three wires each)
- Three wires from the three binding posts (regulated +DC, regulated −DC, and circuit common)

Mounting Q1 and Q2

Under worst-case conditions, Q1 and Q2 might have to dissipate close to 60 watts of power each. To keep them from overheating, some form of

Figure 6-12
Suggested physical layout for the finished lab power supply.

heatsinking must be provided. The chassis metal bottom will work nicely for this purpose, unless you try to operate the power supply in the current limit mode at low voltage, or if short-circuit conditions exist for extended periods of time. This worst-case condition could cause the series-pass transistors (Q1 and Q2) to overheat, even with the chassis heatsinking. Specially designed aluminum heatsinks can be purchased for the mounting of Q1 and Q2 to eliminate the problem, if desired.

When mounting Q1 and Q2 to the chassis, you will have to use an insulator (mica insulators are the most common for this purpose) to keep the transistors *electrically isolated* from the chassis. To aid in heat conductivity, the mounting surface of the transistor and the insulator is coated with a "thin" layer of thermal joint compound, commonly called "heatsink grease." *Thermal joint compound* is a white, silicon oil-based grease used to fill the minute air gaps between the transistor, insulator, and heatsink material. Some power transistors require the use of an *insulating sleeve* to keep the mounting bolt from making electrical contact with the transistor collector. After mounting Q1 and Q2, use your DVM (in the "ohms" setting) to check for any continuity between the transistor's collector lead and the chassis. You should obtain an infinite resistance.

You might be wondering how you could check Q1 and Q2 during the power supply operation to know if they are getting too hot. Without the aid of expensive temperature measurement equipment, there is a very unscientific way to "guesstimate" possible temperature problems with semiconductors. *Be sure that the power is off.* Then, touch the transistors immediately after a period of operation. If they are too hot to comfortably hold your finger on them for more than 1 second, you should probably acquire some better heatsinking.

Circuit Description

The negative regulator functions identically to the positive regulator, except for the difference in voltage polarity. Therefore, I will detail only the operation of the positive regulator.

When power is first applied to the circuit, C1 charges to about 37 volts (as discussed in Chapter 5). R1 supplies sufficient base current to allow uninhibited current flow through Q1. As the Q1 emitter voltage approaches the "raw" DC level of C1, diodes D1, D3, and D5 all go into forward conduction through resistor R3. Once conducting, these diodes maintain a relatively constant *reference voltage* of about 2.1 volts, which is the sum of their individual forward threshold voltages.

As the Q1 emitter voltage continues to rise, it is applied to the R7 and P1 voltage divider. The P1 wiper is connected directly to the Q3 base. When the wiper voltage reaches about 2.8 volts, Q3 will begin to conduct. This occurs because the Q3 emitter is being held at 2.1 volts by the diode assembly, and its base has to be about 0.7 volt more positive than the emitter to overcome the forward threshold voltage requirement of the base to emitter junction. When Q3 starts to conduct, it begins to divert some of the current, being supplied by R1, away from the Q1 base. Consequently, Q1 begins to restrict some of its emitter-collector current flow. This *leveling effect* continues until the output is stabilized at a voltage output relative to the P1 setting.

When a load of some kind is placed on the output, the *loading effect* will try to decrease the output voltage amplitude. As the voltage starts to decrease, this decrease is seen at the P1 wiper and transferred to the Q3 base. Consequently, Q3 decreases its current conduction by an amount relative to the output voltage decrease. This allows more current to flow through the Q1 base, causing Q1 to become more conductive than it was previously. The end result is that the load is compensated for, and the output voltage remains stable (at the same amplitude as before the load was applied).

If a load is placed on the output, causing the current flow to exceed about 1.5 amps, the resultant current flow will cause the voltage drop across R5 to exceed the forward threshold voltage of the base-emitter junction of Q5. When Q5 begins to conduct, it diverts current flow away from the Q1 base, causing Q1 to decrease its current flow, and thus reduce the output voltage. The Q1-and-Q5 combination will not allow the output current to exceed about 1.5 amps, even under a *short-circuit condition.*

Testing the Power Supply

Before applying power to the power supply, set your DVM to measure ohms, and check for continuity of virtually every component to chassis ground. There is not a single part of this circuit that should be electrically connected to the chassis. If you measure resistance from any component to chassis ground, recheck your wiring and connections. Remember that you have a *"floating" circuit common* that is the power supply ground, but not connected to the chassis ground.

If you have double-checked your wiring, and are certain that everything is correct, set P1 and P2 to their approximate center positions, and "pulse" the line power to the power supply on and off quickly. If there were no blown fuses, or other signs of an obvious problem (smoke, the smell of burning plastic insulation, or the "pop" of exploding plastic semiconductor cases), reapply the line power and measure the output voltages at the binding posts with your DVM. The positive and negative output voltages should be measured in reference to circuit common (the center binding post as illustrated in Fig. 6-12), not chassis ground. You should be able to smoothly adjust the positive output voltage by rotating P1 from a minimum of about 3.8 volts, to a maximum of about 36 volts. The negative regulated output should perform identically by rotating P2, except for the difference in voltage polarity.

In Case of Difficulty

If the power supply is blowing fuses on application of line power, you probably have a *short-circuit condition* on one of the raw DC supplies. *Be sure that the line power is off,* and disconnect the raw DC wires from the circuit board. Replace any blown fuses and apply line power. If you still blow fuses, the problem is most likely in the bridge rectifier (BR1). You learned how to check a bridge rectifier in Chapter 4.

It is most likely that the fuses won't blow after disconnecting the raw DC lines because the complete raw DC power supply was previously tested in Chapter 5. If this is the case, *be sure that the line power is off,* and use your DVM to measure the resistance from the collector of Q1 to circuit common. This should be a very high indication (my circuit measures over 2 Mohms). Make the same measurement from the Q2 collector. If either one of these resistance measurements are low, you have a wiring error or a defective component.

The most likely components to be defective in this circuit are the transistors and diodes. You have already learned how to test diodes with a DVM set to the "diode test" function. Transistors can be tested in this manner also.

Checking Transistors with a DVM

If a need arises to functionally test a bipolar transistor, it can be thought of as two back-to-back diodes. With your DVM set to the "diode test" mode, check the base-emitter junction just like any diode junction. The DVM should indicate infinity with one test lead orientation, while reversing the test leads will cause a low indication, or vice versa. Because a transistor is a two-junction device, you must perform the same test on the base-collector junction also. If both junctions appear to be functioning normally, you should measure the resistance from the emitter to the collector. This resistance should be high with both orientations of the DVM test leads.

To check transistors that have already been installed in a circuit, it is usually necessary to consider the parallel effects of other associated components that might influence the measurement. For example, in Fig. 6-11, the base to emitter measurement of Q5 will be predominantly influenced by R5. Because R5 is only 0.4 ohm, it will not be possible to measure the forward or reverse resistance of the base-emitter junction because 0.4 ohm is much less than either. In this case, you would need to temporarily disconnect the base lead before checking the transistor with a DVM. In some cases, it might be necessary to disconnect two of the transistor's leads to obtain a reliable test.

If you discover a defective transistor, there is a good chance that it was destroyed by an error in the wiring. After replacing the defective transistor, double-check its associated wiring for mistakes.

7

Special-Purpose Diodes and Optoelectronic Devices

To this point, you have examined the application of diodes in only two general areas: rectification and voltage referencing (the three-diode voltage reference in the lab power supply project). There are many other applications, and numerous types of special-purpose diodes. *Optoelectronic devices* are electronic devices used in applications involving visual indicators, visible light, infrared radiation, and laser technology. Special-purpose diodes make up a large part of this family.

Zener Diodes

Zener diodes are used primarily as voltage regulator devices. They are specially manufactured diodes designed to be operated in the reverse-breakdown region. Every zener diode is manufactured for a specific reverse-breakdown voltage called the *zener voltage* (abbreviated V_z in most data books).

To understand the operational aspects of zener diodes, refer to Figure 7-1. Note the symbol used to represent a zener diode. For this illustration, a 5.6-volt zener has been chosen. Assume that you can apply a variable DC voltage to the positive and negative terminals of the circuit, ranging from 0 to 10 volts. Beginning at 0 volt, as the voltage is increased up to 5.6 volts, the zener diode behaves like any other reverse-biased diode. It totally blocks any significant current flow. And, because it represents an almost infinite resistance, the entire applied voltage will be dropped across it.

This all holds true up to the point when the applied voltage exceeds the rated zener voltage of the diode. When this *avalanche voltage* occurs, the zener diode will abruptly start to freely conduct current. The point at which this abrupt operational change occurs is called the *avalanche point*. The minimum amount of current flow through the zener diode required to keep it in an avalanche mode of operation is called the *holding current*. The voltage across the zener diode does not decrease when the avalanche point is obtained, but it does not increase by a very significant amount as the applied circuit voltage is increased substantially. Hence, the voltage across the zener diode is *regulated,* or held constant.

To clarify the previous statements, consider the circuit operation in Fig. 7-1 at specific input voltage levels. If 5 volts is applied to the input (observing the polarity as illustrated), the zener diode will block any significant current flow, because its avalanche point will not occur until the voltage across it reaches a level of at least 5.6 volts. If 6 volts is applied

Figure 7-1
Circuit demonstrating zener diode operation.

to the input, about 5.6 volts will be dropped across the zener, and about 0.4 volt will be dropped across R1. Increasing the applied input voltage to 7 volts causes the voltage across R1 to increase to about 1.4 volts, but the voltage across the zener diode will remain at about 5.6 volts. If the applied input voltage is increased all the way up to 10 volts, about 4.4 volts will be dropped by R1, but the zener diode will continue to maintain its zener voltage of about 5.6 volts. In other words, as the applied circuit voltage is increased "above" the rated voltage of the zener diode, the voltage across the zener diode will remain relatively constant and the excess voltage will be dropped by its associated series resistor, R1.

If a load of some kind were to be placed in parallel with the zener diode of Fig. 7-1, the zener diode would hold the voltage applied to the load at a relatively constant level, as long as the applied circuit voltage did not drop below the rated zener voltage of the zener diode.

The two most important parameters relating to zener diodes are the zener voltage and the rated power dissipation. Zener diodes are commonly available in voltages ranging from about 3 volts to over 50 volts. If a higher zener voltage is needed, two or more zener diodes can simply be placed in series. For example, if an application required the use of a 90-volt zener, this could be accomplished by placing a 51-volt zener in series with a 39-volt zener. Unusual zener voltages can be obtained in the same manner. Another method of obtaining an odd (nonstandardized) zener voltage value is to incorporate the 0.7-volt *forward threshold* voltage drop of a general-purpose silicon diode. When using this method, the general-purpose diode is placed in series with the zener diode, but it is oriented in the forward-biased direction, and the zener diode is reverse-biased.

The standardized *power dissipation ratings* for zener diodes are $\frac{1}{2}$, 1, 5, 10, and 50 watts. Zeners rated at 10 and 50 watts are manufactured in stud-mount casings, and must be mounted into appropriately sized heat sinks for maximum power dissipation.

Designing Simple Zener-Regulated Power Supplies

Referring to Fig. 7-2, your design problem here is to build a 12-volt, 500-milliamp zener-regulated power supply to operate the load, designated as R_L in the schematic diagram. Much of the "front end" of the circuit should be rather familiar to you by now. T1, the bridge rectifier, and C1 make up a raw DC power supply. (The methods of calculating the values and characteristics of these components will not be discussed in this context because this was covered in previous chapters.)

The design of a well-functioning zener-regulated power supply is a little tricky because of variations in the raw DC power supply. The output voltage of a raw DC power supply may decrease by as much as 25% when placed under a full load. To roughly estimate the no-load voltage that C1 would charge to, the T1 secondary voltage should be multiplied by 1.414 to calculate the peak secondary voltage. In this case, the peak secondary voltage is about 17 volts. After subtracting about 1 volt to compensate for the loss in the bridge rectifier, you are left with about 16 volts DC across C1. Unfortunately, this calculation will be in error as soon as the power supply is loaded.

The primary component affecting the full-load voltage decrease of a raw DC power supply is the transformer. The percentage of full-load voltage decrease will depend on how close the full-load requirement comes to the maximum secondary current rating of the transformer. For example, our hypothetical load in Fig. 7-2, as stated previously, will require up to 500 milliamps. If T1's secondary rating is 600 milliamps, the secondary voltage will decrease substantially when fully loaded. However, if T1's secondary rating is 2 amps, the 500-milliamps full-load requirement of R_L will have much less effect. In addition, even transformers with similar ratings can behave somewhat differently, depending on certain manufacturing techniques.

A further complication, although not as dramatic, relates to the value of C1. When a load is placed on a raw DC power supply, the ripple content increases. A high ripple content has the effect of reducing the usable DC level.

The easiest solution to overcoming all of these unknown variables is to simply build the raw DC power supply, and place a dummy load across C1 that will closely approximate the full-load requirement of R_L. In this case, you would start with the no-load voltage across C1, which is about

16 volts. Knowing that R_L might require as much as 500 milliamps, the resistance value of the dummy load can be calculated using Ohm's law:

$$R = \frac{E}{I} = \frac{16 \text{ volts}}{0.500 \text{ A}} = 32 \text{ ohms}$$

A 33-ohm resistor would be close enough for calculation purposes. But don't forget the power rating! This dummy resistor must be capable of dissipating about 8 watts.

Assume you built the raw DC power supply, placed the 33-ohm dummy load across C1, and measured the "loaded" DC voltage to be 14 volts. This gives you all the information you need to design the rest of the power supply. (You might have realized that when the raw DC voltage decreased under load, the 33-ohm dummy load no longer represented a full-load condition. Experience has shown that this "secondary" error, which is the difference between the "almost fully loaded" voltage and the "fully loaded" voltage, is not significant in the vast majority of design situations.)

There are three variables you must calculate to complete your design problem: the power rating of the zener, the resistance value of R1, and the power rating of R1.

To calculate these variables, you need to understand how the circuit should function under extreme variations of R_L. When R_L requires the full load of 500 milliamps, the current flow through the zener diode should be as close to the minimum holding current as possible. Assuming the holding current is about 2 milliamps, that means about 502 milliamps must flow through R1; 2 milliamps through ZD, plus 500 milliamps through R_L. R1 is in series with the parallel network of ZD and R_L. The applied voltage to the series-parallel circuit of R1, ZD, and R_L is the voltage developed across C1, which you are assuming to be 14 volts, under loaded conditions. Because 12 volts is being dropped across the parallel network of ZD and R_L, the remaining 2 volts must be dropped by R1 (the sum of all of the series voltage drops in a circuit must equal the source voltage). You now know the voltage across R1, and the current flow through it. Therefore, Ohm's law can be used to calculate the resistance value:

$$R = \frac{E}{I} = \frac{2 \text{ volts}}{0.502 \text{ amps}} = 3.98 \text{ ohms}$$

Of course, 3.98 ohms is not a standard resistance value. You don't want to go to the nearest standard value above 3.98 ohms because this would risk "starving" the zener diode from its holding current when the current flow through R_L was maximum. The nearest standard value below 3.98 ohms is 3.9 ohms, which is the best choice. By using any of the familiar power equations, the power dissipated by R1 comes out to be about 1 watt. A 2-watt resistor should be used to provide a good safety margin.

The worst-case power dissipation condition for ZD occurs when there is no current flow through R_L. If all current flow through R_L ceases, the full 502 milliamps must flow through ZD. Actually, the maximum current flow through ZD could be as high as 513 milliamps because you chose a 3.9-ohm resistor for R1 instead of the calculated 3.98 ohms. The power dissipated by ZD is the voltage across it (12 volts), multiplied by the current flow through it (the worst case is 513 milliamps). This is the familiar power equation $P = IE$. The answer is 6.15 watts. Therefore, ZD would need to be a 12-volt, 10-watt zener with an appropriate heatsink. Another option would be to use two 5-watt, 6-volt zeners in series. The latter option eliminates the need for a heatsink, but care must be exercised to assure plenty of "air space" around the zener diodes for adequate convection cooling.

As the previous design example illustrates, zener-regulated power supplies are not extremely efficient because the zener diode wastes a significant amount of power when the current flow through the load is small. For this reason, zener-regulated power supplies are typically restricted to low-power applications. However, zener diodes are commonly used as voltage references in high-power circuits, as is illustrated later in this chapter.

Varactor Diodes

Going back to diode fundamentals, you might recall that when a diode is reverse-biased, a "depletion region" of current carriers is formed around the junction area. This depletion region acts as an insulator resulting in the restriction of any appreciable current flow. A side effect of this depletion region is to look like the dielectric of a capacitor, with the anode and cathode ends of the diode acting like capacitor plates. As the reverse-bias voltage across a capacitor is varied, the depletion region will also vary in size. This gives the effect of varying the distance between the plates of a capacitor, which varies the capacitance value. A diode that is specifically designed to take advantage of this capacitive effect is called a *varactor diode*. In essence, a varactor is a *voltage-controlled capacitor.*

Varactor diodes are manufactured to exhibit up to 450 pF of capacitance for AM (MW) radio tuning applications; but they are more commonly found in VHF (very-high-frequency) and UHF (ultra-high-frequency) applications, with capacitance values ranging from 2 to 6 pF. Virtually every modern television and radio receiver incorporates varactor diodes for tuning purposes, to reduce costs and to improve long-term performance by eliminating mechanical wear problems.

The schematic symbol used to represent varactor diodes is the same symbol used for general-purpose diodes, but with the addition of a small capacitor symbol placed beside it. Figure 7-3 illustrates the commonly used electronic symbols for special-purpose diodes and optoelectronic devices.

Schottky Diodes

Schottky diodes are sometimes called "four-layer diodes" because their construction includes two layers of each type of semiconductor material. The result is a NPNP device. As with all other types of diodes, they are two-lead devices.

The Schottky family of diodes includes the *PIN diode,* sometimes referred to as the *silicon hot-carrier diode.*

These are *breakdown devices,* meaning that their useful function is to become highly conductive when the reverse voltage across them exceeds an inherent *trigger voltage.* Unlike zener diodes, they do not maintain their avalanche voltage across them after the trigger voltage has been

Figure 7-3
Commonly used electronic symbols for special-purpose diodes and photoelectronic devices.

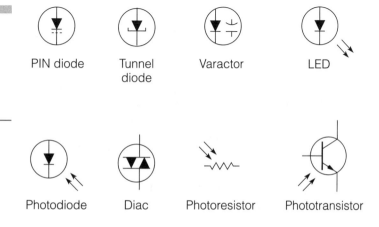

PIN diode   Tunnel diode   Varactor   LED

Photodiode   Diac   Photoresistor   Phototransistor

reached. In contrast, their internal resistance drops to an extremely low value of only a few ohms, and remains there as long as a *minimum holding current* is maintained.

Schottky diodes are commonly used in high-frequency switching, detecting, and oscillator circuits.

Tunnel Diodes

Tunnel diodes are constructed similarly to ordinary diodes, with the exception of heavier impurity doping in the semiconductor material. This results in an extremely thin depletion region under reverse-bias conditions and causes a tunnel diode to be a reasonably good conductor when reverse-biased. The unique characteristic of tunnel diode behavior, however, is in the forward-biased mode.

As the forward-biased voltage across a tunnel diode is increased, there will be three specific voltage levels where the tunnel diode will exhibit *negative resistance* characteristics. This means that the current through the tunnel diode will decrease as the voltage across it increases. As the forward voltage across a tunnel diode is smoothly increased from minimum to maximum, the current response will show a series of three peaks and valleys as the tunnel diode vacillates between positive and negative resistance responses.

Tunnel diodes are used in oscillators and high-speed switching applications for digital circuitry.

Diacs

Diacs are three-layer bilateral trigger diodes. Like PIN diodes, diacs are breakdown devices. However, unlike PIN diodes, diacs are triggered from a blocking-to-conduction state in either polarity of applied voltage. Consequently, there are no band encirclings around the device body to indicate the cathode end, because the orientation of the device is irrelevant.

For example, if the rated breakover voltage (the breakdown voltage, or avalanche point) for a specific diac is 30 volts, it will present an extremely high resistance until the voltage drop across it equals about 30 volts. At that point, it will become highly conductive; and it will remain in this state until the voltage across it reaches a minimum level. At that point, it becomes highly resistive again. It will react this way regardless of the voltage polarity; hence, it is *bilateral* in operation.

Diacs are most commonly used in high-power control circuitry to provide the *turn-on* pulses for silicon controlled rectifiers (SCRs) and triacs. (SCR and triac operation will be discussed in a later chapter.) Diacs were designed to be a solid-state replacement for neon tubes.

Fast-Recovery Diodes

All diodes possess a characteristic called their *recovery time*, which is the amount of time required for the diode to turn off after being in the forward conduction state. This is usually a very short time period; for 60-hertz rectification applications, recovery time is irrelevant. However, in high-frequency rectification applications, the recovery time becomes critical. For these applications, specially manufactured diodes with very fast recovery times are implemented. Logically enough, they are called *fast-recovery diodes.*

Noise and Transient Suppression Diodes

Common household and commercial AC power is fine for powering motors, heaters, and most electromechanical devices. However, in applications where AC power is used to provide the operational power for sensitive and high-speed solid-state circuitry, it can create many problems resulting from noise, voltage spikes, lighting surges, and other undesirable interference signals, which might be conducted into the home or industrial facility by the power lines. To help in eliminating these problems, a whole family of noise and transient suppression diodes have been developed.

Common names for such devices are *unidirectional surge clamping diodes, varistor diodes, unidirectional transient suppression diodes, bidirectional transient suppression diodes, transorbers,* and many others. All of these devices utilize the nonlinear resistive effect or the avalanche effect of semiconductor materials to reduce voltage spikes or overvoltage surges. Their uses are primarily in solid-state power supplies and AC line filters.

A Basic Course in Quantum Physics

Light is a form of radiated energy. As such, it makes up a small part of the total range of radiated energies called the *electromagnetic spectrum*. Radiated energy is composed of extremely small particles of wavelike energy called *quantums* (technically speaking, the plural form of quantum is *quanta*, but the effect of science fiction and media inaccuracy has brought *quantums* into the colloquial language). In reference to the visible and near-visible light frequencies, the older term *quantum* has been replaced with the newer term *photon*.

In the early days of solid-state electronics, it was discovered, quite by accident, that a solid-state diode would emit a small quantity of light as a side effect of the *recombination process* occurring in the PN junction, while forward-biased. This led to the development of the modern *light-emitting diode* (LED). Further research led to the discovery that if an "outside" light source was focused on the junction area of a solid-state diode, the light photons had the tendency to "dislodge" some electrons from their atomic shell positions, resulting in an increase of "minority" carriers. In other words, the "leakage" current in a reverse-biased diode would increase proportionally to the light intensity falling on the junction. This photoconductive property resulted in the development of the photodiode, the phototransistor, and the photoconductive cell. It is also possible to directly convert the energy of photons (light energy) into electrical energy. Devices capable of performing this energy conversion are called *photovoltaic cells*, or, more commonly, *solar cells*.

Further developments, involving a wider diversity of materials and manufacturing processes, led to the more recent member in the optoelectronic field: the laser diode. The term *laser* is actually an acronym for "light amplification (through) stimulated emission of radiation." As stated earlier, normal light consists of small "packets" of energy called *photons*. The typical light emitted all around us consists of photons all traveling in random fashion and random frequencies. Laser light differs from normal light in several ways. First, it is "coherent," meaning that the photons are all traveling in the same direction. To understand this difference, consider a typical flashlight. As you shine a flashlight beam into a distance, the diameter of the beam of light will increase with distance, becoming very broad after only a hundred feet or so. In contrast, a laser light beam will not broaden with distance because all of the photons constituting the beam are going in the same direction. A high-coherency laser beam can easily be bounced off of the moon!

The second radical difference, between laser light and standard "white" light, relates to frequency. Solid-state semiconductor light-emitting devices, such as LEDs and laser diodes, typically emit only a narrow wavelength of light. Therefore, the emitted light consists of only one "pure" color. Common white light, on the other hand, contains all of the colors (meaning all of the frequencies) in the visible light spectrum.

Now that most of the basic principles relating to optoelectronics have been defined, it is appropriate to discuss these devices individually, in more detail.

Light-emitting Diodes

A *light-emitting diode* (LED) is a specially manufactured diode that is designed to glow, or emit light, when forward-biased. When reverse-biased, it will act like any common diode; it will neither emit light nor allow substantial current flow. LEDs can be manufactured to emit any color of the visible light spectrum desired, including "white" light. Red is, by far, the most common color. For certain physical reasons, semiconductor material is especially efficient and sensitive to near-visible light in the infrared region. Consequently, many *photoelectric eyes* used for presence detection and industrial control functions operate in the infrared region.

LEDs are used primarily as indicator devices. The brightness, or intensity, of an LED is relative to the forward current flow through it. Most LEDs are low-voltage, low-current devices, but more recent developments in optoelectronics have led to a family of high-intensity LEDs that approach the light intensity levels of incandescent bulbs.

Most commonly available LEDs operate in the 5- to 50-milliamp range and drop about 1.4 to 2 volts in the forward-biased mode. In most applications, LEDs require the use of a series resistor to limit the maximum current flow.

LEDs have far too many available case styles, shapes, and colors to describe in detail within this context, but each type will use some physical method to indicate the cathode lead. You will learn many of the indication methods through experience, but when in doubt, simply use your DVM in the "diode test" mode to check lead identification. (Most DVMs, in diode test mode, will cause an LED to glow very faintly when checked in forward-biased orientation.)

A common alphanumeric type of LED indicator device is the seven-segment display. (The term *alphanumeric* refers to display devices capable of displaying some, or all, of the characters of the alphabet, as well as numbers.) Seven-segment LED displays are actually seven individual elongated LEDs arranged in a "block 8" pattern. Seven-segment LEDs will have a common connection point to all seven diodes. This common connection might connect all of the cathodes together (making a common-cathode display), or all of the anodes together (making a common-anode display). The choice of using a common-cathode, or a common-anode display, is simply a convenience choice, depending on the circuit configuration and the polarity of voltages used. In addition, many types of "decoder" integrated circuits (integrated circuits designed to convert logic signals into seven-segment outputs), will specify the use of either common-cathode or common-anode displays.

A seven-segment LED will have eight connection pins to the case. One pin is the common connection point to all of the cathodes or anodes. The other seven pins connect to each individual diode within the package. Thus, by connecting the common pin to the appropriate polarity, and forward-biasing various combinations of the LEDs with the remaining seven pins, any seven-segment alphanumeric character can be displayed.

Optoisolators, Optocouplers, and Photoeyes

Regarding their principles of operation, optoisolators, optocouplers, and photoeyes are equivalent; but their sizes, construction, and intended applications can vary dramatically. In essence, all of these devices consist of a light emitter (LED) and a light receiver (photodiode, phototransistor, photoresistor, photo-SCR, or photo-TRIAC).

The intensity of the light emitter can be varied proportionally to an electrical signal. The light receiver can convert the varying light intensity back into the original electrical signal. This process completely eliminates any electrical connection between emitter and receiver, resulting in total isolation between the two. Total electrical isolation is very desirable in circuits that could malfunction from electrical noise, or other interference signals, "feeding back" to the more susceptible areas. Light emitter-receiver pairs used in this manner are called *optical isolators* (*optoisolators*) or *optical couplers* (*optocouplers*). The electrical signal being transmitted to the receiver might be either analog (linear) or digital (pulses).

These same basic components are often used in a photoeye mode. When used in this manner, the light emitter is held constant and sends a continuous beam of light to the receiver. The intended application requires an external object to come between the emitter and the receiver, breaking the beam, and thereby causing the receiver to produce a loss-of-light signal. This type of "presence detection" is used extensively in VCRs, industrial control applications, and security systems.

Optoisolators and optocouplers utilizing a SCR or triac as the receiver are designed for AC power-control applications. The primary advantage in this configuration is the complete isolation from any noise or voltage spikes present on the AC line.

Photodiodes, Phototransistors, and Photoresistive Cells

Photodiodes are manufactured with a clear window in the case to allow external light to reach the junction area. When photodiodes are reverse-biased in a circuit, the amount of "leakage" current allowed to flow through the diode will be proportional to the light intensity reaching the junction. In effect, it becomes a light-controlled variable resistor.

Photoresistors function much like photodiodes, but with a few differences. Unlike photodiodes, photoresistors are junctionless devices. Therefore, like resistors, they do not have fixed orientation in respect to voltage polarity. Also, photoresistors react to light intensity with a very broad resistance range, typically 10,000 to 1. The typical "dark" resistance value of a photoresistor is about 1 Mohm; this resistance then decreases proportionally with exposure to increasing light intensity.

Phototransistors, like photodiodes, incorporate a clear window in the casing to allow ambient light to reach the junction area. The external light affects the transistor operation much like a base signal voltage, so in most cases, the base lead is left unconnected (some phototransistors don't even have base leads). Phototransistors are especially useful in some applications, because they can be used as amplifiers with external light either substituting for, or adding to (modulating), the base signal.

Laser Diodes

The widespread common use of coherent light in the average home has been made a reality by the solid-state laser diode. Every compact-disk (CD) player or CD-ROM (read-only memory) system utilizes a laser diode as the light source for reading the disk data. The commonly seen "laser pointers," familiar to office environments, are little more than a laser diode and a couple of batteries enclosed in a case.

Laser diodes are actually a type of LED. Their operation is similar; the primary difference is in the type of light emitted. Laser diodes emit coherent light.

Laser diodes are available in power ranges from about 0.5 to 5 mW. They are also available as visible red or infrared emitters.

NOTE *Please use caution if you plan to use or experiment with laser diodes. Laser light is dangerous to the eyes. Always follow the manufacturer's recommended safety precautions.*

Liquid-Crystal Displays

Liquid crystal displays (LCDs) have rapidly replaced LED systems in many indicator applications because of several advantages. First, LCDs require much less operational power than do comparable LED systems. This is because LCDs do not actually produce any light of their own; LCD operation depends on ambient light for character display. The second LCD advantage relates to the first. Because LCDs depend on ambient light for operation, they are the most visible in the strongest light where LEDs often appear faint.

An LCD is an optically transparent sandwich, often including an opaque backing. The inner surfaces of the panels making up the sandwich have a thin metallic film deposited on them. On one of the panels, this film is deposited in the form of the desired characters or symbols to be displayed. The space between the two panels contains a fluid called *nematic liquid*. This liquid is normally transparent. When an electric field is placed between the back panel and the desired character to be displayed, the liquid turns black and is displayed, provided that the ambient light is strong enough to see it. This is really no different than using a black magic marker to write a character on a piece of white paper; such a character is clearly visible in normal light, but you couldn't see it in the dark.

Although research is continuing in the LCD field, to date, there are some severe disadvantages. For one, LCDs are much slower than LEDs, making their use in high-speed display applications (such as television) limited. Their speed of operation is greatly affected by temperature; operation becomes visibly slow in cold temperatures. Another disadvantage is versatility. LCD displays must be manufactured for specific applications. For example, an LCD intended for use as a clock display could not be used as a counter display because the colon, which normally appears between the hour and minute characters, places an undesired space between the numerals. After an LCD is manufactured, its character display cannot be modified for another application. A third disadvantage, relating to the home hobbyist or experimenter, is the decoding required for correctly displaying the characters. It could be very complex, requiring ICs that might not be readily available. If you plan on ordering LCD displays from any of the surplus electronics suppliers, be sure that it includes all of the necessary interface documentation.

Charge-Coupled Devices

Charge-coupled devices (CCDs) are actually digital circuits used primarily to replace the older "vidicon" tubes in videocameras. They require less power to operate and provide a much sharper and clearer picture.

Although CCDs are currently used exclusively for video "reception," research toward using CCD technology for solid-state display applications is very promising.

Circuit Potpourri!

As an electrical engineer, instructor, and amateur scientist, I have come to fully appreciate the electronics field as being one of infinite creativity and infinite possibilities. As I write the text for this book, I am surrounded by devices that would have been considered incomprehensible miracles only a few decades ago. Currently, scientists are taking the first infantile steps toward uncovering the mysteries of subatomic structure and quantum physics. These areas of research could lead to gravity and antigravitational generators, total annihilation fusion reactors, and deep-space travel by means of bending the time-space continuum! Virtual reality systems are available today that can positively knock your socks off! Does all this sound a little more than mildly interesting?

Although I can't show you how to build a time machine, this section, together with the concurrent *Circuit Potpourri* sections, contains a collection of projects and circuit building blocks that can be practical, fascinating, and fun (with the emphasis on "fun"). This particular section allows you to start taking the first steps toward discovering all of the creative and ingenious facets within your own self. I'm hoping someday you'll be able to show me how to build a time machine.

I suggest that you read through the description of each circuit even if you do not intend to build or experiment with it. The practical aspects of much of the previous theory is illustrated within them.

Preliminary Steps to Project Building

At this point, I am assuming that you have a lab power supply, DVM, electronic data books, soldering iron, hand tools, suppliers for electronic parts, and miscellaneous supplies needed for project building. In addition, I highly recommend that you purchase a solderless breadboard for testing the following projects before permanently building them. A *solderless breadboard* is a plastic rectangular block with hundreds, or thousands, of contact points internally mounted. Electronic components and interconnection wiring (ordinary no. 22 solid-conductor insulated wire) are simply inserted into the breadboard (without soldering), and the completed circuit can be tested in a matter of minutes.

The circuit can then be modified by simply unplugging the original components and inserting new ones, until the operation is satisfactory. At this point, the user can then remove the components of the perfected circuit, and permanently install them in a universal perfboard or PC board. The solderless breadboard is not damaged in this process, and it can be used repeatedly for designing thousands of additional circuits. An illustration of some excellent-quality solderless breadboards is given in Fig. 7-4. You can also buy prestripped, prebent hook-up wire intended for use with solderless breadboards. For the modest cost involved, I believe this is a good investment.

Flashing Lights, Anyone?

Figure 7-5 illustrates a good basic circuit to cause two LEDs to spontaneously blink on and off. The frequency will be about 1 Hz, depending on component tolerances and the type of transistors used.

The basic circuit is called an *astable multivibrator*. Multivibrators are covered in more detail in successive chapters, but for now you can think of it as a free-running oscillator.

Figure 7-4
Examples of solderless breadboards. (*Photograph courtesy of Interplex Electronics Inc., New Haven, CT.*)

Figure 7-5
A dual-LED flasher.

When power is first applied to this circuit, one transistor will *saturate* (the state of being turned on fully) before the other because of slight component variations. For discussion, assume that Q1 saturates first. In the saturation state, Q1 conducts the maximum collector current lighting LED1. At the same time, this condition makes the positive side of C1 appear to be connected to ground, and it begins to charge to the supply potential through R3. When the charge across C1, which is also the base voltage of Q2, charges to a high enough potential, it causes Q2 to turn on and lights LED2.

At the same time, the positive side of C2 is now placed at ground potential, causing the base voltage of Q1 to go low, forcing Q1 into cut-off (the state of being fully turned off). With Q1 at cutoff, LED1 is dark and C1 begins to discharge back through R3. In the meantime, C2 is now charging and applying a rising voltage to the base of Q1. This continues until Q1 saturates again, and the whole process starts over again. (You might have to slowly reread this functional description and study Fig. 7-5 several times to fully grasp the operation.)

The astable multivibrator shown in Fig. 7-5 is certainly a basic building block for many future applications. Here are a few examples of how this circuit could be modified for a variety of projects. The frequency and on-time/off-time relationship (called the *duty cycle*) can be changed by changing the values of C1, C2, R2, and R3. Experiment with changing the value of each of these components, one at a time, and observe the results. R2 and R3 can be replaced with rheostats (potentiometers connected as variable resistors) for continuously variable frequencies and duty cycles. C1 and C2 can be replaced with smaller values of capacitance making the circuit useful as a simple *square-wave frequency generator.* With the correct choice of transistors and capacitors, this circuit is usable well into the megahertz region. If you wanted to flash brighter lights, LED1 and LED2 could be replaced with small 6-volt relay coils. The relay contacts, in turn, could be connected to the line voltage (120 volts AC), and incandescent lamps for high-brightness flashing (be careful not to exceed the contact current and voltage ratings of the relays). There are many more applications. The fun is in using your imagination.

Three Lights Are Better than Two!

Figure 7-6 illustrates a variation of the same circuit illustrated in Fig. 7-5. The primary differences are that the LEDs will light when their associated transistor is in cutoff, rather than saturation, and an extra transistor circuit has been added for a sequential *three-light effect.*

Only one transistor will be in the cutoff state at any one time. Assume that transistors Q1 and Q3 are saturated, and that Q2 is cutoff. In this state, LED2 is bright from the current flow through Q2's 560-ohm collector resistor. Meanwhile, C1 is charging through Q2's 33-Kohm

Figure 7-6
A triple-LED flasher circuit.

All capacitors rated at 16 WVdc or higher

base resistor. When the voltage across C1 reaches a high enough potential, it turns on Q2 and causes LED2 to become dark (the saturation of Q2 effectively short-circuits the voltage drop across LED2).

Here is where the circuit operation becomes a little tricky. Going back to the prior condition when Q2 was in cutoff, the voltage on the negative side of C2 was actually a little more positive than the voltage on the positive side. This is because the voltage drop across LED2 was a little higher than the base-to-emitter voltage drop of Q3. Therefore, C2 actually takes on a slight reverse voltage charge. When Q2 saturates, this has the effect of forcing Q3 into "hard" cutoff, because a slight negative voltage is applied to its base, before C2 has the chance to start charging in the positive direction through Q3's 33kΩ base resistor. With Q3 in the cutoff state, C3 begins to take on a small reverse charge, while C2 begins to charge toward the point where it will drive Q3 back into saturation. The cycle continues to progress in a sequential manner, with LED1 lighting next, and so on. Even though C1, C2, and C3 are electrolytic capacitors, the small reverse charge is not damaging because the charging current and voltage are very low.

The frequency of operation is a function of the time constant of the capacitors and base resistors. Increasing the value of either component will slow down the sequence. Any general purpose NPN transistor should operate satisfactorily in this circuit. The LED type is not critical, either, although you would have to adjust the resistor values somewhat to accommodate the newer "high brightness" LEDs.

Many of the applications that applied to Fig. 7-5 will also apply to this circuit. One of the advantages of this circuit design is that additional transistor stages can be added on for a longer sequential flashing string.

A Mouse in the House

Figure 7-7 is definitely a "fun" circuit. It consists of two astable multivibrator circuits, very similar to the circuit in Fig. 7-5. The first multivibrator circuit will oscillate very slowly, because of the component values chosen. However, the second multivibrator circuit, consisting of Q3 and Q4, will not oscillate until Q2 saturates, which will occur every few seconds. The component values for the second multivibrator are chosen so that it will oscillate very rapidly and produce a "chirping" or "squeaking" sound if a small speaker is connected to it.

This circuit can be assembled on a very small universal perfboard and enclosed in a small plastic project box together with the speaker, an on-off switch, and a 9-volt transistor battery (which powers it nicely). The two LEDs are optional, but their effect is dramatic.

Once completed, it should be about the size of a pack of cigarettes. It can easily be put in your pocket and carried to a friend's house. When the opportunity arises, turn it on and hide it somewhere inconspicuous. Then, wait for the fun to start! The chirping sounds like a mouse, or some type of large insect. It is not loud enough to cause instant attention, but everyone in the room will notice it in a few minutes. The frequencies and harmonics produced by the multivibrator have the effect

Figure 7-7
An electronic
chirping circuit.

All capacitors rated at 16 WVdc or higher

of making the sound omnidirectional, so it will be difficult to locate. In the meantime, everyone who is a little squeamish toward mice or large insects will get seriously nervous.

It is important to use a very small speaker for this project to achieve the desired effect. Virtually any type of general-purpose NPN transistors will perform well.

If you would like to try a variation on this circuit to produce some really weird sounds, try replacing Q1 and Q3 with a couple of three-lead phototransistors. Connect the phototransistors into the circuit exactly like the original transistors. This causes the changes in ambient light to "sum" with the original base voltages. Various capacitor and resistor combinations will produce some remarkable sounds in conjunction with changing light intensities.

To carry this idea one step further, you can mount this circuit in the center of a bull's-eye target and convert a laser pointer into a "gun" (put a dummy handle on it, and fabricate the on-off switch into a trigger). Using various component values, the target can be made to produce any number of strange sounds, when the laser beam hits the bull's-eye. Including a small power amplifier into the circuit, to boost the output volume, will improve the effect. If you built multiple circuits, adjusted them for individual sound effects, and mounted them in a variety of targets, you could have a high-tech shooting gallery in your own home!

A Sound Improvement

Figure 7-8 is a Hi-Z (high-impedance) audio amplifier circuit that will greatly increase the volume level of a high-impedance headphone (two of these circuits will be needed for stereo headphones). This circuit can come in handy if you want to use your headphones to listen (loudly!) to some of the sounds that you can create with these multivibrator circuits. The input impedance is high enough to keep it from loading down most circuits. You can also use this circuit with most types of speakers that have an impedance-matching transformer connected to the speaker frame. Don't try to use a standard 8- or 4-ohm speaker; you'll destroy the transistor, or the speaker, or both!

This circuit is a modified form of the common-emitter transistor amplifier discussed in Chapter 6. VR2 should be adjusted for the best quality of sound, and VR1 is the volume control.

Figure 7-8
High-impedance headphone amplifier.

Figure 7-9
Time-off delay relay driver.

A Delay Is Sometimes Beneficial

In the field of electronics, there are many control applications that require a *time-delay relay* (TDR). Commercial TDRs are very expensive. Figure 7-9 is a *time-off TDR*, which is both useful and inexpensive. It can also be modified for a variety of functions.

Q1 and Q2 are connected in a configuration called a Darlington pair. The *Darlington pair configuration* is essentially a *beta multiplier*, causing the beta value of Q1 to be multiplied by the beta value of Q2. For example, if both transistors had a beta value of 100, the overall beta value for the pair would be 10,000. The high beta value is particularly useful in this circuit, because only an extremely small Q1 base current is needed to saturate the pair. A discussion of the circuit operation will illustrate why this is important.

When the momentary switch is closed, C1 will appear to charge instantly, because there is no significant series resistance to limit the charge rate. At the same time, Q1 is supplied with more than enough base current to saturate the transistor pair, and the relay is energized. When the switch is released, opening the charge path to C1, C1 begins to "slowly" discharge through the Q1–Q2 base-emitter circuits. Because very little base current is needed to keep the transistor pair saturated, VR1 and R1 can be of a high resistance value, causing a very slow discharge of C1. The majority of C1's discharge cycle will maintain the saturated condition of Q1 and Q2, causing the relay to remain energized for a substantial time period after the switch is released. Theoretically speaking, if you tried to perform this same function with only a single transistor, the resistance values of VR1 and R1 would have to be about 100 times smaller to maintain a base current adequate for saturation (assuming both transistors to have a beta value of 100), and the discharge rate of C1 would be very rapid. You could accomplish the same operation if you increased the value of C1 by a factor of 100, but large electrolytic capacitors are both expensive and bulky.

Figure 7-9 is a *time-off TDR*, meaning that after the control action is instigated (closing and releasing the momentary switch), there is a time delay before the relay deenergizes. The length of the time delay depends on the setting of VR1, which largely controls the discharge rate of C1.

This circuit can be easily modified to provide a time-on/time-off delay. Remove VR1 and connect the opened end of R1 to the positive side of C1. Then connect VR1 into the C1 charge path, between the momentary switch and the applied power source. In this configuration, when the momentary switch is closed, C1 must charge through VR1, causing a time-on delay, until C1 charges to a high enough potential to cause the transistor pair to saturate. The length of this delay would depend on the setting of VR1. On releasing the momentary switch, a time-off delay would occur, while C1 is discharged through R1 and the base-emitter junctions of Q1 and Q2. This delay would be largely controlled by the value of R1.

If you wanted a time-on TDR (without the time-off function), a reasonably good facsimile can be made by simply removing R1, and by connecting the base of Q1 directly to the positive side of C1. Connect R1 in series with VR1 in the C1 charge path. By experimenting with different values of VR1 and C1, a significant time-on delay can be achieved with a fairly rapid turn-off.

Q1 and Q2 are general-purpose NPN transistors. For best results, use low-leakage-type transistors. The voltage amplitude of the circuit power source should be about equal with the relay coil voltage. For experimentation purposes, start with a C1 value of about 100 μF, and an R1 and VR1 value of 100 Kohms and 1 Mohm, respectively. These values can then be adjusted to meet your requirements. Whenever relay coils are incorporated into DC-powered solid-state circuitry, they should always be paralleled with a reverse-biased general-purpose diode. Note the orientation and connection of D1 in Fig. 7-9. The purpose of D1 is to suppress the inductive kickback, transient voltage spike, which will occur when the relay is deenergized. This kickback voltage spike can easily damage solid-state devices. It is generated by the stored energy in the electromagnetic field surrounding the relay coil. Fortunately, these voltage spikes will always be in the opposite polarity of the applied power source. Therefore, D1 will short out the spike, and render it harmless.

A Long-Running Series

Earlier in this chapter, during the discussion of zener diode regulators, it was shown why zeners are not very power-efficient as high-current regulators. A circuit designed to greatly improve the efficiency and operation of *voltage regulation* is illustrated in Fig. 7-10.

Figure 7-10

A series-pass regulator using a zener diode as a voltage reference.

Much of this circuit should already be familiar to you. D1 is needed only if you plan on using this regulator circuit with a battery as the unregulated power source. C1 is the filter capacitor(s) for the raw DC power supply. This raw DC power supply can be of any design you choose. R1 and ZD1 form a simple zener regulator, as discussed previously in this chapter. However, in this circuit, the zener serves as a *voltage reference* for the *series-pass* transistor Q1. Transistor Q1 serves as a *current multiplier* for the zener. For example, if Q1 has a *beta* value of 100, and the current requirement for the load is 1 amp, the zener would only have to supply 10 milliamps of current to the base of Q1 for a 1-amp output. This means that the value of R1 can be chosen so that the current flow through ZD1 is only slightly above its minimum holding current. Therefore, ZD1 is only required to dissipate a small quantity of power, and a much higher load can be regulated (remember, a "high" load means that the load resistance is "small," and vice versa). C3 serves as an additional filter for smoothing the regulated DC.

For regulating most low-voltage loads requiring up to about 1 amp, ZD can be a 1-watt zener. Its zener voltage value should be 0.6 volt above the desired regulated output voltage. Q1 will drop this 0.6-volt excess across the base to emitter junction. For example, if you wanted a 5-volt regulated output, ZD should be a 5.6-volt zener. The value of R1 should be chosen to place the zener diode at about 15 to 20 milliamps above its rated holding current. Commonly used transistors for Q1 are the TIP31, TIP3055, and 2N3055 types.

Capacitor C2 serves a unique purpose in this circuit. Connected as shown, transistor Q1 serves as a *capacitor multiplier*, multiplying the filtering effect of C2 by its beta value. If capacitor C2 had a value of 1000 μF and Q1 had a beta of 100, the regulated output voltage would be filtered as if a 100,000-μF capacitor had been placed in parallel with the output.

Keep It Steady

You will probably run into many situations where you will want to use an LED as an indicator for a variable-voltage circuit. Trying to use a single resistor for current limiting will prove ineffective for this application because the current flow through the LED will vary proportionally to the voltage, and it is likely to go too high, or too low, for good results. Figure 7-11 is a quick and easy solution to the problem. A low-power

Figure 7-11
Regulating the
current flow
through an LED.

Figure 7-12
A voltage-doubler
rectification circuit.

zener (ZD1) will maintain the voltage across the resistor-LED combination at a relatively constant level, regulating the current flow through
the LED, and the voltage variations will be dropped across R1.

Double Your Pleasure

Do you have a transformer in your junk box that you would like to use
for a circuit application, but the secondary voltage is too low? If so, you
can use the voltage-doubler circuit illustrated in Fig. 7-12 to approximately
double the secondary output during the rectification process.

In Fig. 7-12, a transformer with a 12-volt, 1-amp secondary is used to
illustrate this principle. D1 and D2 are configured as two half-wave rectifiers, and C2 and C1 are their associated filters. C1 and C2 are simply
connected so that their voltages are additive.

Capacitors C1 and C2 filter a half-wave rectified voltage, so they must
have a much higher capacity that do comparable capacitors used for full-
wave filtering. D1 and D2 are common, general-purpose diodes. Also notice
that neither secondary transformer lead is used as the common reference.

Figure 7-13
An electronic fuse
monitor circuit.

Show the Blow

As my last entry into this section of circuit potpourri, I submit the blown fuse alarm circuit illustrated in Fig. 7-13. In most homes, there are situations where it is critical to maintain electrical power to certain devices. For example, a chest freezer or a sump pump located in the basement have critical needs for constant power. A blown fuse (or tripped circuit breaker) to either of these appliances could result in a flooded basement, or the loss of hundreds of dollars' worth of food. Unfortunately, it is likely that the blown fuse will not be discovered until the damage has already occurred. The circuit shown in Fig. 7-13 solves that problem by providing an alarm when a blown-fuse condition occurs.

When a fuse blows, it represents an infinite resistance (like an open switch) within an electric circuit. Therefore, the entire source voltage for the circuit will be dropped across it. This is how the fuse-monitoring circuit obtains its operational power.

Assume that the fuse (F1) is protecting a 120-volt AC circuit (I do not recommend this monitoring circuit for AC voltages higher than 120 volts AC). On blowing, it will apply 120 volts AC to the monitor circuit. R1 limits the current, and drops most of the applied voltage. Diodes D1 through D4 rectify the voltage and apply pulsating DC to the zener diode (ZD1). C1 filters the pulsating DC to apply smooth DC to the load. The load can be a piezo buzzer (such as used in smoke detectors) or any other type of low-power, low-voltage visible or audible indicator.

The resistance value and power rating of R1 will depend on the load requirement. Build the circuit as illustrated using a 1-amp, 200-volt PIV bridge rectifier (or comparable diodes), a 1-watt zener diode, and a 100-Kohm, $\frac{1}{2}$-watt resistor for R1. If the load will not operate when 120 volts AC is applied to the circuit, start decreasing the value of R1 a little at a time until you reach the point of reliable operation. If reliable operation requires going below 12 Kohms, I suggest you try using an indicator requiring less operational power. At R1 values lower than 68 Kohms, the power rating should be increased to 2 watts.

This circuit can also be used to monitor fuses used in DC circuits. For these applications, the bridge rectifier is not required, but be sure to observe the correct polarity.

The enclosure for this circuit will depend on the intended application. For instance, 120-volt AC applications require the use of an "approved" metal enclosure, properly grounded, and with wiring and conduit meeting national and local safety standards. It might also be necessary to fuse-protect the monitor circuit.

A final word of caution: *Please don't try to connect this circuit into a fuse box or breaker box,* unless you're fully qualified to do so. Mistakes resulting from a lack of knowledge or experience in this area can result in property damage, fire hazard, and electrocution (especially if you're working on a damp basement floor)!

8

Linear Electronic Circuits

In a general sense, the entire massive field of electronics can be classified into two very broad categories: digital and linear. *Digital* pertains to those circuits and devices that operate on the basis of *switching action,* representing numbers or data by means of on-off pulses. The fundamentals of digital electronics will be covered in later chapters.

In this chapter, you will examine the fundamentals of *linear* circuits. The term *linear* pertains to circuits that operate in a *proportional* manner, accepting inputs and providing outputs that are continuously variable (i.e., analog).

Although the field of linear electronics is very diversified, the basics of linear action are common to almost all of its facets. For example, the same techniques used to *linearize* (i.e., increase proportional accuracy in) an audio amplifier are used to linearize servo systems and operational amplifiers. An accurate understanding of the basic building blocks utilized in linear systems will aid you in understanding a great variety of electronic systems.

Because of a variety of factors, "discrete" (i.e., nonintegrated) circuitry is still utilized in a significant portion of the field of audio electronics. Therefore, this chapter focuses primarily on audio amplification circuits, since they provide a good beginning point to study the fundamentals of linear circuitry. In addition, the associative discussions make it a convenient point to detail printed circuit board construction (in conjunction with a few more "advanced" projects) and the newer computer-automated (also -aided or -assisted) design (CAD) techniques for designing and constructing PC boards.

Transistor Biasing and Load Considerations

The circuit illustrated in Fig. 8-1 should already be familiar to you from the previous discussions of transistor amplifiers. It is a common-emitter amplifier because the output (to the speaker) is taken off from the collector, and the input signal to be amplified is coupled to the base. C1 is a coupling capacitor (blocking the DC bias voltage, but passing the AC audio signal). R1 and R2 form a voltage-divider network to apply the proper DC bias to the base. The emitter resistor (RE) increases the input impedance, and it improves temperature and voltage stability.

Transformer T1 is an audio transformer. It serves two important functions in this circuit. First, it isolates the DC quiescent (steady-state) current flow from the speaker coil (speaker coils can be damaged by even relatively small DC currents). Secondly, it provides a more appropriate load impedance for a transistor collector than would a low-impedance 8-ohm speaker. A transistor amplifier of this configuration could not operate very well with an extremely low collector impedance. A typical audio transformer might have a primary impedance of 100 ohms, for connection into the transistor circuit, and a secondary impedance of 8 ohms for connection to the speaker. Generally speaking, an audio ampli-

Figure 8-1
Basic audio amplifier
circuit.

fier of this type performs satisfactorily for low-power applications. However, based on the basis of modern standards, it has severe problems and limitations.

To begin, examine the real-life problems relating to efficiency and biasing considerations. Choosing some simple numbers for discussion purposes, assume the source voltage in Fig. 8-1 to be 35 volts; T1's primary impedance, 100 ohms; and RE, 10 ohms. As you might recall, the voltage gain of this circuit is approximately equal to the collector resistor (or impedance) divided by the emitter resistor. Therefore, the 100-ohm collector impedance (T1) divided by the 10-ohm emitter resistor (RE) places the voltage gain (A_c) at 10.

R1 and R2 are chosen so that the base voltage is about 2.1 volts. If Q1 drops about 0.6 volt across the base-emitter junction, this leaves 1.5 volts across RE. The 1.5-volt drop across the 10-ohm emitter resistor (RE) indicates the emitter current is at 150 milliamps. Because the collector current is approximately equal to the emitter current, the collector current is also about 150 milliamps. (The 150 milliamps is the "quiescent" collector-emitter current flow. The term *quiescent* refers to a steady-state voltage or current established by a bias.) Now, 150 milliamps of current flow through the 100-ohm T1 primary causes it to drop 15 volts. If 15 volts is being dropped across T1's primary, the collector voltage must be 20 volts (in reference to ground). Then 15 volts plus 20 volts adds up to the source voltage of 35 volts.

If a 500-millivolt rms signal voltage were applied to the input of this amplifier, a 5-volt rms voltage would be applied to the primary of T1 (A_c = 10). If T1 happened to be a "perfect" transformer, it would transfer the total AC power of the primary to the secondary load. In this example, the total power being supplied to the primary is 250 mW rms. Even with no T1 losses, 250 mW of power would not produce a very loud sound out of the speaker.

In contrast, examine the power being dissipated by Q1. As stated earlier, in its quiescent state, the collector voltage of this circuit is 20 volts. The emitter resistor is dropping 1.5 volts; therefore, 18.5 volts is being dropped across the transistor (the emitter-to-collector voltage). With a 150-milliamp collector-emitter current flow, that comes out to 2.775 watts of power dissipation by Q1. In other words, about 2.775 watts of power is being wasted (in the form of heat) to supply 250 mW of power to the speaker. That translates to an efficiency of about 9%.

With a better choice of component values, and a more optimum bias setting, the efficiency of this amplifier design could be improved. However, about the best real-life efficiency that can be hoped for is about 25% at maximum output, in this class A amplifier.

There are actually two purposes to this efficiency discussion. The first, of course, is to demonstrate why a simple common-emitter amplifier makes a poor high-power amplifier. Second, this is a refresher course in transistor amplifier basics. If you had some trouble understanding the circuit description, you might want to review Chapter 6 before proceeding.

Amplifier Classes

Although some audio purists still insist on wasting enormous quantities of power to obtain the high linearity characteristics of class A amplifiers, such as the one shown in Fig. 8-1, most people who specialize in audio electronics recognize the impracticability of such circuits. For this reason, audio power amplifiers have been developed, using different modes of operation, that are much more efficient. These differing operational techniques are arranged into general groups, or "classes." The class categorization is based on the way the output "drivers" [transistors, field-effect transistors (FETs), or vacuum tubes] are *biased*.

The amplifier circuit illustrated in Fig. 8-1 is a *class A audio amplifier* because the output driver (Q1) is biased to amplify the full, peak-to-peak audio signal. This is also referred to as *biasing in the linear mode*.

Again referring to Fig. 8-1, assume that the bias to Q1 were modified to provide only 0.6 volt to the base. Assuming that Q1 will drop about 0.6 volt across the base-emitter junction, this leaves zero voltage across RE. In other words, Q1 is biased just below the point of conduction. In this quiescent state, Q1 would not dissipate any significant power (a little power would be dissipated because of leakage current) because there essentially is no current flow through it. If an audio signal voltage were applied to the input of the circuit in this bias condition, the positive half-cycles would be amplified (because the positive voltage cycles from the audio signal on the base would "push" Q1 into the conductive region, above 0.6 volt), but the negative half-cycles would only drive Q1 further into the cutoff region and would not be amplified. Naturally, this results in severe distortion of the original audio signal, but the efficiency of the circuit, in reference to transferring power to the speaker, would be greatly improved. This mode of amplification is referred to as *class B*, where conduction occurs for about 50% of the cycle.

Of course, the circuit shown in Fig. 8-1 (biased for class B operation) is not very practical for amplifying audio signals because of the *high distortion* that occurs at the output. But if a second transistor were incorporated in the output stage, also biased for class B operation, but configured to amplify only the "negative" half-cycles of the audio signal, it would be possible to re-create the complete original amplified audio signal at the output. This is the basic principle behind the operation of a *class B audio amplifier.*

There is still one drawback with class B amplification. At the point where one transistor goes into cutoff and the other transistor begins to conduct (the *zero reference point* of the AC audio signal), a little distortion will occur. This is referred to as *crossover distortion.* Crossover distortion can be reduced by biasing both output transistors just slightly into the conductive region in the quiescent state. Consequently, each output transistor will begin to conduct at a point slightly in advance of the other transistor going into cutoff. By this method, crossover distortion is essentially eliminated, without degrading the amplifier's efficiency by a significant factor. This mode of operation is referred to as *class AB.* An efficiency factor of up to 78.5% applies to class AB amplifiers.

Figure 8-2 is an example of a hypothetical class-AB audio amplifier. C1 is the input coupling capacitor, R1 and R2 form the familiar voltage divider bias network for biasing Q1, RE is Q1's emitter resistor, and RC is Q1's collector resistor. These components make up a typical common-emitter transistor amplifier. Q2 and potentiometer P1 are configured in a circuit arrangement called an *amplified diode.* The purpose of this circuit is to provide the slight forward bias required on both output driver transistors to eliminate crossover distortion. Q3 and Q4 are the output drivers; with Q3 amplifying the positive half-cycles of the audio signal, and Q4 amplifying the negative half-cycles. C2 is an output-coupling capacitor; it serves to block the DC quiescent voltage from reaching the speaker, while allowing the amplified AC output voltage to pass.

The amplified diode circuit of Q2 and P1 could be replaced with two forward-biased diodes. In theory, each diode would drop about the same

Figure 8-2
A hypothetical class
AB audio amplifier.

voltage as the forward biased base-emitter junction of each output transistor. The problem with this method is a lack of adjustment. If the forward threshold voltage of each diode is not exactly equal to the base-emitter junction voltage of each transistor, some crossover distortion can occur. If three diodes are used, the quiescent conduction current of each output transistor might be too high, resulting in excessive heating of the output transistors.

The amplified diode circuit could also be replaced with an adjustable biasing resistor for biasing purposes. Although this system will function well and eliminate crossover distortion, the adjustable resistor will not thermally "track" with the output transistors. As you might recall, bipolar transistors have a negative temperature coefficient, meaning that they exhibit a decrease in resistance with an increase in temperature. In reference to transistors, a decrease in resistance actually means an increase in leakage current. In other words, bipolar transistors become more "leaky" when they get hot. In bipolar transistor amplifiers, this is a major problem. As output transistors begin to heat up, the leakage current also increases, causing an increase in heat, causing an increase in leakage current, causing an additional increase in heat, causing an additional increase in leakage current, and so forth. This condition will continue to degrade until the output transistors break down. A breakdown of this nature is called *thermal runaway*.

A means of automatic thermal compensation is needed to correct the problem. An adjustable resistor cannot do this (most resistors have a positive temperature coefficient), but that is the beauty of an amplifier diode circuit. Referring to Fig. 8-2, if Q2 is placed on the same heatsink as the output transistors, its temperature rise will closely approximate that of the output drivers. As the leakage current increases with a temperature rise in the output transistors, the leakage current through Q2 also increases. The increase of leakage current through Q2 causes the voltage drop across it to decrease, resulting in a decrease of forward bias to the output transistors. The decrease in forward bias compensates for the increase in leakage current, thus resulting in good temperature stabilization.

Additional Amplifier Classification

There are additional classes of amplifier operation, but they are not typically used for audio amplifiers. *Class C amplifiers* are biased to amplify only a small portion of a half-cycle. They are used primarily in RF (radio-frequency) applications, and their efficiency factors are usually about 80%.

Class D amplifiers are designed to amplify "pulses," or square waves. A class D amplifier is strictly a "switching device," amplifying no part of an input signal in a linear fashion. Strangely enough, class D amplifiers are available (although rarely) as audio amplifiers through a technique called *pulse-width modulation* (PWM). A *PWM audio amplifier* outputs a high-frequency (about 100 to 200 kHz) square wave to the audio speaker. Because this is well above human hearing, the speaker cannot respond. But the *duty cycle* (on-time/off-time ratio) of the square-wave output is varied according to the audio input signal. In effect, this creates a proportional "power signal," which the speaker does respond to, and the audio input signal is amplified. Class D audio amplifiers boast extremely high efficiencies, but they are expensive, and they have drawbacks in other areas. Class D audio amplifiers are sometimes called *digital audio amplifiers*. Most class D amplifiers are more commonly used for high-power switching and power conversion applications.

Audio Amplifier Output Configurations

The circuit illustrated in Fig. 8-2 has a *complementary symmetry output stage*. This term means that the output drivers are of opposite types (one is NPN, and the other is PNP) but have symmetric characteristics (same beta value, base-emitter forward voltage drop, voltage ratings, etc.). Generally speaking, most audiophiles consider this to be the best type of output driver design. Transistor manufacturers offer a large variety of *matched pair,* or *complementary pair,* transistor sets designed for this purpose.

Another common type of output design is called the *quasi-complementary symmetry configuration.* It requires a complementary symmetry "predriver" set, but the actual output transistors are of the same type (either both NPN, or both PNP; NPN outputs are vastly more popular). This type of output design used to be a lot more popular than it is now. The current availability of a large variety of high-power, high-quality complementary transistor pairs has overshadowed this older design. However, it produces good-quality sound with only slightly higher-distortion characteristics than complementary symmetry.

Audio Amplifier Definitions

The field of audio electronics is an entertainment-oriented field. The close association between audio systems and the arts has led to a kind of semiartistic aura surrounding the electronic and electromechanical systems themselves. As with any artform, personal preference and taste play a major role. This is the reason why there are so many disputes among audiophiles regarding amplifier and speaker design. My advice is to simply accept what sounds good to you, without falling prey to current trends and fads.

Unfortunately, there have been many scams and sly stigmas perpetrated by unethical, get-rich-quick manufacturers over the years. This has led to much misunderstanding and confusion regarding the various terms used to define audio amplifier performance.

The most heavily abused characteristic of amplifier performance is "power." *Power,* of course, is measured in watts. The only standardized method of designating AC wattage, for comparison purposes, is by using the rms value. Any other method of rating an amplifier's power output should be subject to suspicion.

Output power is also rated according to the speaker load. For example, an amplifier specification might rate the output power as being 120 watts rms into a 4-ohm load, and 80 watts rms into an 8-ohm load. You might expect the power output to double when going from an 8-ohm load to a 4-ohm load, but there are certain physical reasons why this will not happen. However, when comparing amplifiers, be sure to compare "apples with apples"; an amplifier rated at 100 watts rms into an 8-ohm load is more powerful than an amplifier rated at 120 watts rms into a 4-ohm load.

The human ear does not respond in a linear fashion to differing amplitudes of sound. It is very fortunate for you that you are made this way, because the nonlinear ear response allows you to hear a full range of sounds; from the soft rustling of leaves to a jackhammer pounding on the pavement. For example, a loud sound that is right on the threshold of causing pain to a normal ear is about 1,000,000,000,000 times louder than the softest sound that can be heard. Our ears tend to "compress" louder sounds, and amplify smaller ones. In this way, we are able to hear the extremely broad spectrum of audible sound levels.

When one tries to express differing sound levels, power ratios, noise content, and various other audio parameters, the nonlinear characteristic of human hearing presents a problem. It was necessary to develop a term to relate linear mathematical ratios with nonlinear hearing response. That term is the *decibel*. The prefix *deci* means $\frac{1}{10}$, so the term decibel actually means "one-tenth of a bel."

The bel is based on a logarithmic scale. Although I can't thoroughly explain the concepts of logarithms within this context, I can give a basic feel for how they operate. *Logarithms* are trigonometric functions, and are based on the number of decimal "columns" contained within a number, rather than the decimal values themselves. Another way of putting this is to say that a *logarithmic scale* is linearized according to *powers of ten*. For example, the log of 10 is 1; the log of 100 is 2; the log of 1000 is 3. Notice, in each case, that the log of a number is actually the number of weighted columns within the number minus the "units" column. The *bel* is a ratio of a "reference" value, to an "expressed" value, stated logarithmically. A decibel is simply the bel value multiplied by 10 (bels are a little too large to conveniently work with).

In this case, I believe a good example is worth a thousand words. Assume you have a small radio with a power output of 100 mW rms. During a party, you connect the speaker output of this radio into a power amplifier which boosts the output to 100 watts rms. You would like to express, in decibels, the power increase. The power level that you started with, 100 mW (0.1 watt), is your reference value. Dividing this number into 100 watts gives you your ratio, which is 1000. The log of 1000 is 3 (bel value). Finally, multiply 3 by 10 (to convert bels to decibels), and the answer is 30 decibels.

Each 3-dB increase means a doubling of power: 6 dB gives 4 times the power [3 dB + 3 dB equates to $2x$ power times $2x$ power; $2x(2x) = 4x$]. A

9-dB power increase converts to an $8x$ power increase [6 dB + 3 dB, or $4x(2x) = 8x$].

Each 10-dB increase equals a 10-fold increase; for instance, 20 dB yields $100x$ [10 dB + 10 dB, or $10x$ times $10x$]. And finally, as per the example above, a 30-dB power gain means a 1000-fold increase [10 dB + 10 dB + 10 dB = $10x^3 = 1000x$].

Also, please be aware that negative values of decibels represent negative gain, or *attenuation*. A -3-dB gain means that the power has been halved. Similarly, a -10-dB gain represents a 10-fold attenuation, or a $0.1x$ change in power output.

These figures represent *power logs*. *Voltage* and *current decibel logs* are somewhat different: the square root of the power logs. This is because power is voltage times current, $P = IE$; 30-dB volts is 31.620, and 30-dB amps is also 31.620. Thus, $_{log}P = {_{log}I} \times {_{log}E} = 31.62 \times 31.62 = 999.8x$. Most electronic reference books have decibel log tables for easy reference. Just be aware that there is a difference between the power logs, and the voltage or current logs.

I recognize that if you have not been exposed to the concept of logarithms, or exponential numbering systems, this entire discussion of decibels is probably rather abstract. If you would like to research it further, most good electronics math books should be able to help you understand it in more detail.

Dynamic range is a term used to describe the difference (in decibels) between the softest and loudest passages in audio program material. In a practical sense, it means that if you are listening to an audio system at a 10-watt rms level, then for optimum performance, you would probably want about a 40-watt rms amplifier to handle the instantaneous high-volume passages that might be contained within the program material (a cymbal crash, for example). Compact-disk and "hi-fi" (high-fidelity) videotaperecorders offer the widest dynamic range commonly available in today's market.

Frequency response defines the frequency spectrum that an amplifier can reproduce. The normal range of human hearing is from 20 to 20,000 Hz (if you're a newborn baby, and had Superman as a father). In theory, there are situations occurring in music where ultrasonic frequencies are produced which are not audible, but without them, the audible frequencies are "colored" to some degree, causing a variance from the original sound. For this reason, many high-quality power amplifiers have frequency responses up to 100,000 hertz. The high-end and low-end *frequency response limits* are specified from the point where the amplifier power output drops to 50% (-3 dB) of its rated output.

Distortion is a specification defining how much an amplifier changes, or "colors," the original sound. A perfect amplifier would be perfectly "linear," meaning that the output would be "exactly" like the input, only amplified. However, all amplifiers distort the original signal by some percentage. In the mid-1970s, it was a commonly accepted fact that the human ear could not distinguish distortion levels below 1%. That has since been proved wrong. It is a commonly accepted rule of thumb today that even a trained ear has difficulty detecting distortion below 0.3%, although this figure is often disputed as being too high among many audiophiles. In any case, the lower the distortion specifications, the better.

Distortion is subdivided down into two more specific categories in modern audio amplifiers: harmonic distortion and intermodulation distortion. *Harmonic distortion* describes the nonlinear qualities of an amplifier. In contrast, *intermodulation distortion* defines how well an amplifier can amplify two specific frequencies simultaneously, while preventing the frequencies from interfering with each other in a nonlinear fashion. Typical ratings for both of these distortion types is 0.1% or lower in modern high-quality audio amplifiers.

Load impedance defines the recommended speaker system impedance for use with the amplifier. For example, if the amplifier specification indicates the load impedance as 4 or 8 ohms, you might use either a 4- or 8-ohm speaker system (or any impedance in between) as the output load for the amplifier.

Input impedance describes the impedance "seen" by the audio input signal. This value should be moderately high: 10 Kohms or higher.

Sensitivity defines the rms voltage level of the input audio signal required to drive the amplifier to full output power. Typical values for this specification are 1 to 2 volts rms.

Signal-to-noise ratio is a specification given to compare the inherent noise level of the amplifier with the amplified output signal. Random noise is produced in semiconductor devices by the recombination process occurring in the junction areas, as well as other sources. High-quality audio amplifiers incorporate various noise reduction techniques to reduce this undesirable effect, but a certain amount of noise will still exist and be amplified right along with the audio signal. Typical signal-to-noise ratios are −70 to −92 dB, meaning that the noise level is 70 to 92 dB below the maximum amplifier output.

There are additional specifications that might or might not be given in conjunction with audio amplifiers, but the previous terms are the most common and the most important.

Before proceeding, here is a note of caution. Research has proved that continued exposure to high-volume noise (meaning music or any other audio program material) causes degradation of human hearing response. It saddens me to hear young people driving by in their cars with expensive audio systems blasting out internal sound pressure levels at 120 dB. Even relatively short exposures to this level of sound can cause them to develop serious nerve-deafness problems by the time they're middle-aged. Of course, exposure to high-volume levels at any age is destructive. It isn't worth it. *Keep the volume down* to reasonable levels for your ear's sake.

Power Amplifier Operational Basics

Now that some of the basic audio terms have been established, this section will concentrate on the "front end" of modern audio amplifier design. In addition, this section establishes some of the basics relating to integrated circuit *operational amplifiers*, which will be discussed later in this book.

Figure 8-3 is a kind of semiblock diagram illustrating the input stage of most high-power audio amplifiers. Q1, D1, D2, and P1 form a circuit called a *constant-current source*. For discussion purposes, assume that D1 drops the same voltage as the base-emitter junction of Q1 (which should be a close assumption). That would mean that the voltage drop across D2 would also be the voltage drop across P1. If the voltage drop across D2/P1 is 0.7 volt, and P1 is adjusted to be 700 ohms, the emitter current flow will be 1 milliamp. Because the collector current of Q1 will approximately equal the emitter current, the collector current is also "held" at 1 milliamp. The important point to note here is that the collector current is regulated; it is not dependent on the collector load or the amplitude of the source voltage. The only variables controlling the collector current are D2's forward threshold voltage and the setting of P1. Therefore, it is appropriately named a *constant-current source*.

Transistors Q2 and Q3 form a differential amplifier. Think of a differential amplifier as being like a seesaw in a school playground. As long as everything is balanced on a seesaw, it stays in a horizontal position. If something unbalances it, it tilts, causing one end to go up proportionally, as the other end goes down. This is exactly how a differential amplifier operates with the current flow. As discussed previously, it is assumed that the constant current source will provide a regulated 1 milliamp of current flow to the emitters of Q2 and Q3. If Q2 and Q3 are in a balanced condition, the 1 milliamp of current will divide evenly between

Figure 8-3
Power amplifier oper-
ational basics.

each transistor, providing 0.5 milliamp of current flow through each
collector. If an input voltage is applied between the two base inputs (*A*
and *B*) so that point *A* is at a different potential than point *B*, the balance will be upset. But as the collector current rises through one transistor, it must decrease by the same amount through the other, because the
constant-current source will not allow a varying "total" current. For
example, if the differential voltage between the inputs caused the collector current through Q2 to rise to 0.6 milliamp, the collector current
through Q3 will fall to 0.4 milliamp. The total current through both
transistors still adds up to 1 milliamp.

Notice that the output of the differential amplifier is not taken off of
one transistor in reference to ground; the output of a differential ampli-

fier is the difference between the two collectors. Now let's discuss the advantages of such a circuit.

Assume that the source voltage (supplied externally) increases. In a common-transistor amplifier, an increase in the source voltage will cause a corresponding change throughout the entire transistor circuit. In Fig. 8-3, an increase in the source voltage ($+ V$) does not cause an increase in current flow from the constant current source, because it is regulated by the forward voltage drop across D2, which doesn't change (by practical amounts) with an increase in current. Q2 and Q3 would still have a combined total current flow of 1 milliamp. The collector voltages of Q2 and Q3 would increase, but they would increase by the same amount, even if the circuit were in an unbalanced state.

Therefore, the voltage differential between the two collectors would not change. For example, assume that Q2's collector voltage is 6 volts and Q3's collector voltage is 4 volts. If you used a voltmeter to measure the difference in voltage between the two collectors, it would measure 2 volts ($6 - 4 = 2$). Now assume the source voltage increased by an amount sufficient to cause the collector voltages of Q2 and Q3 to increase by 1 volt. Q2's collector voltage would rise to 7 volts, and Q3's would rise to 5 volts. This didn't change the voltage differential between the two collectors at all; it still remained at 2 volts. In other words, the output of a differential amplifier is immune to power supply fluctuations. Not only does this apply to gradual changes in DC levels; the effect works just as well with power supply ripple and other sources of undesirable noise signals that might enter through the power supply.

One of the most common problems with high-gain amplifiers is noise and *interference signals* being applied to the input through the input wires. Input wires and cables can pick up a variety of unwanted signals, just as an antenna is receptive to radio waves. If you have ever touched an uninsulated input to an amplifier, you undoubtably heard a loud 60-hertz roar (called "hum") through the speaker. This is because your body picks up electromagnetically radiated 60-hertz signals from power lines all around you. Fluorescent lights are especially bad electromagnetic radiators. Figure 8-3 illustrates an example of some electromagnetic radiation causing noise pulses on the *A* and *B* inputs to the differential amplifier. Because electromagnetic radiation travels at the speed of light (186,000 miles per second), the noise pulses will occur at the same time, and in the same polarity. This is called *common-mode interference*. A very desirable attribute of differential amplifiers is that they exhibit *common-mode rejection*. The noise pulses illustrated in Fig. 8-3 would not be amplified.

To understand the principle behind common-mode rejection, assume that the positive-going noise pulse on the *A* input is of sufficient amplitude to try to cause a 1-volt decrease in Q2's collector voltage. Because the noise is common mode (as is all externally generated noise), an identical noise pulse on the *B* input is trying to cause Q3's collector voltage to decrease by 1 volt also. If both collectors decreased in voltage at the same time, it would require an increase in the combined total current flow through both transistors. This can't happen because the constant-current source is maintaining that value at 1 milliamp.

Therefore, neither transistor can react to the noise pulse, and it is totally rejected. (Even if both transistors did react slightly, they would react by the same amount. Because the output of the pair is the difference across their collectors, a slight reaction by both at the same time would not affect their differential output.) This goes back to the analogy of the seesaw I made earlier. If a seesaw is balanced and you placed equal weights on both ends at the same time, it would simply remain stationary. In contrast, the desired signal voltage to be amplified is not common mode. For example, the *B* input might be at signal ground while the *A* input is at 1 volt rms. Differential amplifiers respond very well to differential signals. That is why they are called *differential amplifiers*.

One final consideration of Fig. 8-3 is in reference to R_F and C_F. Notice that this resistor/capacitor combination is connected from the output back to one of the inputs. The process of applying a percentage of the output back into the input is called *feedback*. In high-gain amplifiers, this feedback is almost always in the form of *negative feedback*, meaning the feedback acts to reduce the overall gain. Negative feedback is necessary to temperature stabilize the amplifier, flatten out the gain, increase the frequency response, and eliminate oscillations. Various combinations of resistors and capacitors are chosen to tailor the frequency response, and to provide the best overall performance. Feedback will be discussed further in later chapters.

Building High-Quality Audio Systems

You don't have to be an electrical engineer to build much of your own high-quality audio equipment. Even if your interests don't lie in the audio field, you are almost certain to need a lab-quality audio amplifier for many related fields. In this section, I have provided a selection of audio circuits that are time-proven, and which provide excellent performance.

Figure 8-4 is a block diagram of a typical audio amplification system. It is mostly self-explanatory, with the exception of the volume control and the two 10-μF capacitors. The volume control potentiometer should have an audio taper (logarithmic response). A typical value is 100 Kohms. For stereo applications, this is usually a "two-ganged" pot, with one pot controlling the right channel and one pot controlling the left. The two 10-μF capacitors are used to block unwanted DC shifts that might occur if the volume control is rotated too fast.

Figure 8-5 is a simple preamplifier circuit for use with high-impedance signal sources, such as crystal or ceramic microphones. It is merely a common-collector amplifier with a few refinements. R5 and C4 are used to "decouple" the circuit's power source. The simple RC filter formed by R5 and C4 serves to isolate this circuit from any effects of other circuits sharing the same power supply source. R3 and C2 provide some positive feedback (in phase with the input) called "bootstrapping." Bootstrapping has the effect of raising the input impedance of this circuit to several Mohms.

Figure 8-6 is a *high-gain preamplifier circuit* for use with very low input signal sources. Dynamic microphones, and some types of musical instruments (such as electric guitars), work well with this type of circuit. R1 provides negative feedback for stabilization and temperature compensation purposes. Notice that this circuit is also decoupled by R5 and C2.

Figure 8-7 is an *active tone control circuit* for use with the outputs of the previous preamplifier circuits, or any *line-level* output. Active tone controls incorporate the use of an active device (transistor, FET, operational amplifier, etc.), and can provide better overall response with gain. *Passive tone controls*, in contrast, do not use any active devices within their circuits, and will always *attenuate* (reduce) the input signal. *Line-level outputs*

Figure 8-4
Block diagram of a typical audio amplification system.

Figure 8-5
Preamplifier for use
with high-Z signal
sources.

Figure 8-6
High-gain
preamplifier.

are signal voltages that have already been preamplified. The audio outputs from CD players, VCRs, tape players, and other types of consumer
electronic equipment are usually line-level outputs.

Figure 8-8 is a 12-watt rms audio power amplifier (the term *power
amplifier* implies that the primary function of the amplifier is to provide low-impedance, high-current *driving power* to a typical loudspeaker
system). It is relatively easy to construct, provides good linearity performance, and operates from a single DC supply (most audio power amplifiers require a dual-polarity power supply). A simple "raw" DC power

Figure 8-7
Active tone control
circuit.

supply, similar to the one illustrated in Fig. 5-4, constructed from a 2-amp, 24-volt transformer, a 2-amp bridge rectifier, and a 1000 μF filter capacitor at 50 working volts DC (WVDC) should power this circuit nicely (the rectified and filtered DC voltage produced from a 24-volt transformer will come out to *about* 36 volts DC). The complete circuit can be assembled on a small universal perfboard, or it can be constructed on its own PC board (PC circuit board fabrication is described later in this chapter; this project is mentioned here because it is ideal to test your first attempts at PC board making.) The parts list for this amplifier is given in Table 8-1.

A functional description of Fig. 8-8 is as follows. Beginning at the left-hand side of the illustration, the "audio signal input" is a *line-level signal* from a preceding audio device, such as an FM receiver or cassette tape deck. C1 is a coupling capacitor, blocking the DC bias on the base of Q1 from being applied to the preceding audio device.

Transistors Q1 and Q2 make up a *differential amplifier,* which functions as the first amplification stage of the amplifier. Differential amplifiers are often chosen as the first stage of an audio amplifier because they provide a convenient point of applying negative feedback (i.e., the inverting input; the base of Q2), and because they are capable of high current gain and high input impedance and are relatively insensitive to power supply fluctuations. A differential amplifier's unique quality of

Figure 8.8
A 12-watt rms audio
power amplifier.

common-mode rejection is a paramount issue pertaining to their use in operational amplifiers (discussed in Chapter 12), but it is not used at all within the context of most audio power amplifiers. A single resistor, R3, is used in place of a constant-current source, which is typically adequate for many medium-quality audio power amplifiers. R4 is the load resistor for the differential amplifier. Note that Q2 doesn't have a load resistor; this is because an output signal is not needed from Q2.

Resistors R1 and R2 form a voltage divider that splits the power supply in half, biasing the base of Q1 at half of the power supply voltage, or approximately 18 volts. Resistors R5 and R6 provide the identical function for the base of Q2. The combined voltage-dividing effect of R1, R2, R5, and R6 is to cause the entire amplifier circuit to operate from a reference of half of the power supply voltage, thereby providing the maximum peak-to-peak voltage output signal.

The second amplification stage of Fig. 8-8 is made up of Q3 and its associated components. This is a simple *common-emitter* amplifier stage, in which the audio signal is applied to the base of Q3 and the amplified output is taken from its collector. R8 serves as the collector load for Q3. Diodes D1, D2, and D3 provide a small forward bias to Q4, Q5, Q6, and Q7, to reduce the effects of *crossover distortion*. Capacitor C2 functions as a *compensation capacitor. Compensation* is a common term used with linear circuits (especially operational amplifiers) referring to a reduction of gain

TABLE 8-1

Parts List for
12-Watt Audio
Amplifier Project

Part Designation	Definition	Description
R1, R2, R4	Resistor	22 Kohm, $^1/_2$ watt
R3	Resistor	56 Kohm, $^1/_2$ watt
R5, R6	Resistor	2.2 Kohm, $^1/_2$ watt
R7	Resistor	10 Kohm, $^1/_2$ watt
R8	Resistor	5.6 Kohm, $^1/_2$ watt
R9, R10	Resistor	100 ohm, $^1/_2$ watt
R11, R12	Resistor	0.5 ohm, 1 watt
R13	Resistor	10 ohm, $^1/_2$ watt
C1	Electrolytic capacitor	10 µF, 25 WVDC
C2	Ceramic capacitor	20 picofarads
C3	Ceramic capacitor	0.1 µF
C4	Electrolytic capacitor	1000 µF, 50 WVDC
D1, D2, D3	Diodes	1N4148
Q1, Q2, Q5	PNP transistors	2N3906
Q3, Q4	NPN transistors	2N3904
Q6	NPN transistor	TIP 31C
Q7	PNP transistor	TIP 32C
F1	Fuse	2-amp, GMA type
Miscellaneous		GMA-type fuse clips (PC board mount type), PC board, optional heatsink for TO-220 devices, or metal enclosure

as the signal frequency increases. High-gain linear circuitry will always incorporate some amount of negative feedback (as discussed previously, to improve the overall performance). At higher frequencies, because of the internal capacitive characteristics of semiconductors and other devices, the negative-feedback signal will increasingly *lag* the input signal (remember, voltage lags the current in capacitive circuits). At very high frequencies, this voltage lag will increase by more than 180 degrees, causing the negative-feedback signal to be "in phase" with the input signal. If the voltage gain of the amplifier is greater than unity at this frequency, it will oscillate. Therefore, *compensation* is the general technique of forcing the voltage gain of a linear circuit to drop below unity before the phase-shifted negative feedback signal can lag by more than 180 degrees. In gist, compensation ensures *stability* in a high-gain amplifier or linear circuit.

To understand how C2 provides compensation in the circuit of Fig. 8-8, note that it connects from the collector of Q3 to the base of Q3. As you recall, the output (collector signal) of a common-emitter amplifier is 180 degrees out of phase with the input (base signal). Therefore, as the signal frequency increases, causing the impedance of C2 to drop, it begins to apply the collector signal of Q3 to the base of Q3. The 180-degree out-of-phase collector signal is negative feedback to the base signal, so as the frequency increases, the *voltage gain* of Q3 decreases. The result is a falling-off (generally called *rolloff*) of gain at higher frequencies, promoting good audio frequency stability of the Fig. 8-8 amplifier.

All of the voltage gain in the Fig. 8-8 amplifier occurs in the first two stages. Therefore, when an audio signal is applied to the input of the amplifier, the signal voltage at the collector of Q3 will be the "maximum" output signal voltage that the amplifier is capable of producing. There are two other points regarding the signal voltage at the collector of Q3 that should be understood. First, it is in phase with the input signal. This is because the original audio signal was inverted once at the collector of Q1, and it is inverted again at the collector of Q3, placing it back in phase with the input. Second, the audio input signal was *superimposed* on the DC quiescent level at the base of Q1, which is set to half of the power supply voltage, or about 18 volts. Therefore, a *positive half-cycle* of the audio signal on the collector of Q3 will vary from 18 to 36 volts. In contrast, a *negative half-cycle* of the audio signal will vary from 18 volts down to 0 volt (i.e., the amplified AC signal voltage is superimposed on a quiescent DC level of half of the power supply voltage).

The third stage of Fig. 8-8 consists of Q4, Q5, Q6, Q7, and their associated components. To begin, consider the operation of Q4 and Q6 only. Note that they are both NPN transistors, and the emitter of Q4 connects directly to the base of Q6. This configuration is simply a type of Darlington pair with a few "stabilizing" resistors added. As you recall, the purpose of a Darlington pair is to increase the current gain parameter, or *beta*, of a transistor circuit. Transistors Q4 and Q6 serve as a high-gain, current amplifier in this circuit. They will *current-amplify* the collector signal of Q3.

Under normal operation, the quiescent DC bias of this amplifier is such that the righthand side of R7 will be at half of the power supply voltage, or about 18 volts (this condition is due to the DC bias placed on the input stage, as explained earlier). This also means that the emitters of Q4 and Q6 will be at about 18 volts also (minus a small drop across their associated emitter resistors). If a positive-going AC signal is applied to the input of this amplifier, transistors Q4 and Q6 will current-amplify this signal. However, as soon as the AC signal goes into the *negative* region, the signal applied to the base of Q4 will drop below 18 volts, causing Q4 and Q6 to go into cutoff. Therefore, Q4 and Q6 are only amplifying about half of the audio signal (i.e., the *positive* half-cycles of the audio AC signal). Since transistors Q5 and Q7 are PNP devices, with their associated collectors tied to circuit common, they are current-amplifying the negative half-cycles of the audio AC signal, in a directly inverse fashion as Q4 and Q6. Simply stated, all audio signal voltages above 18 volts are amplified by Q4 and Q6, while all audio signal voltages below 18 volts are amplified by Q5 and Q7. Consequently, the entire amplified audio signal is summed at the positive plate of C4.

Note that resistor R7 is connected from the amplifier's output back to the "inverting" side of the input differential amplifier (i.e., the base of Q2). R7 is a *negative-feedback* resistor. As you recall, the audio signal voltage at the collector of Q3 is in phase with the audio input signal. Q4, Q5, Q6, and Q7 are all connected as common-collector amplifiers (i.e., the input is applied to the bases with the output taken from the emitters). Since common-collector amplifiers are noninverting, the audio signal at the amplifier's output remains in phase with the input signal. Therefore, the noninverted output signal that R7 applies back to the inverting input of the differential amplifier is negative feedback. Negative feedback in this circuit establishes the voltage gain, increases linearity (i.e., decreases distortion), and helps stabilize the quiescent voltage levels. The voltage gain (A_e) of this amplifier is approximately equal to R7 divided by the parallel resistance of R5 and R6 (i.e., about 10).

Capacitor C4, like C1, is a coupling capacitor, serving to block the quiescent 18-volt DC level from being applied to the speaker. Note, however, that the value of C4 is very large compared to the value of C1. This is necessary because the impedance of most speakers is only about 8 ohms. Therefore, in order to provide a time constant long enough to pass low-frequency audio signals, the capacity of C4 must be much greater.

Finally, resistor R13 and capacitor C3 form an output circuit that is commonly referred to as a *Zobel network*. The purpose of a Zobel network is to "counteract" the effect of typical speaker coil inductances, which could have a destabilizing effect on the amplifier circuitry.

As you may have noted, the Fig. 8-8 amplifier consists of three basic stages, commonly referred to as the *input stage* (Q1 and Q2), the *voltage amplifier stage* (Q3), and the *output stage* (Q4, Q4, Q6, and Q7). Virtually all modern solid-state audio power amplifiers are designed with this same basic three-stage architecture, commonly referred to as the *Lin three-stage topology.*

Constructing the 12-Watt RMS Amplifier of Fig. 8-8

If you decide to construct this amplifier circuit on a universal breadboard or solderless breadboard, the construction is rather simple and straightforward. You will need to provide some heatsinking for output transistors Q6 and Q7. If you mount the amplifier in a small metal enclosure, adequate heatsinking can be obtained by simply mounting Q6 and Q7 to the enclosure (remember to ensure that Q6 and Q7 are electrically isolated from the enclosure).

For optimum performance, diodes D1 and D3 should thermally track the temperature of transistors Q6 and Q7. This can be accomplished in several ways. The diodes could be glued to the case of the output devices with a small drop of epoxy, or you can bend a small solderless "ring" terminal into a makeshift clamp, with the ring held in place by the transistor's mounting bolt. If you construct the amplifier circuit similar to the Fig. 8-9a layout, you can simply bend the diodes into a touching position with the output transistors (remember to solder the diodes high above the board surface if you want to use this technique).

If you plan to use a current-limited power supply to test this amplifier circuit (such as the "lab-quality power supply" detailed in Chapters 3 through 6), fuse F1 isn't necessary. If you provide operational power to this amplifier circuit with a simple raw DC power supply as mentioned earlier, F1 must be included for safety purposes. Also, keep in mind that if you accidentally "short" (short-circuit) the speaker output leads together, you will probably destroy one or both of the output transistors (i.e., Q6 and Q7).

Making Printed Circuit Boards by Hand

Making PC boards is not as difficult as you may have been led to believe, or as your past experiences may have indicated. Circuit board

Figure 8-9
12-watt audio ampli-
fier: (*a*) and (*b*) top
views silkscreen lay-
out; (*c*) bottom view
copper artwork.

a

b

c

manufacture is a learned technique, and like any technique, there are several "correct" ways of going about it and many "wrong" ways of accomplishing disaster. In this section, I'm going to detail several methods that can be successfully used by the hobbyist. With a little practice, either of these methods should provide excellent results. The 12-watt audio amplifier described in the previous section is an excellent "first" project for getting acquainted with PC board fabrication, so I included a set of PC board layout illustrations in Fig. 8-9. See Appendix C for the full-size set to be used in your project.

To begin, you will need to acquire some basic tools and materials. If you are a complete novice regarding PC board construction, I recommend that you start by purchasing a "PC board kit," such as the one illustrated in Fig. 8-10. Typically, such kits will contain a bottle of *etchant* (an acid solution used to dissolve any exposed copper areas on the PC board), a *resist ink pen* (a pen used to draw a protective ink coating over any copper areas that you don't want dissolved by the etchant), several "blank" (i.e., unetched) pieces of PC board material, and some miscellaneous supplies that you may or may not need. In addition, you will need a few very small drill bits (no. 61 is a good size), an electric hand drill, some fine-grained emery paper, a small pin punch, a tack hammer, and a glass tray.

If you want to make a PC board of the Fig. 8-9 *artwork* by hand, the following procedure can be used. Begin by observing the illustrations provided in Fig. 8-9. Figure 8-9a is commonly called the *silkscreen* layout. It illustrates a top view of the placement and orientation of the components after they are correctly mounted to the PC board. Figure 8-9b is another top view of the silkscreen, illustrating how the bottom-side

Figure 8-10

Illustration of a typical PC board fabrication kit.

copper artwork will connect to the top-mounted components—the PC board is imagined to be transparent. Figure 8-9c is a *reflected* view of the bottom-side copper artwork. In other words, this is exactly how the copper artwork should look if you turn the board upside down and look at it from the bottom. The reflected view of the copper artwork, Fig. 8-9c, is what you will be concerned with in the next phase of your PC board construction.

Make a good copy of Fig. 8-9c on any copy machine. Cut a piece of PC board material to the same size as the illustration. Cut out the illustration from the copy and tape it securely on the "foil" side of the PC board material. Using a small pin punch and tack hammer, make a dimple in the PC board copper at each spot where a hole is to be drilled. When finished, you should be able to remove the copy and find a dimple in the copper corresponding to every hole shown in the artwork diagram. Next, drill the component lead holes through the PC board at each dimple position. When finished, hold the PC board up to a light, with the diagram placed over the foil side, to make certain you haven't missed any holes and that all of the holes are drilled in the right positions. If everything looks good, lightly sand the entire surface of the copper foil with 600-grit emery paper to remove any burrs and surface corrosion.

Using the resist ink pen, draw a "pad" area around every hole. Make these pads very small for now; you can always go back and make selected ones larger, if needed. Using single lines, connect the pads as shown in the artwork diagram in the following manner. Being sure you have the board turned correctly to match the diagram, start at one end and connect the simplest points first. Using these first points as a reference, eventually proceed on to the more difficult connections. When finished, you'll have a diagram that looks like a "connect the dots" picture in a coloring book. Finally, go back and "color" in the wide foil areas (if applicable) and fill in the wider tracks. The process is actually easier than it appears at first glance. If you happen to make a major mistake, just remove all of the ink with ink solvent or a steel wool pad and start over again—nothing is lost but a little time. You'll be surprised at how accomplished you will become at this after only a few experiences.

When you're satisfied that the ink pattern on the PC board corresponds "electrically" with the reflected artwork of Fig. 8-9c, place it in a glass or plastic tray (not metal!), and pour about an inch of etchant solution over it. Be very careful with this etchant solution; it permanently stains everything it comes in contact with, including skin. Wear goggles to protect your eyes, and don't breathe the fumes. After about 15 to 20 minutes, check the board using a pair of tongs to lift it out of the etchant solution. Continue checking it every few minutes until all of the unwanted copper has been removed. When this is accomplished, wash the board under cold water, and remove the ink with solvent or steel wool. When finished, the PC board will be ready for component installation and testing.

If you construct any PC boards using the aforementioned procedure, you will discover that the copper artwork on the finished PC board is slightly "pitted" in areas where the resist ink did not adequately protect the copper from the etchant. If you want to improve the finished quality of your PC boards, use inexpensive fingernail polish instead of resist ink. Obtain a dried-out felt-tipped marking pen (one with an extrafine point), repeatedly dip the pen in the fingernail polish (like an old quill ink pen), and use it to draw your pads and traces in the same way that you would use the resist ink pen as described above. After etching, the fingernail polish can be removed with ordinary fingernail polish remover, and the copper foil surface of your PC board will be totally free of any pits or corrosion from the etchant.

Making Printed Circuit Boards by the Photographic Method

The previously described method of fabricating PC boards by hand is acceptable for some surprisingly complex PC patterns, but it is very time-consuming and nonrepeatable (i.e., it is difficult to make exact replications). The serious hobbyist will most certainly develop the need to make PC boards in a more rapid and repeatable manner, and the best system that I have discovered for accomplishing this is by means of a photographic process. As in the case of the method discussed above, the photographic method of PC board fabrication is a learned technique, requiring a minimal amount of trial-and-error learning experience for optimum results. Don't get discouraged if your first attempt turns out less than ideal.

In addition to the previously listed materials, the photographic method of making PC boards will require some additional tools and materials. The PC board "blanks" will need to be *presensitized with a positive-acting photoresist*. Such presensitized PC boards are available from a wide variety of electronics suppliers, and the price is very reasonable. You will need a fluorescent light source, consisting of two ordinary 18-inch fluorescent lights positioned side by side, that can be placed approximately 2 inches above the surface of the PC board material. You will also need a supply of office transparencies that can be used with typical copy machines, available from office supply stores. And finally, you will need some *positive PC board developer* (available from electronics suppliers; can be purchased as a dry powder or liquid) and a pane of glass somewhat larger than any PC board size that you intend to make.

The process of making a PC board with the positive photographic method is as follows. Begin by making a *photopositive* of the "reflected" PC board artwork that you intend to fabricate. This is accomplished by making a copy of the black-and-white reflected artwork illustration onto a sheet of clear transparency with a copy machine. The finished photopositive should look identical to the reflected artwork illustration (i.e., all of the copper areas to be *saved* are black; the copper areas to be *etched* are transparent), except that the white areas on the original illustration will be *transparent* on the photopositive. This dark area on this photopositive must be as opaque as possible. If you can see "leakage light" through the dark areas when holding the photopositive up to a light source, you should adjust the copier for a darker copy. If the copy is still too light, you may need to make two copies, and tape one copy over the other for increased opacity.

Once the photopositive is finished, you are ready to expose and develop the PC board artwork. Adjust the ambient lighting of your work area so that it is as dark as possible, but light enough to allow you to see what you're doing. Remove the protective plastic film from the presensitized PC board material, and lay the PC board down (sensitized side up) onto a flat work surface. Place the photopositive on top of the presensitized surface of the PC board, and then place the pane of glass on top of the photopositive (i.e., the photopositive is sandwiched flat between the presensitized surface and the pane of glass). At this point, make sure that the placement of the photopositive is correct on the PC board surface and that you haven't accidentally turned the photopositive upside down (getting the photopositive upside down is a very easy mistake to make— I've done it many times!). If all looks good, adjust your fluorescent lights to about 2 inches above the surface of the pane of glass. Turn on the fluorescent lights and expose the PC board for about 8 minutes.

After the exposure time is completed, remove the glass pane and photopositive, and store them for future use. Place the exposed PC board in a plastic or glass tray containing the "positive PC board developer solution." The recommended quantities of developer solution and mixing instructions will vary with different manufacturers, so follow the directions for preparing the developer solution as indicated on the specific type that you purchase. When the exposed PC board is placed in the developer solution, the photoresist coating that has been exposed to the fluorescent light will begin to dissolve, leaving a clear pattern of the desired PC board artwork. When all of the exposed photoresist has dissolved away, leaving a clean and bright copper surface underneath, remove the PC board from the developer solution and rinse under cold running water.

After the developed PC board has dried, examine the photoresist pattern carefully. If there are numerous "pits" or "light-colored areas" in the photoresist surface, this means you suffered some bleed-through of fluorescent light through your photopositive. In this case, you need to increase the opacity (i.e., darkness) of the foil pattern on your photopositive. If the pitting and/or lightness of the photoresist is not excessive, you can touch up the photoresist pattern with a little fingernail polish. Otherwise, you should scrap the PC board, correct your photopositive, and try again. In contrast, if some of your exposed copper areas (i.e., the areas intended to be etched) have a little remaining photoresist on them, this means that you didn't leave the PC board in the developer solution long enough. The only remedy is to scrape the excess photoresist off of the copper areas intended to be etched, but if the PC board artwork is more than moderately complex, it will probably be more practical to scrap the PC board and try again.

Assuming that you examine the developed PC board and everything looks good (i.e., the copper areas to be etched are bright and clean, the photoresist areas are smooth and pit-free, and all of the artwork edges are neat and sharp), the PC board is now ready for etching. This is accomplished in the same manner as described previously for making PC boards by hand. After the etching is completed, the photoresist can be removed with a fine steel wool pad, fingernail polish remover, or a variety of solvents. Personally, I like to use a spray can of "flux remover" for removing the photoresist.

If this whole photographic fabrication process sounds a little formidable, it is probably due to the lengthy and detailed description of the process that must be contained in a textbook of this sort. You might be a little shaky during your first attempt, but after a few experiences you'll easily get the hang of it. My wife has observed me making PC boards on many occasions, and she considers the whole process much less complicated than her recipe for good chicken gravy!

Other Methods of Fabricating PC Boards

As your experience in the electrical and electronics fields continues to grow, you will doubtless encounter various other methods of fabricating PC boards. Depending on your talents and resources, you may want to try a few. Generally speaking, the following is what I have discovered regarding some other popular methods.

■ The various methods used by large manufacturers and professional PC board fabrication companies can certainly produce professional-quality PC boards. However, the equipment and techniques required are typically too expensive for the hobbyist. Although the photographic method, as previously described, will take more time per unit, the end results can be just as good as the professional manufacturing techniques.

■ The various types of "stick-on transfers" are adequate for small, uncomplicated, one-of-a-kind PC boards. However, for PC boards of medium complexity, the "fabrication by hand technique," as first described in this section, is typically faster and provides better results (it's also less frustrating than trying to work with those little sticky lines that seem to want to stick to everything except your PC board!).

■ The *negative-acting photographic method* is equal to the positive-acting method as described above. However, the materials are not as readily available, and, in some cases, it is more difficult to obtain a good "photonegative" of the PC board artwork.

■ The *iron-on image-transfer system* has become popular. With this method, you must purchase a special film intended for this technique from one of several manufacturers. The PC board artwork is copied onto the film using a copy machine or laser printer (the *toner* causes a chemical change on the film, making the transfer possible). The PC board artwork image is then directly ironed onto the PC board using a clothing iron or some other suitable heat source and pressure. Some hobbyists appear to favor this method, but I have not found the quality of the results to be equal to those using the photographic method. Also, the results are not as repeatable.

CAD Methods of PC Board Development

Before leaving this discussion on PC board fabrication, I would like to briefly describe the newer generation of CAD (i.e., computer-automated design) programs available for PC board development. In bygone days, an electrical engineer was forced to sit down with a schematic diagram, a tablet of graph paper, and a sordid collection of templates, rulers, and other drawing instruments, in the attempt to design a PC board *layout* for a new circuit design. This was usually an aggravating and time-consuming project, requiring the engineer to draw out trial-and-error track runs, reposition component placement, and redefine board dimensions, and all the while the designer had to think in an upside-down perspective as to how the track runs on the bottom of the board would connect to the component leads on top. As PC board complexities increased, this process became proportionally more difficult—in the case of modern multilayer PC boards, it became virtually impossible.

During the 1970s and 1980s, industry began developing various CAD programs to facilitate the difficult task of PC board layout design. Later, these CAD programs were "humanized" to make them more "user friendly," and marketed to electronic hobbyists for the purpose of designing "home-brew" PC boards. Currently, these *PC board layout* programs are available at modest pricing, and they are so easy to use that almost anyone can design PC boards at home. If you anticipate that you will be involved with electronics as an ongoing hobby or profession, I highly recommend that you purchase a good PC board layout program. It will be well worth the investment.

If you are new to CAD PC board layout programs, you may be wondering how you would use such a tool in a practical way. Allow me to provide a hypothetical example. Suppose you wanted to construct a small, portable guitar amplifier with tone control. You could combine the preamplifier circuit of Fig. 8-6 with the tone control circuit of Fig. 8-7 and the 12-watt power amplifier of Fig. 8-8. Obviously, you could construct all of these circuits using perfboard or a universal breadboard, but the process is time-consuming and the finished product doesn't look very professional. In contrast, it would probably take you about an hour and a half to combine all three of these circuits into a single circuit design, download to a layout program, and come up with a single, finished, professional-quality PC board design. You would then use your inkjet printer to print out a full-size reflected artwork diagram directly onto a transparency. It may take you another hour to expose, develop, and etch the PC board for your project using this transparency. If you continued to work consistently, it might take you another 2 hours to drill the holes into the PC board and construct the finished project (provided you had all of the components on hand). In other words, you took a moderately complex project from the "idea" stage to the "professional-quality, completed project" stage in less than 5 hours! Your materials cost for this project was the cost of the presensitized PC board blank (possibly two or three dollars, depending on size), less than one dollar's worth of chemicals, the transparency used in your inkjet printer (about 60 cents), and the cost of the electronic components.

If you are intimidated by the thoughts of "designing with computers" or "running complex programs," you shouldn't be. Most of these types of user-friendly CAD programs are self-explanatory and very easy to become comfortable with. Personally, I use the Electronics Workbench–brand Multisim and Ultiboard design programs, and I can testify that on the very first time I tried using these programs, I was designing circuits in less time than it took me to figure out what the opening icons meant in Windows 95. And don't get the idea that I'm some sort of genius computer hacker—I still haven't made it through *Windows '95 for Dummies.*

A Professional-Quality Audio Power Amplifier

Figure 8-11 is a schematic illustration of a very high-quality professional audio power amplifier. Modern high-quality, solid-state audio power amplifiers are a culmination of almost all of the commonly used linearization techniques applied to the entire field of linear electronic circuits. Therefore, a detailed dissection of the building blocks incorporated into a modern power amplifier is a very good method of understanding much of the theory and physics involved with the broad spectrum of linear electronics, including such subjects as operational amplifiers, servo systems, and analog signal processing. Also, it should be remembered that the humble audio amplifier is probably the most common electronic subsystem to be found in the vast field of consumer electronics, including such common devices as televisions, stereo systems, radios, musical instrument amplifiers, multimedia computer systems, and public address systems.

Figure 8-11 Schematic diagram of a professional-quality audio power amplifier.

Preconstruction and Preliminary Design

The amplifier design of Fig. 8-11 is not a typical commercial-quality design; it is significantly superior. I chose this particular design to provide a baseline for discussion and illustration of various discrete building blocks that are commonly used in linear circuits. Also, it is a good project if you wish to construct something a little more advanced, and I have provided the PC board artwork if you would like to practice your PC board construction skills. (Unfortunately, space restrictions will not allow me the opportunity to provide PC board artwork for the majority of projects in this textbook—this is where a good CAD PC board design program will come in very handy!)

Beginning with C1 and C2 in Fig. 8-11, note how they are connected in series, oriented with opposing polarities. This is a common method of creating a *nonpolarized electrolytic capacitor* from two typical electrolytic capacitors. The combination of C1 and C2 will look like a single nonpolarized capacitor with a capacitance value equal to half of either individual capacitor and a voltage rating equal to either one of the capacitors individually. In other words, if C1 and C2 are both rated at 22 μF at 35 WVDC, the equivalent nonpolarized capacitor will perform like an 11-μF capacitor at 35 WVDC. A nonpolarized coupling capacitor is needed for this amplifier because the quiescent voltage on the base of Q1 will be very close to circuit common potential, so the AC input signal current flow through C1 and C2 will be in both directions.

Degeneration (Negative Feedback) R1 establishes the input impedance of the amplifier at approximately 10 kohms (the input impedance at the base of Q1 is much higher than R1). C3 provides some high-frequency filtering to protect the amplifier from ultrasonic signals that could be superimposed on the audio signal. R2 and R3 are *degeneration resistors* for the differential amplifier, consisting of Q1 and Q2. The term *degeneration* is used to describe a type of negative feedback and usually applies to resistors placed in the emitter circuits of transistors, but it can be applied to other techniques in which a single resistor is placed in the signal path of a gain stage, providing a voltage drop that opposes the gain action of an active device, thereby reducing gain and improving linearity. In simple terms, degeneration resistors flatten out the gain response of transistor circuits and improve their linearity.

Constant Current Sources   The circuit consisting of Q3, Q4, and R4 is an upside-down version of the circuit illustrated in Fig. 6-8d of the *Transistor Workshop* section of Chapter 6. If you refer back to Fig. 6-8d and the corresponding description, you will note that the Fig. 6-8d circuit serves the function of providing current-limit protection. In other words, it allows current to flow up to a maximum level, and then restricts it at that maximum level regardless of increases in the load or input signal. Referring back to Fig. 8-11, note that R5 has been added to the basic circuit of Fig. 6-8d, which provides a constant maximum base signal to Q4, which forces Q3 into a constant state of current limiting. Since the collector current of Q4 is constantly at the current-limit level (determined by the resistance value of R4), the circuit of Q3, Q4, R4, and R5 makes up a *constant current source*. Actually, this type of constant-current source provides better performance than do the more conventional types of constant-current sources, such as illustrated in Fig. 6-8a. This is due to the fact that you have the combined effect of both transistors in regulating the current, instead of a single transistor referencing a voltage source. Note that the circuit consisting of Q10, Q11, R9, and R12 makes up an identical constant-current source for the voltage amplifier stage; however, the level of regulated current is set differently because of the resistance value of R12.

Current Mirrors   Now, turn your attention to the peculiar-looking circuit consisting of Q5, Q6, R6, and R7 in Fig. 8-11. If you ever study the equivalent schematic diagrams of linear integrated circuits, you will see tons of these circuits represented. They are called *current mirrors*. Current mirrors "divide" current flows into equal portions. For example, if 4 milliamps of current is flowing through the collector of Q4, the current mirror of Q5 and Q6 will force 2 milliamps to flow through Q1 and 2 milliamps to flow through Q2—it will divide the total current flowing through the differential amplifier equally through both legs. This is important because a differential amplifier will be most linear when the current flows through both legs are "balanced" (i.e., equal to each other). Therefore, by incorporating the current mirror of Q5 and Q6, the linearity of the differential amplifier (i.e., Q1 and Q2) is optimized. The operational physics of Q5 and Q6 is rather simple. Since the collector of Q6 is connected directly to its base, the current flow through the collector of Q2 will be directly proportional to the voltage drop from the base to emitter of Q6. Since Q5's base-emitter junction is connected in parallel with Q6's base-emitter junction, Q5 is forced to "imitate" the collector current flow of Q6. Therefore, whatever the total current flow through the differential amplifier happens to be, it will be forced into a balanced state. R6 and R7 are degeneration resistors, serving to flatten out the small parameter differences between Q5 and Q6.

Voltage Amplification   Q8 provides the same function in this amplifier design as Q3 provides in the Fig. 8-8 amplifier. Namely, it provides the majority of "voltage gain" inherent to the amplifier. Likewise, C7 is the *compensation capacitor* for this amplifier, serving the same function as C2 does in the Fig. 8-8 amplifier.

Instead of utilizing three forward-biased diodes (to provide a small forward voltage drop, reducing the effects of crossover distortion) as in the design of Fig. 8-8, the *amplified diode* circuit of Q9, P1, and C8 is incorporated into the Fig. 8-11 design. Amplified diode circuits are often called V_{be} *multiplier* circuits. An amplified diode circuit has the advantages of a more precise control of the continuous forward bias applied to the output transistors, as well as improved thermal tracking characteristics when the V_{bias} transistor (Q9 in this case) is physically mounted to the same heatsink as the output transistors. C8 provides smoothing of voltage variations that would normally appear across the amplified diode circuit during signal variations.

Active Loading   If you compare the amplifier design of Fig. 8-11 with the design of Fig. 8-8, you'll note that the collector load for the voltage amplifier transistor (R8 in Fig. 8-8) has been replaced with the Q10/Q11 constant-current source in Fig. 8-11. The technique of using a resistor (a passive device) for a collector load in a transistor amplifier is appropriately called *passive loading*. In contrast, the technique of using an active circuit for a collector load, such as the Q10/Q11 constant-current source, is called *active loading*.

The principles and reasons for active loading relate back to the basic operational fundamentals of a simple common emitter amplifier. As you may recall, the voltage gain of a common emitter amplifier is determined by the ratio of the emitter resistance to the collector resistance. In theory, if you wanted a common-emitter amplifier with a very high gain, you would be forced to make the collector resistor a very high resistance value. Unfortunately, as you may also recall, the output impedance of a common-emitter amplifier is approximated as the value of the collector resistor. Therefore, if you increase the resistance of the collector resistor to extremely high values, the amplifier's output impedance also increases to such high values that the amplifier is no longer capable of "matching" to any kind of practical load. In other words, it can't "drive" anything. The optimum collector load for a common-emitter amplifier would be a load that appeared to represent extremely high resistance (i.e., a high internal impedance), while simultaneously providing its own "drive current" to an external load. This is the effect that a constant-current source provides to a common emitter amplifier.

The easiest way to understand the desirable effects of active loading is to imagine Q8 in Fig. 8-11 to be a *variable resistor* (signal voltages applied to the base of Q8 will cause the collector-to-emitter impedance of Q8 to vary proportionally, so this imaginary perspective is not in error). Also, consider Q7 and the *amplified diode* circuitry to be out of the circuit for purposes of discussion. Imagine that the variable collector-to-emitter impedance provided by Q8 can range from the extremes of 100 ohms to 6 Kohms, which is a reasonable estimation. Assume that throughout this entire drastic range of collector-to-emitter impedance variations, the constant current source of Q10/Q11 maintains a constant, regulated current flow of 6.7 milliamps. The only way that the same "constant current" effect could be accomplished using "passive" loading would be to make the collector load resistance for Q8 extraordinarily high, so that Q8's collector-to-emitter impedance variations would be negligible. In other words, if the Q10/Q11 constant-current source were replaced with a 1-Mohm resistor (i.e., 1,000,000 ohms), the 100-ohm to 6-Kohm collector-to-emitter impedance variations of Q8 would have a negligible effect on the circuit current (i.e., the dominant current-controlling factor would be the very high resistance of the collector resistor). However, to accomplish a 6.7-milliamp current flow through a 1-Mohm resistor, you would need a 6700-volt power supply! Consequently, the "active load" provided by the constant current source of Q10/Q11 looks like a very high-resistance collector load to Q8, which improves the gain and linearity factors of this voltage amplifier stage.

Output Stage Considerations   Note that the collector circuit of Q8 must provide the *signal drive current* to the bases of Q14 and Q15 (through resistors R13 and R14). The technique of active loading, as previously described, also aids in providing some of the signal drive current to Q14 and Q15, which makes the output impedance of the Q8 voltage amplifier stage appear reasonably low. This desirable condition also improves the linearity of the voltage amplifier stage.

Transistors Q14, Q15, Q16, and Q17 in Fig. 8-11 perform the same function as described previously for transistors Q4, Q5, Q6, and Q7 in the 12-watt amplifier of Fig. 8-8. Collectively, these transistors constitute a "near-unity voltage gain, current amplifier." They are biased in a class

B mode, meaning that Q14 and Q16 provide current gain for the *positive* half-cycles of the AC signal, while Q15 and Q17 provide current gain for the *negative* half-cycles of AC signal.

Capacitor C9 in Fig. 8-11 aids in eliminating a type of distortion known as switching distortion. High-power bipolar transistors have to be manufactured with a rather large internal crystal geometry to be able to handle the higher current flows. The larger crystal area increases the inherent "junction capacitance" of power transistors. The higher junction capacitance can cause power transistors to be reluctant to turn off rapidly (i.e., when they are driven into cutoff). If one output transistor is so slow in turning off that it is still in conduction when its complement transistor turns on, a very undesirable condition occurs which is referred to as cross conduction (i.e., both complementary output transistors are on at the same time). Cross-conduction creates switching distortion. Note how C9 is connected across both base terminals of the output transistors in Fig. 8-11. Under dynamic operating conditions, C9 acts to neutralize charge carriers in the base circuits of the output transistors, thereby increasing their turn-off speed and eliminating switching distortion. Although cross-conduction creates undesirable switching distortion in audio power amplifiers, it can be destructive in various types of motor control circuitry utilizing class B outputs. Consequently, it is common to see similar capacitors across the base circuits of high-current class B transistor outputs utilized in any type of high-power control circuitry.

Diodes D4 and D5 in Fig. 8-11 can be called by many common names. They are most commonly known as *freewheeling, catching, kickback,* or *suppression* diodes, although I have heard them called by other names. Their function is to suppress reverse-polarity inductive "kickback" transients (i.e., reverse electromotive force pulses) that can occur when the output is driving an inductive load. As you may recall, this is the same function performed by D1 in Fig. 7-9 of Chapter 7.

Most high-quality audio power amplifiers will incorporate a low-inductance-value air-core coil on the output, such as L1 in Fig. 8-11. L1 has the tendency to negate the effects load capacitance, which can exist in certain speakers or speaker crossover networks. Even small load capacitances can affect an audio power amplifier in detrimental ways, because it has a tendency to change the high-frequency phase relationships between the input and negative-feedback signals. Compensation is incorporated into an amplifier to ensure that its gain drops below unity before the feedback phase shift can exceed 180 degrees, resulting in oscillation. However, if the amplifier's speaker load is somewhat capacitive, it can cause a more extreme condition of phase lag at lower frequencies, leading to a loss of stability in the amplifier. At higher frequencies, the *inductive reactance* of L1 provides an opposing force to any *capacitive reactance* that may exist in the speaker load, thereby maintaining the stability of the amplifier under adverse loading conditions. The inductance value of L1 is too low to have any effect on frequencies in the audio bandwidth. R26, which is in parallel with L1, is called a "damping" resistor. R26 reduces "ringing oscillations" that can occur at resonant frequencies of the speaker and L1 (resonance will be discussed in a later chapter).

Short-Circuit Protection   The circuit consisting of Q12, D2, R16, R15, and R19 provides short-circuit protection during *positive* signal excursions for output transistor Q16. Likewise, Q13, D3, R17, R18, and R20 provide short-circuit protection for Q17 during *negative* signal excursions. Since these two circuits operate in an identical fashion but in opposite polarities, I will discuss only the positive protection circuit, and the same operational principles will apply to the negative circuitry.

Referring to Fig. 8-11, imagine that some sort of adverse condition arose, creating a direct "short" from the amplifier's output to circuit common. Note that all of the speaker load current flowing through Q16's emitter must also flow through R22. As the Q16 emitter current tries to increase above 3 amps, the voltage drop across R22 (which is applied to the base of Q12 by R19) exceeds the typical 0.7-volt base-emitter voltage of Q12, causing Q12 to turn on. When Q12 turns on, it begins to divert the drive current away from the base of Q14, which, in turn, decreases the base drive current to Q16. The more the emitter current of Q16 tries to increase, the harder Q12 is turned on, thereby limiting the output current to a maximum of about 3 amps, even under a worst-case short-circuit condition. In other words, under an output short-circuit condition, this protection circuit behaves identically to the other current-limit protection circuits you have examined thus far. The only difference is the inclusion of Q14, which serves only to "beta-enhance" the output transistor Q16.

In addition to protecting the amplifier of Fig. 8-11 from short-circuit conditions at the output, the protection circuit also provides a more complex action than does simple current limiting. The voltage divider of R15 and R16, in conjunction with the dynamic action of the output rail (which is connected to the emitter of Q12), causes the current-limit action of the protection network to be "sloped." In other words, it will limit the maximum current to about 3 amps when the output rail voltage (i.e., the amplifier's output voltage) is 0 volt. However, as the output rail voltage increases in the positive direction, the protection circuitry will allow a higher maximum current flow. For example, if the amplifier's output voltage is at 20 volts, the protection circuit may allow a maximum of 5 amps of current flow. This type of sloped protection response allows the maximum output power to be delivered to a speaker system, while still providing complete protection for the output transistors. And finally, diode D2 is placed in the collector circuit of Q12 to keep Q12 from going into a state of conduction during negative signal excursions.

Now that the output protection circuitry has been explained, it becomes easier to understand the function of Q7. Q7 provides current-limit protection for transistor Q8. If you refer back to Fig. 6-8d, you will note that the Q7/Q8 combination of Fig. 8-11 is nearly identical; the only difference is in the value of the emitter resistors. To understand why Q8 in Fig. 8-11 needs to be current-limited, imagine that the amplifier output is short-circuited. If a positive signal excursion occurs, Q16 will be protected by Q12 turning on and essentially shorting (short-circuiting) Q14's base current to the output rail. In this situation, the drive current to Q14's base is actually originating in the constant-current source of Q10/Q11. Since Q10 and Q11 are already in a state of current limit, Q12's shorting action is of no consequence. However, if a negative signal excursion occurs, Q13 will turn on to protect Q17, shorting Q15's drive current to the output rail. During negative signal excursions, the drive current for Q15 is the collector current of Q8 through the negative power supply rail. Without the current-limit protection provided by Q7, Q8 would be destroyed as soon as Q13 turned on, because there would be nothing in the collector circuit of Q8 to limit the maximum current flow.

Negative Feedback   Negative feedback in the Fig. 8-11 amplifier is provided by R8, R10, and C6. As you recall from previous studies of the common emitter amplifier, a "bypass" capacitor could be placed across the emitter resistor, causing the DC gain response to be different than the AC gain response. C6 serves the same purpose in this design of Fig. 8-11. The AC voltage gain response of this design is determined by the ratio of R10 and R8, since C6 will look like a short to circuit common to AC signal voltages. With the values shown, it will be about 31 [(R10 + R8) divided by R8 = 31.3]. In contrast, C6 will look like an infinite impedance to DC voltages, placing the DC voltage gain at unity (i.e., 100% negative feedback). Simply stated, the negative-feedback arrangement of Fig. 8-11 provides the maximum negative DC feedback to maintain the maximum accuracy and stability of all DC quiescent voltages and currents. Regarding AC signal voltages, it provides the necessary voltage gain to provide the maximum output power to the speaker load.

Note that C6 is in parallel with D1. If a component failure happened to occur that caused a high negative DC voltage to appear on the output of the amplifier, a relatively high *reverse-polarity* voltage could be applied across C6, resulting in its destruction. Diode D1 will short any significantly high negative DC levels appearing across C6, thereby protecting it.

Other Concerns Finally, a few loose ends regarding Fig. 8-11 have not been discussed. R24 and C10 make up the Zobel network, which serves the same purpose as the Zobel network previously explained in reference to Fig. 8-8. C11, C12, C4, and C5 are all *decoupling capacitors* for the power supply rails. You'll note that the "signal common" connection at the input of the amplifier is not connected to circuit common. This signal common point must be connected to circuit common, but it is better to run its own "dedicated" circuit common wire back to the power supply, thereby eliminating electrical noise that might exist on the "main" circuit common wire connecting to all of the other components.

Now that you have completed the theoretical analysis of the complex audio power amplifier of Fig. 8-11, as well as the other circuits illustrated in this chapter, you may be wondering how an in-depth understanding of these circuits will be beneficial in your continued involvement in electronics. Allow me to explain. If you happen to be interested in audio electronics, obviously this chapter is right down your alley. But if your interest lies with robotics or industrial control systems, don't despair! If you look at a schematic diagram of a servo motor controller, you'll say, "Hey! This is nothing but a customized audio power amplifier." If you look at the schematic of an analog proportional loop controller, as used extensively in industry, you'll say, "Hey! This is nothing but an adjustable gain amplifier with the means of obtaining negative feedback from an external source." If you look an the "internal" schematic of an operation amplifier IC, you'll say, "Hey! This is exactly like an audio power amplifier with external pins for customizing the feedback and compensation." (In reality, modern audio power amplifiers are nothing more than large "discrete" versions of operational amplifiers—you'll examine operational amplifiers in more detail in Chapter 12.)

Expectedly, there will be times when you look at the schematic of an electronic circuit and say, "Hey! I have no idea what this is." However, if the circuit is functioning with analog voltages, currents, and/or signals, it will consist of some, or all, of the following circuit building blocks: single-stage or multistage amplifiers, differential amplifiers, current mirrors, constant-current sources, voltage reference sources, protection circuits, power supplies, and power supply regulation and protection circuits. It will also probably utilize some or all of the following techniques: beta enhancement, negative feedback, active loading, compensation, filtering, decoupling, and variations of gain manipulation. The only exception to these generalities will be in circumstances where the electronic circuitry consists primarily of linear integrated circuits performing the aforementioned functions and techniques. The main point is that if you have a pretty good understanding of the linear circuitry discussed thus far, you have a good foundation for comprehending the functions and purpose of most linear circuitry.

Constructing the Professional-Quality Audio Power Amplifier

If you would like to construct the professional-quality audio power amplifier as described above, I have provided a professional layout design (Figs. 8-12 and 8-13) and the PC board artwork (Fig. 8-14) for you to copy. See Appendix C for a full-size copy of Fig. 8-14, to be used in your pro-

Figure 8-12

Top-view silkscreen of the 50-watt professional-quality amplifier.

Top-view silkscreen of
the 50-watt amplifier
showing top view of
the PC board
artwork.

Bottom-view reflected
artwork for the 50-
watt professional-
quality amplifier.

ject. I have constructed amplifiers more involved than this using the PC board "fabrication by hand" process described earlier, so it is not overly difficult to accomplish if you have a little time and patience. Also, I have constructed similar amplifiers on perfboard, but this process is really time-consuming, and you are more prone to make mistakes. Naturally, the preferred method of construction is to fabricate the PC board as illustrated using the photographic technique. Regardless of the method you choose, the amplifier circuit is reasonably forgiving of wire or trace size, and component placement. If you would like to avoid the task of PC board fabrication all together, a complete kit (including etched and drilled PC board) is available from SEAL Electronics (the contact information is provided in Appendix B of this textbook).

The parts list for the professional-quality amplifier is provided in Table 8-2. Most of these components are available at almost any electronics supply store, but a few details need to be highlighted. If you want to go first class on this project, you can use 1% metal film resistors for all of the $\frac{1}{2}$-watt resistors listed. Metal film resistors will provide a little better signal-to-noise performance, but it is probable that you will not be able to hear the difference—standard 5% carbon film resistors will provide excellent performance. Output inductor L1 is easily fabricated by winding about 18 turns of either 16 or 18 AWG "magnet wire" around a $\frac{1}{2}$-inch form of some type (I use an old plastic ink pen that happens to be $\frac{1}{2}$ inch in diameter). The exact inductance of L1 is not critical, so don't become unduly concerned with getting it perfect.

All of the transistors used in this amp are readily available from MCM Electronics or Parts Express (contact information is provided in Appendix B of this textbook), as well as many other electronic component suppliers. The fuse clips used in the PC board design are GMA types (four needed) with two solder pins that insert through the holes in the PC board. Pay particular attention to the orientation of transistors Q14 and Q15 when soldering them into the PC board—it is easy to install them backward. You should mount a small TO-220-type heatsink to Q14 and Q15. Almost any type of TO-220 heatsink will be adequate, since the power dissipation of Q14 and Q15 is very low. A common style of heatsink that will be ideal is U-shaped, measuring about $\frac{3}{4} \times 1$ inch ($w \times h$) (width $\times$ height), constructed from a single piece of thin, stamped aluminum.

You will need one medium-sized heatsink for the output transistors (i.e., Q16 and Q17) and the V_{bias} transistor (Q9). Just to provide you with a rough idea of the size of this heatsink, the commercial models of this amplifier use a heatsink that measures $6 \times 4 \times 1\frac{1}{4}$ ($l \times w \times d$) (length $\times$ width $\times$ depth) (if you're lucky enough to find a heatsink accompanied

TABLE 8-2

Parts List for
50-Watt
Professional-Quality
Audio Amplifier

Part Designation	Definition	Description
R1, R9, R10	Resistor	10 Kohm, $\frac{1}{2}$ watt
R2, R3, R6, R7, R12, R13, R14	Resistor	100 ohm, $\frac{1}{2}$ watt
R4	Resistor	180 ohm, $\frac{1}{2}$ watt
R5	Resistor	22 Kohm, $\frac{1}{2}$ watt
R8	Resistor	330 ohm, $\frac{1}{2}$ watt
R11	Resistor	33 ohm, $\frac{1}{2}$ watt
R15, R18	Resistor	12 Kohm, $\frac{1}{2}$ watt
R16, R17, R21	Resistor	150 ohm, $\frac{1}{2}$ watt
R19, R20	Resistor	270 ohm, $\frac{1}{2}$ watt
R22, R23	Resistor	0.22 ohm, 5 watt
R24, R25	Resistor	8.2 ohm, 5 watt
C1, C2	Tantalum capacitor	22 μf at 16 WVDC
C3	Ceramic capacitor	68 pF
C4, C5	Electrolytic capacitor	100 μF at 50 WVDC
C6	Electrolytic capacitor	220 μF at 25 WVDC
C7	Ceramic capacitor	150 pF
C8	Electrolytic capacitor	10 μF at 16 WVDC
C9	Mylar capacitor	1 μF at 100 volts
C10, C11, C12	Mylar capacitor	0.1 μF at 100 volts
D1, D2, D3	Diode	1N4148
D4, D5	Diode	1N5060
L1	Inductor	1 μH (see text)
Q1, Q2, Q3, Q4, Q10, Q11, Q13	PNP transistor	2N5401
Q5, Q6, Q7, Q8, Q12	NPN transistor	2N5551
Q9, Q14	NPN transistor	2SD669
Q15	PNP transistor	2SB649
Q16	NPN transistor	MJ15003
Q17	PNP transistor	MJ15004
P1	Trim potentiometer	1 Kohm (horizontal mount)
F1, F2	Fuses	3-amp GMA
(Not illustrated)	Fuse clips	GMA type (see text)
(Not illustrated)	Heatsink	Double T0-3 type (see text)
(Not illustrated)	Heatsink	Small TO-220 type (see text)
Miscellaneous		Transistor insulators, PC board, mounting hardware

with the manufacturer's thermal resistance rating, the specified rating is
0.7°C/watt). However, if this is all confusing to you, don't worry about it.
If you can find a heatsink that is close to the same dimensions and is
designed for the mounting of two TO-3 devices, it should suffice very
well. A few examples of larger heatsink styles are provided in Fig. 8-15; the
square-shaped Wakefield type in the forefront of the illustration is a good
choice for this amplifier project. If the heatsink is manufactured so that it
can be mounted vertically, there is space provided for mounting it directly to the PC board. If not, you can run longer connection wires out to an
externally mounted heatsink (the length of the transistor connection
wiring in this design is not critical). Q9 should be mounted in close proximity to the output transistors on the same heatsink so that it will thermally track the temperature of the output transistors. The connection
wiring to Q16, Q17, and Q9 can be made with ordinary 20- to 22-AWG
stranded, insulated hookup wire. Be sure to double-check the accuracy of
your wiring according to the silkscreen diagrams provided—it's very easy
to get the transistor leads confused on the TO-3 devices.

Testing, Setup, and Applications of the Professional-Quality Audio Power Amplifier

Once you have finished the construction of the amplifier, I suggest that
you use the schematic (Fig. 8-11) and accompanying illustrations (Figs. 8-12
to 8-14) to recheck and then double-recheck your work. The amplifier

Figure 8-15
Some examples of
larger heatsink styles.

design of Fig. 8-11, like all modern high-quality audio power amplifier designs, is classified as *direct-coupled*. This term simply means that all of the components in the signal path are directly coupled to each other, without utilizing transformers or capacitors to couple the signal from one stage to the next. Direct-coupled amplifiers provide superior performance, but because of the direct-coupled nature of the interconnecting stages, an error in one circuit can cause component damage in another (a damaged electronic circuit causing subsequent damage in another electronic circuit is referred to as *collateral damage*). Therefore, it is extremely important to make sure that your construction is correct; one minor mistake is all that it takes to destroy a significant number of components.

The amplifier of Fig. 8-11 is designed to operate from a "raw" dual-polarity power supply providing 42-volt rail potentials. Under these conditions, it is very conservatively rated at 60 watts rms of output power into typical 8-ohm speaker loads. The design also performs very well with dual-polarity rail voltages as low as 30 volts, but will provide proportionally lower output power. Figure 8-16 illustrates a good power supply design for this amplifier. As shown, it will provide dual-polarity rail potentials of about 38 volts. This equates to a maximum power output of a little over 50 watts rms.

There are a few details that bear mentioning in the power supply design of Fig. 8-16. The 0.01-μF capacitor connected across the power switch (SW1) is provided to eliminate "pops" from the amplifier when the

Figure 8-16
Recommended power supply for 50-watt audio amplifier.

switch is opened. Note that when SW1 is open, the full 120-volt AC line voltage will be dropped across the 0.01-μF capacitor, so the capacitor should be rated for about 250 volts (the peak voltage of ordinary 120-volt AC residential power is about 170 volts). R1 and R2 are called *bleeder resistors*, and they are incorporated for safety reasons. Referring back to the amplifier schematic of Fig. 8-11, imagine that a failure had occurred causing the rail fuses (F1 and F2) to blow. With the rail fuses blown, the filter capacitors of the power supply (i.e., C3 and C4 in Fig. 8-16) do not have a discharge path. Therefore, they could maintain dangerous electrical charges for weeks, or even months. If you attempted to service the amplifier without recognizing that the filter capacitors were charged, you could suffer physical injury from accidental discharges. R1 and R2 provide a safe discharge path for the filter capacitors to prevent such servicing accidents. R1 and R2 can be 5.6-Kohm, $^1\!/_2$-watt resistors for this power supply. C1 and C2 are typically 0.1-μF ceramic disk capacitors providing an extra measure of high-frequency noise filtering on the power supply lines. They are seldom required, and may be omitted if desired. The capacitance value for the filter capacitors (i.e., C3 and C4) should be about 5000 μF (or higher), with a voltage rating of at least 50 WVDC.

While on the subject of power supply designs, the power supply illustrated in Fig. 5-6 will also function very well with this amplifier design. You may want to consider this approach if you have difficulty finding a 50-volt center-tapped, 3-amp transformer (if you decide to use the Fig. 5-6 circuit for the amplifier power supply, it would be a good idea to install bleeder resistors as illustrated in the Fig. 8-16 design).

After connecting the Fig. 8-11 audio power amplifier to a suitable dual-polarity power supply, adjust P1 to the middle of its adjustment range and apply operational power. If either one of the rail fuses blows (i.e., F1 or F2), turn off the power immediately and correct the problem before attempting to reapply power. If everything looks good, and there are no obvious signs of malfunction, use a DVM to measure the DC voltage at the amplifier's speaker output (i.e., the right side of L1). If the amplifier is functioning properly, this *output offset voltage* should be very low, with typical values ranging between 10 and 20 mV.

If the output offset voltage looks good, measure the DC voltage from the emitter of Q16 to the emitter of Q17, and adjust P1 for a stable DC voltage of 47 mV. Once this procedure is accomplished, the amplifier is tested, adjusted, and ready for use.

From a performance perspective, the input sensitivity of this power amplifier design is approximately 0.9 volts rms. The rms power output is a little above 50 watts using the power supply design of Fig. 8-16. The signal-to-noise ratio (SNR) should be at or better than -90 dB, and the total harmonic distortion (% THD) should be better than 0.01%. The amplifier can drive either 8- or 4-ohm speaker load impedances. Its design is well suited for domestic hi-fi applications, and since it includes excellent short-circuit and overload protection, it is also well suited for a variety of professional applications, such as small public address systems, musical instrument amplifiers, and commercial sound systems.

Speaker Protection Circuit

I decided to include this project as a final entry in this chapter, because it is a good example of how various circuit building blocks can be put together to create a practical and functional design, and also because it represents the last electronic "block" in an audio chain starting from the preamplifier and ending at the speaker system.

There are several problems associated with high-performance direct-coupled amplifier designs, such as the design illustrated in Fig. 8-11. If one of the output transistors (Q16 or Q17) failed in the amplifier of Fig. 8-11, it is probable that one of the DC power supply rails could be shorted directly to the speaker load (bipolar transistors normally fail by developing a short between the collector and emitter). DC currents are very destructive to typical speaker systems, so it is very possible that a failure in an audio power amplifier could also destroy the speaker system that it is driving. Another problem relating to audio power amplifiers is their power-up *settling time* (often called *turn-on transients*). When operational power is first applied to most higher-power audio amplifiers, they will exhibit a temporary period (possibly as long as 100 milliseconds) of radical output behavior while the various DC quiescent voltages and currents are "settling" (balancing out to typical levels). This settling action usually produces a loud "thump" from the speakers during power-up, which is appropriately referred to as "turn-on thumps."

Figure 8-17 illustrates a very useful, practical, and easy-to-construct circuit that can be implemented into almost any audio system to eliminate the two aforementioned problems associated with direct-coupled audio power amplifiers; (1) it automatically disconnects the speaker system from the audio power amplifier if any significant level of DC current is detected at the amplifier's output; (2) it delays connection of the amplifier to the speaker for about 2 seconds on power-up, thereby eliminating any turn-on

Figure 8-17
Speaker protection
circuit.

thumps; and (3) it provides a visual "status" indication of the amplifier's operation utilizing a typical LED indicator.

Operational power for the Fig. 8-17 protection circuit is normally obtained from the secondary of the power transformer used in the power supply for the audio power amplifier (the current drain of this circuit is very low, so it shouldn't represent any significant load to any large power transformer used to provide high currents to an audio power amplifier). The connection terminal extending from the negative side of C3 should be connected to the *center-tap* of the power transformer—the transformer center-tap will always be the *circuit common* point (see Fig. 8-16). The connection point extending from the anode of D1, labeled "ac in" in Fig. 8-17, is connected to either "hot" side of the transformer secondary (e.g., either 25-volt secondary wire coming out of the 50-volt center-tapped transformer of Fig. 8-16).

Referring to Fig. 8-17, D1 and C3 make up a simple half-wave rectifier circuit, providing a rectified and filtered DC voltage from the AC operational power obtained from the secondary of the power transformer. For example, if you connected this circuit to the 50-volt transformer secondary of Fig. 8-16 (as described previously), the DC voltage at the cathode of D1 should be approximately 35 volts (25 volts × 1.414 = 35.35 volts

DC). ZD1 and R1 form a simple zener voltage regulator circuit. Typically, R1 is a 220-ohm, 1-watt resistor and ZD1 is a 24-volt, 1-watt zener diode.

D3, R2, R3, C6, Q4, Q5, D9, and the relay form a time-delay relay circuit. When AC power is first applied to the AC input (i.e., the anode of D1), D3 is forward-biased only during the positive half-cycles. The positive pulses are applied to the *RC* circuit of R2 and C6. Because of the *RC* time constant, it takes several seconds for C6 to charge to a sufficiently high potential to turn on the Darlington pair (Q4 and Q5) and energize the relay. D9 is used to protect the circuit against inductive kickback spikes when the relay coil is deenergized. R3 is incorporated to "bleed" off C6's charge when the circuit power is turned off.

Transistors O2 and Q3 and their associated circuitry form the familiar astable multivibrator (discussed in previous chapters of this textbook). The values of C4 and C5 (typically 1 µf) are chosen to cause the circuit to oscillate at about 2 hertz.

The protection circuit of Fig. 8-17 is designed to accommodate "two" audio power amplifiers, since most domestic hi-fi applications are in stereo. The outputs of the audio power amplifiers are connected to the "right in" and "left in" connection points, while the right and left speaker systems are connected to the "right out" and "left out" connection points, respectively. If you desire to provide protection to only one speaker system (obviously indicating that you will be using only one audio power amplifier), you can delete F2 and R5, and the output relay need only be a *single-pole, double-throw* (SPDT) type. Fuses F1 and F2 are speaker fuses. Their current ratings will depend on the power capabilities of your audio power amplifier and speaker system.

Since the operational power for this protection circuit is obtained from the power transformer in the audio power amplifier's power supply, operational power to this circuit will be applied simultaneously to applying power to the amplifier(s). Therefore, when the operational power is first turned on, the speakers will not be connected to the power amplifier(s) because the relay will not be energized. The relay will not energize until C6 reaches a potential high enough to turn on Q4 and Q5. This will take several seconds. In the meantime, the audio power amplifier(s) will have had sufficient time to stabilize, thereby eliminating any turn-on thumps from being applied to the speaker systems.

While C6 is charging, before Q4 and Q5 have been turned on, the astable multivibrator is oscillating, causing LED1 to flash at about 2 hertz. This gives a visible indication that the protection circuit is working, and has not yet connected the speakers to the power amplifier(s).

When the time delay has ended and Q4 and Q5 turn on, the collector of Q4 pulls the collector of Q3 low, through D8, stopping the oscillation of the multivibrator, and causing LED1 to light continuously. The relay energizes simultaneously. By staying bright continuously, LED1 now gives a visual indication that the circuit is working, and that the speakers are connected to the power amplifier(s). At this point, the protection circuit of Fig. 8-17 has no further effect within the amplification system unless a DC voltage occurs on one or both of the audio power amplifier outputs.

Under normal conditions, when no DC voltage is present on the output of either of the audio power amplifiers, the amplified audio AC signal from both power amplifiers is applied simultaneously to R4, R5, C1, and C2. Because the time constant of the *RC* circuit is relatively long, C1 and C2 cannot charge to either polarity. In effect, they charge to the average value of the AC waveshape, which is (hopefully) zero. This is the same principle as trying to measure an AC voltage level with your DVM (digital voltmeter) set to measure DC volts—pure AC will provide a zero reading.

If a significant DC voltage (i.e., higher than about 1.2 volts) appears on the output of either power amplifier, C1 and C2 will charge to that voltage regardless of the polarity (note that C1 and C2 are connected with both positive ends tied together, forming a nonpolarized electrolytic capacitor—this principle was discussed when describing the Fig. 8-11 amplifier circuit). This DC voltage will be applied to the bridge rectifier (D4 through D7). Although it might seem strange to apply DC to a bridge rectifier, its effect in this circuit is to convert the applied DC to the correct polarity for forward-biasing Q1 (a technique often called *steering*). When Q1 is forward-biased, it pulls the base of the Darlington pair (Q4 and Q5) low, deenergizing the relay, and disconnecting the speaker systems from the power amplifiers before any damage can result. At the same time, the astable multivibrator is enabled once more, causing LED1 to flash, which gives a visual indication that a malfunction has occurred. The circuit will remain in this condition as long as any significant DC level appears on either of the power amplifier outputs. On removal of the DC, the protection circuit will automatically resume normal operation.

Transistors Q1 through Q5 can be almost any general-purpose NPN types; common 2N3904 transistors work very well. The contact ratings of the relay must be chosen according to the maximum output power of the audio power amplifier(s). The voltage ratings for all of the electrolytic capacitors should be 50 WVDC for most applications. However, if the operational AC power obtained from the power amplifier's power supply transformer is higher than 30 volts AC, you will have to adjust most of the component values in the circuit to accommodate the higher operational voltages. The diode bridge consisting of D4 through D7 can be constructed from almost any type of general-purpose discrete diodes (e.g., 1N4148 or 1N4001 diodes), or you can use a small "modular" diode bridge instead.

CHAPTER 9

Power Control

The term thyristor refers to a broad family of semiconductor devices used primarily for power control. Thyristors are basically fast-acting electronic switches. Common members of the thyristor family include silicon controlled rectifiers, unijunction transistors, triacs, diacs, and (strangely enough) neon tubes.

Silicon controlled Rectifiers

Silicon controlled rectifiers (SCRs), which are probably the most common of all thyristors, are three-lead devices resembling transistors. The three leads are referred to as the *gate, cathode,* and *anode.* An illustration of SCR construction, lead designation, and electrical symbol is given in Fig. 9-1.

An SCR will allow current to flow in only one direction. Like a diode, the cathode must be negative, in relation to the anode, for current flow to occur. However, forward current flow will not begin until a positive potential, relative to the cathode, is applied to the gate. Once current flow begins, the gate has no more control of the SCR action until it drops below its designated holding current.

One method of stopping the forward current flow, in a conducting SCR, is to reverse-bias it (cause the cathode to become positive relative to the anode). Another method is to allow the forward current flow to drop below the SCR's holding current. The *holding current* is a manufacturer's specification defining the minimum current required to hold the SCR in a conductive state.

If the forward current flow drops below the specified holding current, the SCR will drop out of conduction. When this happens (as a result of either a voltage polarity reversal or a loss of minimum holding current), control is again returned to the gate, and the SCR will not conduct (even if forward-biased) until another positive potential (or pulse) is applied to the gate.

Like a transistor, the SCR is considered a *current device* because the gate current causes the SCR to begin to conduct (if forward-based). Also, the forward current flow is the variable that maintains conduction, once the SCR is turned on by the gate current.

Because SCRs can be turned off when they are reverse-biased, they are very commonly used in AC power applications. Because AC power reverses polarity periodically, an SCR used in an AC circuit will auto-

Figure 9-1
SCR construction and electronic symbol.

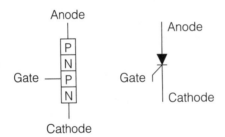

matically be reverse-biased (causing it to turn off) during one-half of each cycle. During the other half of each cycle, it will be forward-biased, but it will not conduct unless a positive gate pulse is applied. By controlling the *coincidental point* at which the gate pulse is applied together with a forward-biasing half-cycle, the SCR can control the amount of given power to a load during the half-cycles it is forward-biased.

Consider the circuit shown in Fig. 9-2. As long as SW1 remains open, R_L will not receive any power from the AC source because the SCR will not conduct during either half-cycle. If SW1 is closed (providing a continuous positive potential to the gate), the load will receive half of the available power from the AC source. In this condition, the SCR acts like a diode, and conducts current only during the half-cycles when it is forward-biased. (Resistor R_G is placed in the gate circuit to keep the gate current from exceeding the specified maximum.)

If it were possible to rapidly turn SW1 on and off, so that the SCR received a gate pulse at the "peak" of each forward-biasing half-cycle, the SCR would only conduct for "half" of the half-cycle. This condition would cause the load to receive one-fourth of the total power available from the AC source. By accurately varying the timing relationship between the gate pulses and the forward-biasing half-cycles, the SCR could be made to supply any percentage of power desired to the load, up to 50%. It cannot supply more than 50% power to the load, because it cannot conduct during the half-cycles when it is reverse-biased.

There are several important points to understand about the operation of this simple circuit. First, once the SCR has been turned on by closing S1, it cannot be turned off again during the remainder of the forward-biasing half-cycle. As long as the SCR is conducting a current flow higher than its minimum holding current, the gate circuit loses all control. Second, before the SCR receives a gate pulse and begins to conduct, there is virtually no power consumption in the circuit; it looks like an open

Figure 9-2
Demonstration of SCR
operating principles.

switch. Once the SCR begins to conduct, virtually all of the power is delivered to the load.

An SCR wastes very little power when controlling power to a load, because it functions in either of two states: ON (looking like a closed switch), or OFF (looking like an open switch). A closed switch might have a high amplitude of current flow through it, but it poses no opposition (resistance). Therefore, the voltage drop across a closed switch is virtually zero. Because power dissipation is equal to current times voltage ($P = IE$), when the voltage is close to zero, so is power dissipation. In contrast, an open switch might drop a high voltage, but does not allow current flow. Again, it becomes irrelevant how high the voltage is, if there is no current flow; power dissipation is still zero.

To understand the importance of efficient power control, consider another method of varying power to a load. Figure 9-3 illustrates a circuit in which a rheostat is used to vary the power delivered to a 50-ohm load (R_L).

The rheostat is adjustable from 0 to 50 ohms. When it is adjusted to 0 ohm, it will appear to be a short and the entire 120-volts AC source will be dropped across R_L. The power dissipated by R_L can be calculated as follows:

$$P = \frac{E^2}{R} = \frac{120^2}{50} = \frac{14400}{50} = 288 \text{ watts}$$

The power dissipated by the rheostat will be

$$P = \frac{E^2}{R} = \frac{(0)^2}{0} = 0$$

If the rheostat is adjusted to present 50 ohms of resistance, the voltage dropped by the rheostat will be equal to the voltage dropped by R_L. Therefore, 60 volts AC will be dropped by both. Because they are both equal in voltage drop and resistance, the power dissipation will also be

Figure 9-3
Rheostat power control circuit.

equal in both. Therefore, under these circumstances, the power dissipation of either is

$$P = \frac{E^2}{R} = \frac{60^2}{50} = \frac{3600}{50} = 72 \text{ watts}$$

Because the purpose of the circuit shown in Fig. 9-3 is to control the power delivered to the load (R_L), all of the power dissipated by the rheostat is wasted. In the previous example, the efficiency of the power control is 50%. At different settings of the rheostat, different efficiency levels occur; but it is obvious that this level of waste is unacceptable in high-power electrical applications.

A disadvantage of using a single SCR for power control, as illustrated in Fig. 9-2, is that it is not possible to obtain full 360-degree control of an AC waveform (only 180 degrees, that is, the forward-biasing half-cycle, can be controlled). To overcome this problem, two SCRs might be incorporated into a circuit for full-wave power control.

The Triac

Another member of the thyristor family, the triac can be used for full-wave power control. A *triac* has three leads designated as the *gate, M1,* and *M2*. Triacs are triggered on by either a positive or negative pulse to the gate lead, in reference to the M1 terminal. Triacs can also conduct current in either direction between the M1 and M2 terminals. Like an SCR, once a triac has been triggered, the gate loses all control until the current flow through the M1 and M2 terminals drops below the manufacturer's specified holding current. Triacs are considered current devices. The electrical symbols used for triacs are illustrated in Fig. 9-4.

The principle of efficient power control is essentially the same for the triac as it is for the SCR. Because a triac operates in only two modes (ON

Figure 9-4
Additional thyristor symbols.

or OFF), full-wave power control can be obtained without appreciable power losses in the triac itself. The primary advantage of a triac is its capability of being triggered in either polarity, and controlling power throughout the entire AC cycle.

Triacs are commonly used for smaller power control applications (light dimmers, small DC motors and power supplies). Unfortunately, triacs have the disadvantage of being somewhat difficult to turn off (especially when used to control inductive loads). Because of this problem, SCRs (rather than triacs) are used almost exclusively in high-power applications.

Unijunction Transistors, Diacs, and Neon Tubes

So far in this chapter, you have examined the theoretical possibility of controlling power with SCRs or triacs. If you could vary the gate *trigger pulse timing* relative to the AC cycle (this is called varying the *phase angle* of the trigger pulse), you would have an efficient electronic power control tool. Obviously, it would be impossible for a human to turn a switch on and off at a 60-hertz rate to provide trigger pulses for power control. Unijunction transistors, diacs, and neon tubes are commonly used to accomplish this function. (The neon tube is not actually a member of the thyristor family, but its function is identical to that of the diac. Some equipment still uses neon tubes for triggering, because they serve a dual purpose as power-on indicators.)

Like the transistor, the *unijunction transistor* (UJT) is a three-lead device. The three leads are referred to as the *emitter, B1,* and *B2.* The schematic symbol for UJTs is given in Fig. 9-4. Unlike the previously discussed SCR and triac, the UJT is a voltage device. When the voltage between the emitter and B1 leads reaches a certain value (a ratio of the applied voltage between the B1 and B2 leads, and the manufactured characteristics of the UJT), the resistance between the emitter and B1 decreases to a very low value.

Contrastingly, if the voltage between the emitter and B1 decreases to a value below the established ratio, the resistance between the emitter and B1 increases to a high value. In other words, a UJT can be thought of as a voltage breakdown device. It will avalanche into a highly conductive state (between the emitter and B1 leads) when a peak voltage level (expressed as

V_p) is reached. It will continue to remain highly conductive until the voltage is reduced to a much lower level called the "valley voltage" (V_v).

Firing circuits are electronic circuits that vary the amplitude and phase of an AC trigger voltage applied to the gate lead of an SCR (or other thyristor). Using a combination of amplitude adjustment and phase-shifting techniques, the V_p level (resulting in UJT voltage breakdown) can be made to occur precisely and repeatedly at any "phase angle relative to an AC power waveform applied to a triac or SCR." The conductive breakdown characteristics of the UJT are used as a means of providing the gate trigger pulses to "fire" the SCRs or triacs. Therefore, the SCRs or triacs can be repeatedly fired at any point on the AC waveform, resulting in full-wave (0 to 100% duty cycle) power control.

Another voltage breakdown device is the diac. Because diacs are special-purpose diodes, their operational description and symbols have already been discussed in Chapter 7. *Diacs* are simply the solid-state replacement for neon tubes. The schematic symbol for neon tubes is given in Fig. 9-4. Neon tubes and diacs are devices which will remain in a nonconductive state until the voltage across them exceeds a breakover voltage (or ionization voltage in the case of neon tubes). At breakover, they will become conductive and remain so until the voltage across them drops below a holding voltage (a much lower voltage than the breakover voltage), at which time they will become nonconductive again.

Building a Soldering Iron Controller

Soldering irons are available in many power ranges. The smallest sizes, around 15 watts, are recommended for very small and precise work, such as surface-mount-component (SMC) work. Medium sizes, about 30 watts, are recommended for most general electronic work, including PC board soldering. Larger sizes, from 60 watts and up, are for large soldering jobs, such as making solder connections to large bus bars or stud-mount diodes.

If you run into situations where you need to do a variety of different types of soldering, there are several solutions. The obvious solution is to buy several different types of soldering irons. Another solution is to buy a soldering station, with an automatic tip temperature regulator (starting at about $100.00). A good middle-of-the-road solution is to buy a 60-watt soldering iron, and use it with the circuit illustrated in Fig. 9-5. In addi-

Figure 9-5
A soldering-iron controller.

tion to being a useful and convenient tool, this circuit will help illustrate most of what has just been discussed concerning thyristors and power control.

Referring to Fig. 9-5, the incoming 120-volt AC power is applied across the triac through the load (lightbulb and soldering iron, in parallel). Assuming SW1 is turned on at the instantaneous point in time that the AC voltage is at zero, the triac is off (nonconductive). As the AC voltage begins to increase through a half-cycle (the polarity is irrelevant because the diac and triac are both bilateral in operation), all of the AC voltage is dropped across the triac because it looks like an open switch.

Similarly, the same voltage is dropped across the firing circuit, or trigger circuit (P1, the diac, and C1), because it is in parallel with the triac. C1 will begin to charge at a rate relative to the setting of P1. As the AC half-cycle continues, C1 will eventually charge to the specified breakover voltage of the diac, causing the diac to avalanche, and a current pulse (trigger) to flow through the gate and M1 terminals of the triac.

This trigger pulse causes the triac to turn on (much like a closed switch), resulting in the remainder of the AC half-cycle being applied to the load (lamp and soldering iron). When the AC power has completed the half-cycle and approaches zero voltage (prior to changing polarity), the current flow through the triac drops below the holding current and the triac returns to a nonconductive state. This entire process continues to repeat with each half-cycle of the incoming AC power.

There are several important points to understand about the operation of this circuit. The diac will reach its breakover voltage and trigger the triac at the same relative point during each half-cycle of the AC waveform. This relative point will depend on the charge rate of C1, which is controlled by the setting of P1. In effect, the setting of P1 controls the average power delivered to the load. P1 can control the majority of the AC half-cycle because C1 also introduces a *voltage-lagging* phase shift. Without the phase shift, control would be lost after the peak of the AC power cycle was reached. Throughout the entire power control range of this circuit, the power wasted by the triac is negligible, compared to the power delivered to the load.

PL1 is a standard 120-volt AC three-prong plug. If you build this project in an aluminum project box, the ground prong (round prong) should be connected to the aluminum box (the chassis in this case). For safety reasons, the 120-volt AC hot lead should also be fuse-protected. P1 is mounted to the front panel of the project box for easy access.

I used a flat, rectangular aluminum project box large enough to set the soldering iron holder on. I also connected the soldering iron internally to a phenolic solder strip with a strain relief to protect the cord. This, of course, is a matter of opinion. You might want to wire the circuit to a standard 120-volt AC socket for use with a variety of soldering irons.

The triac, diac, and C1 can be assembled on a small universal perfboard or wired to a phenolic solder strip.

The 15- to 25-watt lamp is a standard 120-volt AC incandescent lightbulb of any style or design you like (it might also be any wattage you desire, up to 60 watts). It is mounted on the outside front panel of the project box and serves several useful indicator functions. First, it indicates that the power is on and that the circuit is functioning. Second, with a little practice, the brightness of the bulb is a good indicator of about how much power you are applying to the soldering iron. For example, if you're using this circuit with a 60-watt soldering iron, and you adjust P1 until the bulb is about half as bright as normal, you're supplying about 30 watts of power to the tip. Third, the lightbulb makes a good reminder to turn off the soldering iron when you're finished working. (I can't count how many times I have come into my shop and found the soldering iron still turned on from the previous day.)

The best value for C1 will probably be about 0.1 μF. After building the circuit so that it can be tested using the lightbulb as the load, try a few different values for C1, and choose the one giving the smoothest operation throughout the entire power range. C1 must be a nonpolarized capacitor rated for at least 200 volts.

In addition to controlling the power delivered to a soldering iron, this circuit is a basic light-dimmer circuit. You might use it to control the power delivered to any "resistive" load up to about 150 watts. For controlling larger loads, you will need to use a larger triac and, depending on the triac, you might need to use different values for P1 and C1. For controlling large loads, it's also a good idea to place a varistor (MOV) across the incoming 120-volt AC line (such as an NTE2V115).

Circuit Potpourri

The theories and postulations are done. Now's the time for applications and fun!

Watts an Easier Way?

They don't get any simpler than this: a very simple soldering iron power control as shown in Fig. 9-6. This circuit inserts a diode (with SW1 open) in series with the soldering-iron heating element. The diode will block one-half of the incoming AC voltage to the heating element, resulting in a power decrease.

Although it would seem logical that the soldering iron would operate at half-power with the diode in the power circuit (that is, a 60-watt iron would become a 30-watt iron), it is not quite that simple. The type of wire universally used in resistive heating elements is called *nichrome*. Nichrome, like most resistive substances, has a positive temperature coefficient; as it becomes hotter, its resistance value goes up. When a resistive heating element designed for a 120-volt AC application (such as in a soldering iron) experiences a decrease in applied power, its temperature goes down, resulting in a proportionate decrease in resistance. This

Figure 9-6
Simple soldering iron power control.

decrease in resistance has the effect of causing the wire to dissipate more "power per volt" than it did at the rated voltage.

The circuit illustrated in Fig. 9-6, if used in conjunction with a 60-watt soldering iron, would decrease its operational power down to about 40 watts. In reality, this would handle the vast majority of soldering jobs you will encounter. If you used this circuit with another soldering iron rated at 30 watts, it would decrease it down to about 18 watts (just about right for SMC work). In other words, two soldering irons (a 60-watt and a 30-watt), two general-purpose diodes, and two switches will provide you with the full gamut of soldering iron needs. One additional benefit—using a 50- or 60-watt soldering iron with the diode in the power circuit (and placing the wattage at about the optimum point for general-purpose electronic work) will cause the heating element to last about 40 times longer, and will thus increase tip life.

By the way, placing a general-purpose diode in series with incandescent (but *not* fluorescent!) lightbulbs will cause the lightbulbs to last about 40 times longer. The disadvantage resulting from this is the lower quantity and efficiency of the light produced.

Isolation Might Be Good for the Soul

Figure 9-7 shows how to construct a simple isolation transformer for your electronics bench. Typically, isolation transformers are needed when testing line-powered equipment with line-powered *test* equipment.

For example, referring back to Fig. 9-5, suppose you wanted to observe the waveshape across the parallel circuit of the soldering iron and lamp with a line-powered oscilloscope. You connect the ground and scope probe across the lamp, turn on SW1, and "boom"—the fuse blows and the triac might be destroyed. What happened?

For safety reasons, the ground (common) lead connections on most line-powered test equipment are connected to earth ground. Looking at

Figure 9-7

General-purpose isolation circuit.

Fig. 9-5, imagine the results of connecting earth ground to either side of the lamp/soldering iron connections. From a problematic viewpoint, there are several possibilities: two possibilities of how PL1 is wired (correct and incorrect), and many possibilities of how the scope can be connected. Out of these possibilities, three would result in improper operation, or destroyed components.

For example, if the iron controller's power cable were connected with the black wire going to the fuse (which is standard, safe, and correct), and the scope "common" probe was inadvertently touched to that part of the circuit (fuse, SW1, or that side of the lamp), then a direct short to ground would exist! If the probe connection was made to the cable side of the fuse, a catastrophic short would occur, hopefully blowing the fuse or breaker at the main box. If the connection were made beyond the fuse, this short could blow either the main fuse, the controller fuse, or both.

If the controller were isolated from the main power line, by using the isolation transformer, neither of the lines (black or white) going to the controller have any continuity to ground. Remember, in household wiring, the white wire is called the *neutral*, and it is kept at "ground" potential. The black and red wires are hot. Without the isolation provided by the transformer, the common scope probe, which is connected to earth ground and neutral (through the house wiring) and attached to the hot side of the controller, would present a dead short. For this reason, among others, isolation transformers are handy to have around.

In Fig. 9-7, T1 and T2 are any two "identical" power transformers, with appropriate ratings for the desired application. For example, assume that T1 has a secondary rating of 25 volts at 2 amps. It is, therefore, a 50-VA (volt-amp) transformer (25 volts × 2 amps = 50 VA). To calculate the maximum current rating of the primary, simply divide the volt-amp rating by 120 volts (intended primary voltage). This comes out to a little over 410 milliamps. The primary and secondary of any individual power transformer will always have the same volt-amp rating.

If you decided to use two such transformers for the isolation circuit, the primary of T1 would be connected to the line (120 volts AC). The secondary of T1 would be connected to the secondary of T2, whose primary would be the output. In other words, the function of the first transformer is to convert 120 volts AC down to 25 volts AC (step-down application), and the second transformer converts 25 volts AC back up to 120 volts AC (step-up operation). This "conversion from the original" and "conversion back to the original" process is the reason why both transformers must be identical.

As explained earlier, 410 milliamps is the highest current load that can be drawn from T2's "intended primary" under ideal conditions. However, there are losses in power transformers, and, in this case, the losses are doubled. Therefore, it is necessary to derate T2's primary maximum current value by 10%. Consequently, the maximum current output from this hypothetical isolation circuit will be only about 370 milliamps—a little too small for most general-purpose isolation requirements.

If you decide to build this type of isolation circuit for your bench, you'll probably want to use transformers with significantly higher volt-amp ratings. Unfortunately, it soon becomes apparent that power transformers in the 400- to 500-VA range are expensive, but so are isolation transformers. However, some "odd" value industrial transformers with high VA ratings can be purchased very inexpensively from many surplus electronic dealers. Just make sure that they are not ferroresonant transformers, or power transformers intended for use with 400-hertz AC power.

Curiosity Catcher

Figure 9-8 is an SCR latch circuit. When assembled as illustrated, depressing the "normally open" momentary switch (SW1) provides a positive gate pulse (relative to the cathode) and fires the SCR. Once conducting, the gate has no more control over the SCR, and it continues to conduct, powering the piezo buzzer, until the cathode/anode current flow is interrupted by depressing the "normally closed" momentary switch (SW2). Because the cathode/anode current flow drops below the holding current of the SCR (it actually drops to zero), control is returned to the gate, and the SCR will not conduct again until SW1 is depressed.

Figure 9-8
SCR latch circuit.

A circuit of this type is called a *latch circuit*. (The electromechanical counterpart of this circuit is a latching relay.) Just about any commonly available SCR can be used. The piezo buzzer can be substituted for almost any type of load that you need to be latched. There are many, many applications for such a circuit, but I would like to describe one that I had a lot of fun with.

I mounted this circuit in a small black box, using a 9-volt transistor battery as the power source. The piezo buzzer is commonly available in any electronics parts store. The buzzer was mounted close to the front of the box, where *speaker holes* were drilled. SW2 was located *inside* the project box. I mounted it in a lower back corner, so that the only way it could be depressed was to use a straightened paper clip through an almost invisibly small hole drilled through the back of the box. SW1 was mounted to the front plate with a large, red, plastic knob. Underneath the SW1 push button, I arranged stick-on letters to read "Do not push." You can probably guess the rest.

If the box is left in an obvious place, it will drive many people crazy, until they see what happens when the button is pressed. After their curiosity gets the better of them and they press the button, the piezo buzzer sounds off, and there isn't any apparent way to stop it.

After building this box, user discretion is advised! After playing this joke on a person lacking a really good sense of humor (while the buzzer was still sounding), I made the mistake of saying he could never figure out how to stop it. He promptly dropped it on the floor, and stomped on it with a large, silver-tip boot. I stood corrected.

See Ya Later, Oscillator

The circuit illustrated in Fig. 9-9 is a *UJT oscillator.* UJTs make good oscillators for audio projects, and they have some distinct advantages over astable multivibrators. They are easier to construct, and require the use of only one UJT. Also, the output is in the form of a "sawtooth" wave, which provides more pleasing audio harmonics than square waves. (This circuit will be used as a foundation for later projects in this book.)

UJT oscillators do have a drawback. Their usable audio output, which is taken across the capacitor, has a very high impedance. Therefore, it is best to use an impedance-matching circuit, such as the Darlington pair common-collector amplifier shown in Fig. 9-9. (A JFET works very well for this application also. JFETs will be discussed in the next chapter.) Impedance-matching circuits used for this type of application are often called *buffers.*

Figure 9-9
Buffered UJT audio oscillator.

Using the values given in the illustration, a fairly wide range of the audio frequency spectrum can be produced by rotating P1 (frequency adjustment). The value of C1 can be increased for extremely low oscillations, and vice versa. The output of the Darlington pair can drive a small speaker directly, but the sound quality will be greatly improved by coupling the output with an audio transformer or, better yet, using a separate audio amplifier and speaker.

If you would like to experiment with some really unique sounds, build two of these UJT oscillator circuits and connect their outputs into the modulator circuit, also illustrated in Fig. 9-9. One UJT oscillator output will connect to point *A* and the other, to point *B*. The output of the modulator circuit must be connected to the input of a separate amplifier and speaker. The diode can be any general-purpose rectifier diode.

An understanding of this modulator circuit requires an explanation of several new principles; the first involves diodes. When diodes are forward-biased with very low voltages and currents, they react in a highly nonlinear fashion. This area of diode conduction is called the *forward-biased knee* of the diode response.

A second new principle involves the process of modulation. *Modulation* occurs when two different frequencies are mixed in a nonlinear circuit. The effect of modulation causes some additional frequencies to be created. For example, if a 1-kHz frequency and a 10-kHz frequency are applied to the inputs of a linear mixer, the output will be the original frequencies, only mixed together. If the same two frequencies are applied to the input of a nonlinear mixer circuit, the original frequencies will still appear at the output, but two additional frequencies, called *beat fre-*

quencies, will also occur. The beat frequencies will be the sum and difference of the original frequencies. Using 1 kHz and 10 kHz as the originals, the beat frequencies will be 11 kHz (sum) and 9 kHz (difference).

When the two UJT oscillators are connected into the modulator circuit (with its output applied to the input of a power amplifier and speaker), P2 and P3 are adjusted to cause the outputs of the oscillators to fall into the "knee" area of D1's forward conduction response. You might have to add some additional series resistance between each oscillator and each modulator input. If this is the case, try increasing the resistance in 100-Kohm increments. When the two frequencies modulate, the sound produced will be instantly recognizable. This type of sound, or "tonality," has been in the background of most science fiction movies since the 1950s.

Forbidden Planet Revisited

In 1956, MGM released a classic science fiction movie entitled *Forbidden Planet,* starring Walter Pidgeon, Anne Francis, and Leslie Nielsen. This movie was unique in that it contained no conventional music. Rather, the soundtrack was described as "tonalities," which were the original creation of Louis and Bebe Barron. I estimate that about 80% of these tonalities were generated with the fascinating circuit, called a *ring modulator,* illustrated in Fig. 9-10.

The simple modulator circuit shown in Fig. 9-9 can produce interesting modulation effects, but the original frequencies appear at the output along with the beat frequencies. The effect is somewhat muddy because of the various blends and amplitudes of the four frequencies (i.e., two beat frequencies plus two original frequencies). In contrast, the ring modulator circuit of Fig. 9-10 cancels out the two original frequencies, outputting only the beat frequencies. The effect is fascinating and quite impressive. The Fig. 9-10 circuit will function well with two audio signal sources of any type (i.e., two oscillator circuits of Fig. 9-9, two signal generators, two function generators, or any combination thereof). The modulator inputs must be several volts in amplitude for a suitable output.

Almost any two audio transformers will function well for the Fig. 9-10 ring modulator, but they must be matched (i.e., they must be the same model and type). Diodes D1 through D4 should be germanium types, although silicon diodes will work to some degree.

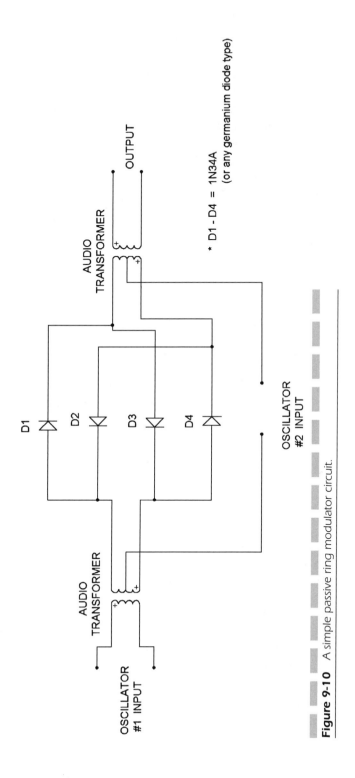

Figure 9-10 A simple passive ring modulator circuit.

The theory of operation of the Fig. 9-10 circuit is fairly simple. Modulation occurs in the nonlinear diode circuitry, but the diodes are arranged so that the rectification process cancels out the original frequencies. Consequently, the only frequencies remaining at the output are the upper and lower modulation sidebands.

Ring modulators are used in many types of electronic music circuits, as well as sound-effect circuits, producing such effects as bells, clangs, and metallic sounds. They also produce a "harmonizing" effect if one of the oscillator inputs is held stable while a musical instrument input is applied to the other input.

Figure 9-10 is personally endorsed by Robby the Robot.

10

Field-Effect Transistors

The field-effect transistor (FET) is an active "voltage" device. Unlike bipolar transistors, FETs are not current amplifiers. Rather, they act much like vacuum tubes in basic operation. FETs are three-lead devices similar in appearance to bipolar transistors. The three leads are referred to as the *gate, source,* and *drain.* These three leads are somewhat analogous to the bipolar transistor's base, emitter, and collector leads, respectively. There are two general types of FETs: junction field-effect transistors (JFETs) and insulated-gate metal oxide semiconductor field-effect transistors (MOSFET or IGFET).

FETs are manufactured as either N-channel or P-channel devices. *N-channel FETs* are used in applications requiring the drain to be positive relative to the source; the opposite is true of *P-channel FETs*. The schematic symbols for N-channel and P-channel JFETs and MOSFETs are shown in Fig. 10-1. Note that the arrow always points toward the channel (the interconnection between the source and drain) when symbolizing N-channel FETs, and away from it in P-channel symbologies.

All types of FETs have very high input impedances (1 Mohm to over 1,000,000 Mohm). This is the primary advantage to using FETs in the majority of applications. The complete independence of FET operation from its input current is the reason for their classification as voltage devices. Because FETs do not need gate current to function, they do not have an appreciable loading effect on preceding stages or transducers. Also, because their operation does not depend on "junction recombination" of majority carriers (as do bipolar transistors), they are inherently low-noise devices.

FET Operational Principles

The basic operational principles of FETs are actually much simpler than those of bipolar transistors. FETs control current flow through a semiconductor "channel" by means of an electrostatic field.

Figure 10-1
JFET and MOSFET symbols.

N-channel JFET

P-channel JFET

N-channel MOSFET

P-channel MOSFET

Referring to Fig. 10-2, examine the construction of a JFET. Notice that there are two junctions, with the P material connected to the gate, and the N material making up the channel. Assume that the source lead is connected to a circuit common, and a positive potential is applied to the drain lead. Current will begin to flow from source to drain with little restriction. The N-channel semiconductor material, although not a good conductor, will conduct a substantial current.

Under these conditions, if a *negative* voltage is applied to the gate, the PN junctions between the gate and channel material will be *reverse-biased* (negative on the P material). The reverse-biased condition will cause a depletion region extending outward from the gate/channel junctions. As you might recall, a depletion region becomes an insulator because of the lack of majority charge carriers. As the depletion region spreads out from the gate/channel junction deeper into the channel region, it begins to restrict some of the current flow between source and drain. In effect, it reduces the conductive area of the channel, acting like a water valve closing on a stream of flowing water. This depletion region will increase outward in proportion to the increase in amplitude of the negative voltage applied to the gate.

If the negative gate voltage is increased to a high enough potential, a point will be reached when the depletion region entirely pinches off the current flow through the channel. At this point, the FET is said to be "pinched off" (this pinch-off region is analogous to cutoff in bipolar transistors), and all current flow through the channel stops. The basic principle involved is controlling the channel current with an electrostatic field. This field effect is the reason for the name field-effect transistor.

Figure 10-2
FET construction.

JFET MOSFET Enhancement mode MOSFET

Continuing to refer to Fig. 10-2, notice the difference in construction between a MOSFET and JFET. Although a JFET's input impedance is very high (because of the reverse-biased gate junction), there can still be a small gate current (because of leakage current through the junction), which translates to a reduced input impedance. However, gate current through a MOSFET is totally restricted by an insulating layer between the gate and channel.

A MOSFET functions in the same basic way as a JFET. If a negative voltage is applied to the gate of an N-channel MOSFET, the negative electrostatic charge round the gate area repels the negative-charge carriers in the N-channel material, forming a resultant depletion region. As the negative gate voltage varies, the depletion region varies proportionally. The variance in this depletion region controls the current flow through the channel and, once again, current flow is controlled by an electrostatic field.

A third type of FET, called an *enhancement-mode MOSFET*, utilizes an electrostatic field to "create a channel," rather than deplete a channel. Referring again to Fig. 10-2, notice the construction of an enhancement-mode MOSFET. The normal N channel is separated by a section of a P-material block, called the *substrate*. N-channel enhancement-mode MOSFETs, such as the one illustrated, require a positive voltage applied to the gate. The positive potential at the gate attracts "minority" carriers out of the P-material substrate, forming a layer of "N material" around the gate area.

This has the effect of connecting the two sections of N material (attached to the source and drain) together to form a continuous channel, and thus allows current to flow. As the positive gate potential increases, the size of the channel increases proportionally, which results in a proportional increase in conductivity. Once more, current flow is controlled by an electrostatic field.

All of the operating principles discussed in this section have been applied to N-channel FETs. P-channel FET devices will operate identically; the only difference is in the reversal of voltage polarities.

FET Parameters

As discussed previously, the primary gain parameter of a standard bipolar transistor is beta. Beta defines the ratio of the current flow through the base relative to the current flow through the collector. In reference to FETs, the primary gain parameter is called transconductance (G_m). The *transconductance* is a ratio defining the effect that a gate-to-source voltage (V_{GS}) change will have on the drain current (I_D). Transconductance is typically defined in terms of *micromhos* (the *mho* is the basic unit for expressing conductance). Typical transconductance values for common FETs range from 2000 to 15,000 micromhos. The equation for calculating transconductance is

$$G_m = \frac{\text{change in drain current}}{\text{change in gate-to-source voltage}}$$

For example, assume that you were testing an unknown FET. A 1-volt change in the gate-to-source voltage caused a 10-milliamp change in the drain current. The calculation for its transconductance value would be

$$G_m = \frac{I_D}{V_{GS}} = \frac{10 \text{ milliamps}}{1 \text{ volt}} = 0.01 \text{ mho} = 10,000 \text{ micromhos}$$

Referring to Fig. 10-3, assume that this illustration has the same transconductance value as calculated in the previous example. A 1-volt change in the gate-to-source voltage (input) will cause a 10-milliamp change in the drain current. According to Ohm's law, a 10-milliamp current change through the 1-Kohm drain resistor (R_D) will cause a 10-volt change across the drain resistor (10 milliamps × 1000 ohms = 10 volts). This 10-volt change will appear at the output. Therefore, because a 1-volt change at the gate resulted in a 10-volt change at the output, this circuit has a voltage gain (A_e) of 10.

In numerous ways, FET circuits can be compared with standard bipolar transistor circuits. The circuit shown in Fig. 10-3 is analogous to the common-emitter configuration, and it is appropriately called a

Figure 10-3
JFET common-source amplifier.

common-source configuration. The output is inverted from the input, and it is capable of voltage gain. If the output were taken from the source, instead of the drain, it would then be a common-drain configuration. The output would not be inverted, and the voltage gain would be approximately 1. Of course, the common-drain FET amplifier is analogous to the common-collector amplifier in bipolar design.

FET Biasing Considerations

Referring again to Fig. 10-3, note that the gate is effectively placed near the same potential as circuit common through resistor R_G. With no input applied, the gate voltage (relative to circuit common) is zero. However, this does not mean the gate-to-source voltage is zero. Assume the source resistor (R_S) is 100 ohms, and that the drain current, which is the same as the source current, is 15 milliamps. This 15-milliamp current flow through RS would cause it to drop 1.5 volts, placing the source lead of the FET at a positive 1.5-volt potential "above circuit common." If the source is 1.5 volts more positive than the gate, it could also be said that the gate is 1.5 volts more negative than the source. (Is an 8-ounce glass, with 4 ounces of water in it, half-full or half-empty?) Therefore, the gate-to-source voltage in this case is *negative* 1.5 volts. This also means that the gate has a −1.5-volt negative bias. If a signal voltage is applied to the input, causing the gate to become more negative, the FET will become less conductive (more resistive), and vice versa. A JFET exhibits maximum conductivity (minimum resistance), from the source to the drain, with no bias voltage applied to the gate.

MOSFETs are biased in similar ways to JFETs, except in the case of enhancement-mode MOSFETs. As explained previously, enhancement-mode MOSFETs are biased with a gate voltage of the opposite polarity to their other FET counterparts. Some enhancement-mode MOSFETs are designed to operate in either mode of operation.

In general, FETs provide a circuit designer with a higher degree of simplicity and flexibility, because of their lack of interstage loading considerations (a transistor stage with a high input impedance will not load down the output of a previous stage). This can also result in the need for fewer stages, and less complexity in many circuit designs.

Static Electricity: An Unseen Danger

The introduction of MOS (metal oxide semiconductor) devices brought on a whole new era in the electronic world. Today, MOS technology has been incorporated into discrete and integrated components, allowing lower power consumption, improved circuit design and operation, higher component densities, and more sophisticated operation. Unfortunately, a major problem exists with all MOS devices. They are very susceptible to destruction by static electricity.

Inadvertent static electricity is usually caused by friction. Under the proper conditions, friction can force electrons to build up on nonconductive surfaces, creating a charge. When a charged substance is brought in contact with a conductive substance of lesser charge, the charged substance will discharge to the other conductor until the potentials are equal.

Everyone is "jolted" by static electricity from time to time. Static electrical charges can be built up on the human body by changing clothes, walking over certain types of carpeting, sliding across a car seat, or even friction from moving air. The actual potential of typical static charges is surprising. A static charge of sufficient potential to provide a small "zap" on touching a conductive object is probably in the range of 2000 to 4000 volts!

Most MOS devices can be destroyed by static discharges as low as 50 volts. The static discharge punctures the oxide insulator (which is almost indescribably thin) and forms a tiny carbon arc path through it. This renders the MOS device useless.

The point is that whenever you work with any type of MOS device, *your body and tools must be free of static charges.* There are many good methods available to do this. The most common is a "grounding strap," made from conductive plastic, that might be worn around the wrist or ankle and attached to a grounded object. Soldering irons should have a grounded tip, and special "antistatic" desoldering tools are available. Conductive work mats are also advisable. MOS devices must be stored in specially manufactured small parts cabinets, antistatic bags, and conductive foam.

NOTE *Do not try to make your own grounding straps out of common wire or conductive cable of any type!*

This is very dangerous. It is like working on electrical equipment while standing in water. Specially designed grounding straps, for the removal of static charges, are made from conductive plastic exhibiting very high resistance. Consequently, static charges can be drained safely, without increasing an electrocution risk in the event of an accident.

The susceptibility to static charges has led many people to believe that MOS devices are somehow "fragile." There is some evidence to support this notion; but in actuality, the problem is usually the result of an inexperienced design engineer incorporating a MOS device into an application where it doesn't belong. If properly implemented, MOS devices are as reliable as any other type of semiconductor device. However, care should be exercised in handling PC boards containing MOS devices, because some designers might extend an open, unprotected MOS device lead to an edge connector where it is susceptible to static voltages once unplugged.

Building a High-Quality MOSFET Audio Amplifier

Many audiophiles today are adamant supporters of the virtues of MOSFETs used as output drivers in audio amplifiers. They claim that MOSFETs provide a softer, richer sound—one more reminiscent of vacuum-tube amplifiers. Although I won't get involved in that dispute, I will say that MOSFETs are more rugged than bipolar transistors and consequently provide a higher degree of reliability.

There are good economic and functional reasons to use MOSFETs as output drivers, however. At lower power levels, power MOSFETs display the same negative temperature coefficient as bipolar transistors. But at higher power levels, they begin to take on the characteristic of devices with a positive temperature coefficient. Because of this highly desirable attribute, temperature compensation circuits are not required and there is no danger of thermal runaway. Also, power MOSFETs, being voltage devices, do not require the high current drive that must be provided for their bipolar counterparts. The result is a simpler, more temperature-stable amplifier circuit. The only disadvantage in using power MOSFETs (that I have discovered), is the lack of availability of high-power, complementary pairs.

Figure 10-4 is a schematic diagram of a professional-quality 120-watt rms MOSFET audio power amplifier. If you compare this schematic with the Fig. 8-11 amplifier design, you will discover that they are very similar. Most of the same operational physics and principles apply to both designs. There are a few differences, however, which will be detailed in this section.

Figure 10-4 Schematic diagram of a professional-quality 120-watt rms MOSFET power amplifier.

To begin, the Fig. 10-4 design incorporates four power MOSFETs in the output stage: two pairs of complementary MOSFETs that are connected in parallel with each other. This was done to increase the power output capability of the amplifier. If only about 50 or 60 watts of output power capability were desired, only a single pair of MOSFETs would have been needed. The conventional method of increasing the output power capability of any audio power amplifier, bipolar or MOSFET, is to add additional output devices in parallel.

A few of the changes incorporated into the Fig. 10-4 amplifier have nothing to do with the use of MOSFET outputs—they are due to the higher rail voltages and subsequent higher output power capability. F1 and F2 have been increased to 5-amp fuses to accommodate the higher rail currents, and Q8 and Q11 have been replaced with higher dissipation transistor devices because of the higher rail voltages. Also, although not shown in the schematics, the rail decoupling capacitors (i.e., C4, C5) must have their voltage ratings increased up to at least 63 WVDC (preferably 100 WVDC). C10, C11, and C12 were already specified at 100 WVDC for the Fig. 8-11 project, so they are suitable for this design also. The voltage gain of this design also had to be increased in order to keep the same approximate input sensitivity. Remember, the peak-to-peak output voltage of this amplifier must be significantly greater than the Fig. 8-11 design in order to deliver about double the output power to the speaker system. Therefore, if the input signal is maintained at about 0.9 volt rms, the voltage gain must be increased. R8 was lowered to 270 ohms, which sets the gain at about 38 [(R10 + R8) divided by R8 = 38.03].

Now, getting into the modifications involving the incorporation of MOSFETs, note that transistor Q9 in the Fig. 8-11 diagram has been removed and replaced with a 500-ohm potentiometer (connected as a rheostat) in the Fig. 10-4 diagram. Since the source-to-drain impedance values of MOSFETs increase with rising temperature, temperature compensation circuitry and dangers of thermal runaway are nonexistent. Therefore, the V_{bias} temperature-tracking transistor of Fig. 8-11 (Q9) can be replaced with a simple potentiometer. Typically, P1 will be adjusted to drop about 0.8 volt to provide the small forward bias on the output devices for the minimization of crossover distortion.

Since MOSFETs are voltage devices, drawing insignificant gate currents for operational purposes, the predriver transistors (Q14 and Q15 in Fig. 8-11) have been removed. Unfortunately, MOSFETs do suffer the disadvantage of fairly high gate capacitance (up to about 1000 pF in many devices), which can create parasitic RF oscillations in the gate circuits (i.e., destructive high-frequency oscillations localized in the gate circuitry and not a function of the overall amplifier stability characteristics). This is especially problematic if you are using paralleled output pairs, such as in the Fig. 10-4 design. The cure for this idiosyncrasy is to install gate resistors, commonly called *gate-stopper resistors,* to provide resistive isolation from one gate to another. This is the function of resistors R19, R20, R25, and R26.

Note that C9 in Fig. 8-11 has been deleted in the Fig. 10-4 design. As you may recall, C9 was implemented to improve the turn-off speed of the bipolar output transistors (eliminating the possibility of switching distortion). Unlike bipolar transistors, MOSFETs do not have a junction capacitance mechanism that can store charge carriers and inhibit their turn-off speed. This is the reason why power MOSFETs are superior to power bipolar transistors for high-frequency applications. Therefore, C9 is not needed.

The MOSFET devices specified for the Fig. 10-4 amplifier (i.e., 2SK1058/2SJ162 pairs) are a special type of MOSFETs commonly called *lateral MOSFETs.* These devices are specifically designed for applications in audio power amplifiers, and they will provide better performance with greater reliability than the more common *HEXFET* or *D-MOSFET* families. However, these other device types are incorporated into quite a few commercial MOSFET amplifiers because of their lower cost—lateral MOSFETs are comparatively expensive.

The performance of the Fig. 10-4 audio power amplifier is quite impressive. Most of the performance specifications are virtually identical to those of the Fig. 8-11 amplifier, but the percent THD (total harmonic distortion) is a little higher, measuring out to about 0.02% at 120 watts rms. One reason for this slightly higher distortion figure is the inherent lower transconductance of MOSFETs in comparison to bipolar transistors. (Bipolar transistors can be evaluated on the basis of transconductance just like MOSFETs, but their gain factor is usually looked at from the perspective of current gain rather than transconductance.) The point is that bipolar transistors have a gain capability much higher than do MOSFETs, so their higher gain can be converted to higher levels of "linearizing" negative feedback, which results in a little better distortion performance.

I included the amplifier design of Fig. 10-4 in this chapter primarily for discussion purposes, but it can be considered an advanced project if you want to invest the time and money into building it. However, I certainly don't recommend it for a *first* project. In case you believe that your construction experience and safety practices are satisfactory for such an endeavor, I have provided the following construction details.

If you design a PC board for this project, don't worry about trying to make it overly compact—the heatsinking for the MOSFETs will take up most of the enclosure space. Make sure that all of the high-current PC board tracks are extra wide. The heatsinking for the lateral MOSFETs will need to be approximately doubled over that needed for the Fig. 8-11 amplifier. (MOSFETs are a little less efficient than bipolar transistors, but they can also tolerate higher temperatures than bipolar transistors.) If you have to run fairly long connection wiring to the MOSFET leads (i.e., over ~ 6 inches), it is best to solder the gate resistors directly to the gate leads of the MOSFETs and insulate them with a small piece of heat-shrink tubing. Two small heatsinks should be mounted to transistors Q8 and Q11. The remaining construction details are essentially the same as for the Fig. 8-11 amplifier design.

The raw dual-polarity power supply that you will need to power this amplifier will be quite hefty. The power transformer must be an 80-volt center-tapped model with a secondary current rating of at least 4 amps (i.e., a 320-VA transformer). I recommend a 25-amp bridge rectifier (over-rated because of the high surge currents involved) and about 10,000 to 15,000 µF of filtering for each power supply rail. These capacitors will need to have a voltage rating of 75 WVDC. The AC line fuse (on the primary of the power transformer) should be a 3-amp, 250-volt slow-blow type, and don't forget to incorporate bleeder resistors.

During the initial testing of the Fig. 10-4 power amplifier, or almost any power amplifier design that you want to test, the "lab-quality power supply" project (the first project in this book) can be used to detect major faults or wiring errors in the amplifier circuitry. The power amplifier under test can be connected up to the dual-polarity 38-volt outputs of the lab supply and functionally tested without a speaker load. Most modern audio power amplifier designs will function at much lower rail voltages than what they might be designed for. Obviously, under these conditions, the amplifier will not operate at maximum performance levels, but it is a less risky means of testing a newly constructed amplifier (or an amplifier that has just undergone major repairs). If you have made any catastrophic mistakes, the current-limited outputs of the lab supply will reduce the risk of destroying a large quantity of expensive components through collateral damage.

Circuit Potpourri

Time for more practical fun.

Sounds Like Fun

Figure 10-5 is my favorite project in this book. It can actually be "played" similar to a musical instrument to produce a variety of pleasing and unusual sounds. It is also capable of running in "automatic mode" for unattended fascination.

The heart of this light-controlled sound generator is the basic UJT oscillator illustrated in Chapter 9, Fig. 9-9. Referring to this illustration, the 4.7-Kohm resistor in series with P1 is replaced with a photoresistor. Three such oscillators are needed for the Fig. 10-5 circuit. Each oscillator should have a different C1 value; the lowest C1 value chosen should be placed in oscillator 1 (to produce the highest audio frequency), the intermediate value in oscillator 2, and the highest capacity value (producing the lowest frequency) in oscillator 3.

The outputs of the three oscillators are capacitor-coupled to the input of a JFET audio mixer. The output of the mixer is connected to a line-level input on any audio amplification system.

The P1 potentiometer in each oscillator is adjusted for a good reference frequency under ambient lighting conditions. By waving your

Figure 10-5
Light-controlled
sound generator.

hand over the photoresistor (causing a shadow), the frequency will decrease accordingly. With all three oscillators running, the waving of both hands over the three photoresistors can produce a wide variety of sounds. By experimenting with different P1 settings in each oscillator, the effects can be extraordinary.

Another feature, added for automatic operation, is two "high-brightness"-type LEDs on the output of oscillator 3. Referring to Fig. 10-5, when SW1 is in the position to connect the oscillator 3 output to the LEDs, the LEDs will flash on and off at the oscillator's frequency. If these LEDs are placed in close proximity to the photoresistors used to control the frequency of oscillators 1 and 2, their frequency shifts will occur at the oscillator 3 frequency. In addition, even subtle changes in ambient light will cause variances. The possibilities are infinite. Although not shown in Fig. 10-5, you will need to add some series resistance between the output of oscillator 3 and the LEDs to limit the current to an appropriate level (depending on the type of LEDs used).

This circuit works very well with the 12-watt amplifier discussed in Chapter 8. Placed in an attractive enclosure, it is truly an impressive project.

The JFET mixer (Fig. 10-5) is a high-quality audio frequency mixer for any audio application. If additional inputs are needed, additional 1-Mohm potentiometers and 100 Kohm-resistor combinations can be added.

Emergency Automobile Flasher

Figure 10-6 illustrates how one HEXFET (a type of MOSFET) can be used to control a high-current automobile headlight for an emergency flasher.

Figure 10-6
Automobile headlight flasher.

A UJT oscillator (Chapter 9, Fig. 9-9) is modified for extremely low-frequency (ELF) operation (about 1 hertz), and its output is applied to the gate of the HEXFET as a switching voltage. The 1-Kohm resistor and 1000-μF capacitor are used to decouple the oscillator from the power circuit.

Any high-current automotive accessory (up to 10 amps) can be controlled with this circuit—even inductive loads, such as winch motors.

Home-Made AC

The circuit illustrated in Fig. 10-7 is used to convert a low-voltage DC power source (usually 12 volts from an automobile battery) to a higher-voltage AC source. Circuits of this type are called *inverters*. The most common application for this type of circuit is the operation of line-powered (120-volt AC) equipment from a car battery. There are, of course, many other applications.

For example, if you wanted to use this circuit for the previously mentioned application, the 12-volt DC source from the car battery would be applied to the DC terminals shown in Fig. 10-7 (observing the correct polarity, and fuse-protecting the 12-volt line from the battery). A standard 12.6-volt ct secondary/120-volt primary power transformer is used. The VA rating of the transformer will depend on the load of the line-powered equipment intended for use with this circuit. If the line-powered device required 120 volts AC at 1 amp (for example), a minimum size of 120 VA is needed (I recommend using at least a 10 to 20% higher VA rating to compensate for certain losses). With the components specified, a 200-VA transformer is the largest transformer that can be used.

Figure 10-7
Power MOSFET
inverter.

Q1 = Q2 = NTE2388

The combination of C1, C2, and the transformer secondary make up a resonant circuit (resonance will be discussed in a later chapter). Used in conjunction with the active components (Q1 and Q2), this circuit becomes a free-running oscillator, with the frequency of which is determined primarily by the value of C1 and C2. The transformer will operate the most efficiently at about a 60-hertz frequency, so the value of C1 and C2 should be chosen to "tune" the oscillator as closely to that frequency as possible.

Capacitors C1 and C2 should be equal in capacitance value. Start with 0.01 μF for C1 and C2, and use a resistance value of 100 Kohms for R1 and R2. These values should bring you close to 60 hertz. If the frequency is too high, decrease the values of the capacitors slightly and vice versa for a lower-frequency condition.

11

Batteries

Because many hobby and commercial electronic products receive their operational power from batteries, I thought it suitable to include this chapter. Before getting into details, you must begin by learning some basics.

The term *battery* actually refers to an electrochemical DC power source containing multiple cells. Nominal voltage levels for individual cells can vary, depending on the type, from 1.25 to 2 volts per cell. The 1.5-volt electrical devices most people refer to as "flashlight batteries" are, technically speaking, cells.

Primary batteries, or "dry cells," are typically thought of as non-rechargeable, although this description is not always true. The electrolyte used in primary batteries is not always "dry," either.

Secondary batteries are rechargeable, and usually do contain a liquid or paste-type electrolyte.

The *ampere-hour* (amp-hour, Ah) *rating* is a term used to define the amount of power that a battery, or cell, can deliver.

Battery Types

All batteries use a chemical reaction to produce electrical current. Battery types are usually defined by the types of chemicals or materials used in their construction.

There are seven main types of commercially available batteries: zinc, alkaline, rechargeable alkaline, nickel-cadmium, lead acid, "gelled" electrolyte, and lithium.

Zinc batteries are the most common type of "flashlight" battery. They are available in regular and heavy-duty types, but neither is very exemplary in regard to performance levels. They are not recommended for the majority of electronic projects.

Regular *alkaline batteries* last 300 to 800% longer than zinc batteries, depending on their application.

Nickel-cadmium batteries, or "nicads," are very popular secondary batteries. These are the type used in the majority of commercially available rechargeable products. A newer type of high-capacity nicad is also being used extensively today. It boasts a much longer service life, and a faster recharge rate (typically 5 to 6 hours). A bothersome peculiarity of nicads is their tendency to develop a "memory." Because of this characteristic, it is recommended that they be fully discharged before recharging.

Lead acid batteries are most commonly used as automobile batteries. They are also available in smaller sizes that are sealed (except for a blow-hole to allow gases to escape). "Motorcycle"-type lead-acid batteries are available in a wide range of amp-hour ratings, and are a good choice for heavy-duty projects.

Gelled electrolyte batteries fall into the same category as lead-acid types. They are most often used in *uninterruptable power supplies* (UPS; power supplies intended to supply 120-volt AC power in the event of a power failure), burglar alarms, and emergency lights.

Lithium batteries are designed to supply a small amount of power for a long period of time. Lithium batteries are most often used to power memory backup systems in computers, and in wristwatches. They are very expensive.

The most recent entry into the common marketplace is called the *rechargeable alkaline battery.* It was developed and patented by Rayovac Corp. under the trade name Renewal. These batteries offer 2 to 3 times the energy storage capacity of nicad batteries, higher terminal voltage (1.5 vs. 1.2 volts), retention of full charge for up to 5 years in storage, and lower operating temperatures, and they are "environmentally friendly." Rechargeable alkaline batteries can be used through well over 25 full charge/discharge cycles with very little degradation in performance, so that can add up to quite a cost savings over buying "throwaway" batteries.

Battery Ratings

As stated previously, the power delivering ability of a battery is rated in amp-hours. However, this does not mean that a 5-Ah battery will deliver a full 5 amps for one hour, and then suddenly quit. The battery manufacturer will calculate this rating over a longer period of time and then *back-calculate* to find the rating. For example, a 5-Ah battery should be capable of providing 500 milliamps for 10 hours. This calculates out to 5 amps for 1 hour, 1 amp for 5 hours, or 500 milliamps for 10 hours.

Trying to operate a battery at its maximum current rating for 1 hour is destructive to the battery. A 5-Ah battery would be a good choice for a project requiring one amp of current, possibly even two amps. Primary battery life and secondary service life (the total number of charge/discharge cycles it can withstand before failure) can be extended (by factors ranging from 100 to 400%) by avoiding excessive current drain.

Battery Care

In addition to excessive-duty operation, the most destructive variable to batteries is heat. Batteries should always be stored in a cool place, even a refrigerator (not a freezer!). Batteries being used in vehicles that are left outside in the sunlight should be removed and brought indoors, if possible.

Secondary batteries should be maintained in a charged state. Lead acid and gelled electrolyte batteries need to be recharged about once every 3 to 6 months if not used. Nicads, on the other hand, have a self-discharge rate of about 1% per day! For optimum performance, they need to be left on a trickle charge continuously when not in use.

A Few Words of Caution

Some types of batteries contain very potent acids, caustic substances, or highly poisonous materials. These materials include mercury, lead, hydrochloric acid, and other substances so toxic that the EPA (U.S. Environmental Protection Agency) classifies them as *toxic waste*. If you want to experiment with chemistry, buy a Gilbert Chemistry Set, but don't attempt to mess around with this stuff.

Don't try to recharge a battery that is not designed for recharging. It could explode. And if the explosion doesn't do enough damage in itself, toxic waste can be sprayed in the eyes and mouth. Even rechargeable batteries can explode if recharged too fast, or if overcharged.

Lead acid batteries can produce enormous current flows. An accidental direct short, with a lead acid battery as the power source, can literally blow up in your face; this could result in eye damage, or fires. *Large batteries should be fuse-protected right at the battery terminals.*

Recharging Batteries

The recharging of secondary batteries is not a complex process, but there are a few rules to follow. Nicad batteries should be completely discharged before trying to recharge them. They should not be discharged too quickly, however. One good way to discharge them properly is to connect them to a small incandescent lamp rated for the same voltage as the battery. When the lamp goes out, they're discharged. Of course, other types of resistive loads will perform this function as well as a lightbulb, but they won't provide a visual indication of when the discharge has been completed. As stated previously, it is destructive to discharge any battery too quickly. Nicads, however, are the only secondary battery type in which mandatory discharge becomes a concern. Other types of batteries can be recharged without having to be fully discharged.

Recharging procedures are generally the same for all types of secondary batteries. The primary rule to keep in mind, in regard to charging rates, is to not try to recharge them too fast. In some cases, trying to recharge a secondary battery too fast may cause it to explode. A good rule of thumb to follow is to not allow the charging current to exceed one-tenth the value of the amp-hour rating. For example, a 5-Ah battery should not receive a charging current any higher than 500 milliamps ($^1/_{10}$th of 5 amps).

In many cases, the recharge voltage applied to a secondary battery is higher than the rated battery voltage. For example, a typical automobile battery is rated at 12 volts (six cells at 2 volts each). The charging voltage applied to an auto battery is 13.8 volts. There are practical reasons for doing this with a car battery, but it is seldom advisable to follow this practice with batteries in other applications. In the majority of situations, the use of higher voltages ends up being a waste of power, and a potential risk toward overheating the battery.

Building a General-Purpose Battery Charger

By now, you may have come to the conclusion that all you need to properly recharge a battery is a variable DC power supply and the properly sized series resistor to limit the current. You're right! For example, if you wanted to recharge a 5-Ah, 12-volt battery, you will want to limit the charging current to 500 milliamps.

When the "dead" battery is first connected to the power supply, assume it to look like a short (it will come close to that if it is totally discharged). That means you need a resistor to drop 12 volts at 500 milliamps. Using Ohm's law, that comes out to 24 ohms. Because 24 ohms is not a standard resistor value, a 27-ohm resistor will do nicely. In the beginning of the charge cycle, this resistor will be required to dissipate almost 6 watts of power, so use one with a 10-watt rating. Alternatively, you could use two 56-ohm, 5-watt resistors in parallel; or, you could use three 8-ohm, 2-watt (or higher, 5-watt is preferable) resistors in series. Set the power supply to 12 volts, put the 27-ohm resistor in series with the battery, and the battery should recharge properly.

There is a small problem with this simple resistor power supply method. As the battery begins to charge, the voltage across it will increase. This causes a subsequent decrease of voltage across the resistor, and the charging current decreases. When the battery gets close to being totally recharged, the charging current drops to a very low value. The result is that the recharging process takes more time. This might or might not be a problem, depending on your needs.

If a more rapid recharge is desired, the circuit illustrated in Fig. 11-1 will provide a constant charge current, regardless of the battery voltage. This charge current will be maintained until the battery is fully charged, then it will automatically drop to a safe level. This circuit can be used with any type of variable power supply (as long as the power supply can be adjusted to a voltage slightly higher than the battery voltage), and it can charge any type of battery up to about a 15-Ah rating.

Referring to Fig. 11-1, the variable power supply is connected to the positive and negative input terminals of the charging circuit, and adjusted to be about 1 volt higher than the battery voltage. (The "lab-quality power supply" project, discussed in earlier chapters, will work well with this circuit.) The rotary switch (RS1) is set to the desired charging current position:

- Position 1 = 15-Ah batteries (or larger)

- Position 2 = 7.5-Ah batteries

- Position 3 = 2-Ah batteries

- Position 4 = 1-Ah batteries

- Position 5 = 0.2-Ah batteries (or smaller)

Figure 11-1
Battery charger
circuit.

RS1 is set to the position rated at, or below, the actual amp-hour rating of the battery. For example, a battery rated at 5 Ah should be recharged in position 3, not in position 2!

Figure 11-1 is a simple current-limiting circuit. It is identical in function to the current-limit circuit used in the "lab power supply" project discussed earlier in this book. However, some component values have been changed to provide different current-limit values. D1 is designed to protect the circuit in the event the battery terminals are connected in the wrong polarity.

If you would like to modify the circuit in Fig. 11-1 to provide different charging currents for special needs, this can be easily accomplished. Simply divide 600 mV by the desired charging current in milliamps. The answer, in ohms, will be the total resistance needed between Q2's base-emitter leads to provide the desired current.

12

Integrated Circuits

The process of miniaturizing multidevice circuits (transistors, resistors, diodes, etc.) is called *integration*. An *integrated circuit* (IC) is a chip that contains (or can perform the function of) many discrete (nonintegrated) devices. In appearance, most common types of ICs are small, rectangular packages with 8 to 16 pins (or legs) extending from the package. This physical configuration is called a *dual in-line package* (DIP). Other common package styles include round metal casings with multiple leads extending from them, and larger rectangular packages with up to 40 pins.

Most ICs are manufactured as general-purpose building blocks within a specific functional area. For example, the common 741 operational amplifier is designed specifically for amplification type functions, but its external component design can modify its performance for literally thousands of different applications. In contrast, many ICs are manufactured for very specific applications, especially in the consumer electronics field. For example, an IC specified for use as a rotational speed control for VCR heads can be used for little else.

It is easy to integrate semiconductor components (or components that can be made from semiconductor materials). Very large-scale integration (VLSI) chips can contain as many as 250,000 transistors. Resistors can be made very accurately from semiconductor material. However, large-value "reactive" components, such as inductors and capacitors, cannot be reduced in size. Also, semiconductor components designed to dissipate large quantities of power cannot be integrated very successfully. For these reasons, many ICs are nothing more than the total low-power semiconductor part of a larger circuit. The reactive and high-power components must be added for a functioning circuit. "Switching regulator control" ICs are a good example of this kind of design.

Because of the vast number of integrated circuits available, it is absolutely necessary to have a good selection of manufacturers' data books in your electronics library. You will have to depend on these data books to obtain pin-out diagrams (illustrations of the functional aspects of each connection lead on an IC), application information, and functional specifications.

ICs can be divided into two main families: *digital* and *linear* (or analog). Digital ICs will be discussed in the next chapter, so in this chapter, we will concentrate primarily on linear devices.

Linear ICs can be further subdivided into several classifications: operational amplifiers, audio amplifiers, voltage regulators, and special-purpose devices.

For very common applications, a manufacturer may build a complete circuit, using ICs and discrete components, and encapsulate the complete circuit into a block. These devices are called *hybrid modules*. Common hybrid modules include high-power audio amplifiers, power supply regulators, motor controls, and certain types of high-power, high-voltage devices.

Operational Amplifiers

Operational amplifiers are basic amplification building blocks. They consist of multiple high-gain differential amplifiers (exhibiting high common-mode rejection) without any kind of feedback loop. Depending on the configuration of the external components, they can be used for voltage amplifiers, transconductance amplifiers (voltage-to-current converters), transimpedance amplifiers (current-to-voltage converters), differentiators, integrators, comparators, oscillators, and regulators.

Virtually all operational amplifiers ("op amps") have two inputs marked with a positive (+) sign and negative (−) sign. These are the non-inverting inputs and inverting inputs, respectively. Don't confuse these with the power supply connections. Many op amps have frequency compensation inputs. Normally, a capacitor or resistor-capacitor combination is connected between these pins for controlling high-frequency characteristics. Offset null inputs are provided on many op amps to bias the output at a desired DC quiescent level.

Op amps are not specified in regard to frequency response, because it will vary depending on the way the op amp is used in a circuit. For example, an op amp used as a voltage amplifier (with the external components setting its gain at 10) will have a much broader frequency response than if the external components set its gain at 1000. Therefore, the frequency response characteristic of op amps is defined by the term *slew rate,* which is the speed (given in microseconds or nanoseconds) in an op amp's output change in accordance with an instantaneous change on the inputs. The higher the slew rate, the higher the maximum usable frequency response.

Most op amps are designed for use with dual-polarity power supplies. The power supply voltages should be equal, but opposite in polarity. Op amps designed for dual-polarity power supplies can work with single polarity supplies for certain applications. Also, some op amps are designed exclusively for single-polarity supplies.

The "perfect" op amp would have infinite gain, zero output impedance, infinite input impedance, and instantaneous slew rate, and be totally immune to common-mode noise on the power supply inputs and signal voltage inputs. If a perfect op amp existed, the circuit designer could use external components to design any feasible type of op amp circuit without any consideration of op amp parameters or limitations. Of course, perfection is not possible, but modern op amps, especially "FET input" op amps, come close to perfection parameters. Therefore, the equations used for designing amplifiers, filters, oscillators, and other circuits, seldom include any op amp variables.

Figure 12-1 illustrates some examples of common op amp circuit configurations. These illustrations do not give pin numbers or power supply connections because they are general in nature. Virtually any general-purpose op amp will work in these circuits, and many high-quality, low-noise

op amps will work better. For these circuits, and almost any type of op
amp circuit, I highly recommend the low-noise NE5532 (NTE778A) dual
op amp ICs, although the industry-standard 741 type will probably perform satisfactorily in most cases. For critical audio circuits (preamplifiers,
mixers, tone controls, etc.), I recommend the TL-074 op amps, which boast
a high slew rate and 0.005% THD (total harmonic distortion).

Referring again to Fig. 12-1, examine some details of each circuit illustrated. Circuit A is an inverting voltage amplifier. The equation for calculating voltage gain shows that it is solely dependent on the ratio of R1 and
R2. For example, if R1 is 10 Kohm and R2 is 100 Kohm, the voltage gain is
10. The negative sign $(-)$ is placed in front of the R1/R2 ratio expression to
show that the output will be inverted. R2 in this circuit is the negative-
feedback resistor. Note that it is connected from the inverting input to the
output. The output portion being "fedback" to the input will be inverted,
or opposing in nature. Consequently, it is negative feedback. R3 is used for
circuit stabilization, and its calculation is included in the illustration.

Circuit B is a noninverting voltage amplifier and, as the gain equation
shows, its gain is also dependent on the ratio of R1 and R2.

Circuit C is a transconductance amplifier, or a voltage-to-current converter. The current output, which is the current flow through $R_{L'}$ will

not be dependent on the resistance value of R_L. Rather, it will be dependent on the ratio of the input voltage to the value of R1.

Circuit D is a transimpedance amplifier, which functions in a manner opposite to that of circuit C. Circuit D converts an input current into a proportional voltage. The associated equation for this function defines the component relationship.

All of the circuits illustrated in Fig. 12-1 must be designed with their power supply limitations in mind. For example, circuit A cannot be designed so that a desired input and voltage gain will drive the output in excess of the power supply voltages used. In the same way, the manufacturer's specified maximum power dissipation cannot be exceeded without destroying the chip. In multiple op amp circuit designs, it may also be necessary to decouple the individual op amp circuits (usually with about a 100-μF electrolytic capacitor and a 0.1-μF nonpolarized capacitor).

IC or Hybrid Audio Amplifiers

Many currently available audio amplifiers are offered as "totally integrated" or hybrid. To the electronic "user," the only major difference is in reference to size; hybrid circuits are typically larger, and might require a larger heatsink.

In general, audio amplifier ICs might be thought of as "simple-to-use power op amps," because that is essentially what they are. Many types are designed for single-polarity power supplies (which make them excellent for battery-powered applications), and they frequently have internal gain and frequency compensation networks. These attributes, together with others, add up to a quick and simple, general-purpose audio amplification system for low- to medium-power applications. Heatsinks are typically required for outputs in excess of 1 watt.

Although great progress has been made in the integration techniques of audio amplifiers, the performance levels of many medium to high-power systems leaves much to be desired. For this reason, most audiophiles still prefer discrete audio power amplifiers in the majority of cases.

IC Voltage Regulators

This is an area where ICs have virtually taken over. Typical IC voltage regulators provide almost unexcelled voltage regulation, overvoltage protection, current limiting, and automatic thermal shutdown (the unit shuts off when it gets too hot), and many are adjustable. In addition, they are commonly available and very inexpensive. It's no wonder they are popular.

In addition to all of their other desirable attributes, voltage regulator ICs are very easy to use and implement for almost any application. Most have only three connection terminals; one is the "raw" DC input, another is the output, and the third simply connects to circuit common, or to a single potentiometer for voltage adjustment. You can't get much simpler than that!

The small size and low cost of regulator ICs has made it practical to use them as onboard regulators for each individual printed circuit board within an entire system. This has reduced the need (and cost) of large, high-current regulated power supplies for larger electronic systems. And, because this puts them in close proximity to each PC board within a system, the regulation and noise immunity is usually better than can be achieved with a large central power supply.

Regulator ICs commonly require heatsinking for maximum output. Most are available in either TO-220 or TO-3 packages. This causes them to look like power transistors, but it provides them the flexibility to be mounted on the many styles of transistor heatsinks already available.

Special-Purpose ICs

There are many more special-purpose integrated circuits than there are pages in this book. The determining factor as to whether any particular circuit will be available in IC form is purely economic. It is expensive for a manufacturer to do the research and development (R&D) required to produce a new type of IC. Consequently, manufacturers do extensive marketing research (to ensure the demand, in great quantities, for a special-purpose IC) before they can justify the cost of the initial investment. If they do not believe that they can regain their R&D costs, the IC is not manufactured.

Skim through several IC data books to acquire a basic knowledge of "what's out there" regarding special-purpose ICs. Before using a special-purpose IC, closely examine its specifications. There are many examples where discrete versions of circuits will perform better than their IC counterparts.

Improvement of Lab-Quality Power Supply

The lab-quality power supply, as discussed in Chapters 3 through 6 of this book, has a minor shortcoming. If you have used this supply for many of the prior projects, you might have noticed that the output voltage will drop slightly when the supply is heavily loaded. Because this power supply was offered as the first project in this book, I felt it was prudent to keep the parts count and complexity to a minimum. Although many commercial power supplies utilize this same basic design, it might not measure up to some of your future requirements in more critical circuits. Therefore, this section will detail two ways of improving the voltage regulation performance; it's up to you to decide on either method. You might also decide to leave this supply as it is, and build the quad power supply detailed in the next section instead.

First, refer back to Fig. 6-11 in Chapter 6. To improve the voltage regulation of this circuit, it is necessary to increase the gain of the Q3 and Q4 stages. One method of increasing gain would be to replace Q3 and Q4 with Darlington pairs. Another method, providing even tighter control, is illustrated in Fig. 12-2.

Although the circuit in Fig. 12-2 shows only the positive regulator, you would simply duplicate the basic theory and design in the negative regulator. (The current-limiting circuit of Fig. 6-11 is not shown in Fig. 12-2 for the sake of clarity.) Q3 of Fig. 6-11 would be replaced by the general-purpose op amp shown in Fig. 12-2. The inverting input of the op amp would connect to the wiper of the voltage control potentiometer, P1

Figure 12-2
Improvement of lab-quality power supply.

$$V_{REG} = V_{REF}\left(1 + \frac{R2}{R3}\right)$$

(Fig. 6-11), and the noninverting input would connect to the anode of D1 (the voltage reference diodes in Fig. 6-11). The output of the op amp connects to the base of Q1 (the series-pass transistor, Fig. 6-11), and the op amp power supply connections are made to the positive and circuit common points as illustrated.

The circuit, shown in Fig. 12-2, is included as a basic reference for all of your future power supply needs. To use this design, start by designing a simple zener-regulated power supply (the transformer, bridge rectifier, filter capacitor, R1, and D1). Be sure that the zener voltage of D1 is somewhat less than the anticipated voltage drop across R3. The equation for calculating the value of R2 and R3 is provided in the illustration. Of course, Q1 is chosen according to the current and power dissipation requirements of the proposed loads to be applied to the circuit.

The best method of improving the lab power supply is illustrated in Fig. 12-3. The full regulator circuit design has been included so that the circuit modifications might be more easily located. These alterations should be self-explanatory by comparing Fig. 6-11 with Fig. 12-3. Note that you will need to mount two additional binding posts for the positive and negative regulated outputs. Keep the original two posts intact, connected to the unregulated current limit circuit, for testing audio power amplifiers (and other projects requiring higher-voltage, dual-polarity supplies). Also note that P1 and P2 (Fig. 6-11) must be replaced with 5-Kohm potentiometers.

Building a Quad Power Supply

Figure 12-4 illustrates a quad-output lab power supply. This is an extremely versatile lab power supply. It provides the most frequently used voltages in dual polarity. Each output is current-limited at 1.5 amps, and all of the regulator ICs are internally protected from overvoltage and overtemperature conditions. The regulator ICs are of the *fixed-voltage type*, meaning that they are not voltage-adjustable, but they provide extremely good voltage regulation. All of the power supply components have been chosen so that all four outputs can be loaded to maximum capacity simultaneously.

All four regulator ICs will require some form of heatsinking. Mounting the ICs to a metal enclosure will probably be sufficient, but the combination of the chassis and the "sinks" is recommended. After constructing the power supply, load down all four outputs, and let it run for a while. If the

Figure 12-3
Improvement of original lab power supply illustrated in Fig. 6-11.

Figure 12-4
A quad-output lab power supply.

enclosure (or the ICs) start to become too hot, you will need to include some additional heatsinking between the enclosure and IC mounting tabs.

Circuit Potpourri

Noise Hangs in the Balance

Most high-quality microphones, those designed for professional entertainment use, are low-impedance devices and have balanced XLR-type connector outputs. Although these "mikes" can be impedance matched for a high-impedance input with a single transistor stage, this will defeat the whole purpose of having a balanced line.

The circuit illustrated in Fig. 12-5 will match the impedance correctly, and utilize the high common-mode rejection characteristic of operational amplifiers to remove the unwanted "hum" picked up by long microphone wires.

Hum Reducer

Furthermore, on the subject of hum (stray 60-hertz noise, inductively coupled to an audio circuit), the circuit illustrated in Fig. 12-6 can do a very effective job of removing almost all hum content, even after it has already been mixed with other audio material. However, it will also reduce some low-bass frequencies (from about 40 to 100 hertz).

Figure 12-6 shows a "twin-tee" notch filter. The component values chosen will cause the output to drastically attenuate 60-hertz frequencies,

Figure 12-5

A balanced input circuit for low-impedance microphones.

▮▮ ▮▮ ▮▮ ▮▮
Figure 12-6
A 60-hertz notch
filter.

$R_1 = R_2$ $R_3 = \dfrac{R_1}{2}$

$C_1 = C_2$ $C_3 = 2(C_1)$

$F_N = \dfrac{1}{2\pi R_1 C_1}$

ICI = NTE778A

but allow most other frequencies to pass unaltered. Attenuation at 60 hertz should be on the order of about 12 dB.

Equations are provided in the illustration to design notch filters for other frequencies, also. Notice the equation for finding the notch frequency (F_N). The little symbol with the two legs and a wavy top is called *pi*. Pi is a mathematical standard, and it is approximately equal to 3.1416.

Let the Band Pass

Figure 12-7 is an example of a bandpass filter. A *bandpass filter* only passes a narrow band of frequencies, and severely attenuates all others. This circuit illustrates how a JFET can be used as a voltage-controlled resistor; to "tune" the bandpass frequencies of the filter by means of a control voltage (about 0 to 2 volts DC). The usable frequency range is from subaudio to about 3 kHz. This circuit is commonly used to produce "wah-wah" effects for electric guitars and other instruments.

When several of these circuits are placed in parallel with component values chosen to provide different ranges of bandpass, the result is a *parametric filter*. If a manual control of the bandpass is desired, Q1 might be replaced with a 5-Kohm resistor.

High-Versatility Filters

Figures 12-8 and 12-9 are both examples of equal-component, *Sallen-Key active filter* circuits. Equal-component filters are easier to build, because

Figure 12-7
A voltage-controlled bandpass filter.

Figure 12-8
Low-pass filter.

$R_1 = R_2$

$C_1 = C_2$

$F_C = \dfrac{1}{2\pi R_1 C_1}$

$A = \dfrac{R_4}{R_3}$

$R_3 = 0.586\,(R_4)$

$ICI = NTE778A$

Figure 12-9
High-pass filter.

$R_1 = R_2$

$C_1 = C_2$

$F_C = \dfrac{1}{2\pi R_1 C_1}$

$A = \dfrac{R_4}{R_3}$

$R_3 = 0.586\,(R_4)$

$ICI = NTE\ 778A$

other filter designs require exact multiples or divisions of the values of key components which might not be standard in value. Figure 12-8 is a low-pass filter, and Fig. 12-9 is a high-pass design. Both designs provide about 12 dB-per octave of signal attenuation (rolloff) for the unwanted frequencies.

To design these filter circuits for your own applications, start with R3 = 12 Kohms and R4 = 22 Kohms. Using R1 and R2 values of 4.7 Kohms, and C1 and C2 values of 0.005 μF, F_c should be a little over 6 kHz. From this point, you can adjust component values, as needed, by following the equations provided in the illustrations.

These circuits can be cascaded (in series) for sharper cutoff responses. For example, if two Fig. 12-8 circuits were built, each having a cutoff frequency of 6 kHz, and they were placed in series, the final output would have a 24-dB/octave rolloff. The Fig. 12-8 and Fig. 12-9 circuits can also be placed in series to provide precise bandpass responses with both upper and lower cutoff frequencies. For example, assume that you need a filter circuit to reject all frequencies except those occurring between 6 and 8 kHz. You would design the low-pass filter for a cutoff frequency of 8 kHz, and the high-pass filter for a cutoff of 6 kHz. By placing the two filter circuits in series, the total response would be that of a bandpass filter with a "passband" of 6 to 8 kHz.

How Was Your Trip?

Figure 12-10 illustrates an adjustable trip-point relay driver. This is an example of how an operational amplifier can be used as a "comparator."

Figure 12-10
Adjustable trip-point relay driver.

A comparator circuit "compares" two voltages. When one exceeds the other, a signal is provided.

The op amp compares an adjustable reference voltage, provided by P1, to an input voltage. When the input voltage exceeds the reference voltage, the op amp output goes high, turning on Q1, which energizes the relay. The relay, of course, can be used to turn on, or provide power to, just about any kind of device. For precise, repeatable performance, the positive rail should be regulated by an LM7812 (NTE966) positive voltage regulator.

Circuits of this nature are used primarily for control applications. For example, different variables such as temperature, humidity, or light could be utilized to control other circuits, appliances, or machinery. For a security light controller, a photoresistor and fixed resistor, in series as a voltage-divider network, could be connected between the positive rail and ground. The input to the op amp could be connected to the junction between the two. If this circuit were to be placed outdoors, the resistance of the photoresistor would increase as the sun started to set. When the voltage drop across the photoresistor became high enough to exceed the reference voltage applied to the op amp, the relay would be energized turning on an outside light.

When two op amps are configured as comparators, one referenced to a low voltage and the other to a higher voltage (called *high and low trip points*), the resulting circuit is called a *window comparator*. Window comparators are used to control a variable within a set of operating parameters.

Watt an Amplifier

The simple circuit of Fig. 12-11 is a 1-watt IC audio amplifier. P1 is the volume control. The rest of the circuit is self-explanatory.

Figure 12-11
A 1-watt audio amplifier.

This circuit works very well in battery-powered applications, and makes an excellent little amplifier for many of the sound-effect circuits covered previously in this book.

Control Your Tone

As a final entry into this section of circuit potpourri, I submit the easy-to-construct and extremely versatile tone control and preamplifier combination illustrated in Fig. 12-12. I've also added some additional circuitry to illustrate an optional method of incorporating this circuit into almost any high-quality audio power amplifier system.

Starting at the left side of the illustration, IC1 and its associated circuitry constitute a typical inverting amplifier, identical to the amplifier circuit illustrated in Fig. 12-1, circuit A. As you will note, the ratio of the feedback resistor (47 Kohm) to the input resistor (4.7 Kohm) is 10:1, so the voltage gain is set at 10. However, you can change this ratio for more or less gain, according to your needs. The 100-pF capacitor across the audio input connections is for filtering out any unwanted high-frequency (HF) interference signals that may be present on the audio input signal. The output of IC1 is connected to a 100-Kohm potentiometer that allows you to "tap off" any percentage of the amplified audio signal desired. In other words, it is your "volume" control.

From the tap of the volume potentiometer, the audio signal must pass through a system of *RC* filters, similar to the tone control circuit illustrated in Fig. 8-7 of Chapter 8. Note that the inverting input of IC2 is coming from the filter network. If you follow the output of IC2 to the junction of the two 5.6-Kohm resistors, you'll note that the output connects back into the filter network. In other words, the filter network actually makes up the negative-feedback path for IC2 (i.e., the output of IC2 connects to the filter network, and the output signal is fed back to the inverting input of IC2 through the filter network). Therefore, this circuit is an "active" tone control circuit, and it provides very good performance.

You will notice that the operational power for the op amps is obtained from two zener-regulated power supplies connected to a dual-polarity 38-volt DC power supply. The zener diodes are 15-volt, $\frac{1}{2}$-watt units, so the voltage at the cathode of ZD1 is approximately 15 volts DC, and the voltage at the anode of ZD2 is approximately −15 volts DC. The two 1.5-Kohm resistors in series with the zeners are the "dropping" resistors. This power supply configuration shows how this circuit can be

Figure 12-12 General-purpose preamplifier and tone control circuit.

connected directly to "rail" power supplies of typical audio power amplifiers. For rail voltages higher or lower than 38 volts, the value of the series dropping resistors would have to be changed accordingly. Naturally, if you wanted to provide operational power to this circuit with typical dual-polarity 12- or 15-volt supplies, you would simply omit the two zener diodes and their associated components (i.e., the 1.5-Kohm dropping resistors and 220-μF filter capacitors).

LM1458 or NTE778A operational amplifiers are specified for this circuit, but almost any general-purpose op amp will provide satisfactory results. If you want to significantly improve performance for a high-quality application, you could use TL-074, OP-176, or AD797 operational amplifiers. In addition to utilizing high-performance operational amplifiers, the signal-to-noise ratio (SNR) can be improved by using 1% metal film resistors instead of the more common carbon film or carbon composition types of resistors.

13

Digital Electronics

Digital electronics is a large branch of the electrical-electronics field relating to those electronic functions used in performing logical functions. It encompasses every type of logical system from simple combinational logic control in a coffee pot to the largest computer systems. The concept of digital control is not new. Even before the first computer appeared, or solid-state electronics came into being, large banks of relays were performing logical control functions in industrial facilities.

The earliest types of computers were called *analog computers*. They were enormous machines made from thousands of vacuum tubes. Computations were performed using voltage levels based on the decimal (base 10) numbering system. Although this method seems natural because you think in terms of tens (because you were created with 10 fingers), the old analog computers soon gave way to the more modern binary computers (base 2). There are very good reasons for this change, which will be explained as this chapter continues.

Although many people seem to have trouble comprehending different numbering systems, it's really quite simple. The key is in understanding the mechanics behind the decimal system, and then applying those principles to any other numbering system.

The term *decimal* means "base 10." If the number 1543 is broken down into decimal column weights, it comes out to 3 units (or ones), 4 tens, 5 hundreds, and 1 thousand. Notice how each succeeding weight is actually the base number (10) times the "weight" of the preceding column. In other words, $1 \times 10 = 10$, $10 \times 10 = 100$, $10 \times 100 = 1000$, and so on.

The *binary numbering system* works exactly the same way, except that it is based on 2 instead of 10. For example, the first weight (or least significant digit) is the units, or ones, column. The weight of the second column is 2×1, or *2*. The next column is 2×2, or *4*. The next column is 2×4, or *8*. Instead of the column weights being ones, tens, hundreds, thousands, ten thousands, and so forth, the binary column weights will be ones, twos, fours, eights, sixteens, and so on.

In the decimal system, there are 10 possible numbers which can be placed in any weight column (0, 1, 2, 3, 4, 5, 6, 7, 8, 9). In the binary system, there are only two possible numbers for any one weight column (a 0 or a 1; a "yes" or a "no"; an ON or an OFF). If a binary number such as 0111 is broken down into weights, it means 1 one, 1 two, 1 four, and 0 eights. By adding the weights together, the binary number can be converted to decimal. In the previous example, $1 + 2 + 4 = 7$. Therefore, 0111 is the binary equivalent to decimal 7.

The following example demonstrates how it is possible to count up to 9 using the binary numbering system:

Column Values	
8421	**8421**
0000 = 0	0101 = 5
0001 = 1	0110 = 6
0010 = 2	0111 = 7
0011 = 3	1000 = 8
0100 = 4	1001 = 9

The binary numbering system is used in digital electronics because the binary digits 1 and 0 can be represented by an electronic device being either ON or OFF. For example, a relay can be energized or deenergized; or a transistor can be saturated or cut off. The advantage to a simple ON/OFF status is that the "absolute value or voltage level is not important." In other words, it is totally irrelevant whether a transistor in cutoff has 4.5 volts, or 5.5 volts, on its collector. The only important data from a binary point of view is that it is OFF.

There are other numbering systems used extensively in digital electronics besides the binary system. The two most common ones are the octal system (base 8) and the hexadecimal system (base 16 system). These different numbering systems come in handy when interfacing with humans, but at the actual component level, everything is performed in binary.

Logic Gates

Various forms of logical building blocks are available in integrated circuit form. These logical building blocks are called "gates," with each gate having a distinct function. *Logic gates* can be further combined into more complex digital building blocks to perform a variety of counting, memory, and timing functions.

Digital ICs are grouped into families, with each family possessing certain desirable traits that make them more or less suited to a variety of applications. One logic family might not be compatible with another family, so it is typical for the designer to use only one family type for each application. The most commonly used *logic families* in the present market are complementary metal oxide silicon (CMOS) and transistor-transistor logic (TTL).

Logic gates respond to "high" or "low" voltage levels. The specific voltage level for a "high" or "low" condition will vary from one logic family to another. For example, a logical one (high) in TTL logic is about 5 volts; in contrast, a possible 12-volt level might be used for CMOS logic. However, the functional operation and symbolic representation is universal throughout all of the families. Figure 13-1 lists some of the more common logic devices and their associated symbols.

Logic gates, and other logic devices, are functionally defined by using truth tables. Figure 13-2 illustrates a variety of truth tables for some common logic gates. Compare the AND-gate illustration in Fig. 13-1 with its corresponding truth table in Fig. 13-2. Because the AND gate has two input leads, there are a total of four possible logic conditions that could

Figure 13-1
Common logic
symbols.

AND	NAND	OR	NOR

Buffer	Inverter	Exclusive-OR	Exclusive-NOR

Set-Reset flip-flop	JK flip-flop	Toggle flip-flop

Figure 13-2
Truth tables for common logic gates.

IN	Out
AB	X
00	0
01	0
10	0
11	1

Truth table
for 2-input
AND gate

IN	Out
ABC	X
000	0
001	0
010	0
011	0
100	0
101	0
110	0
111	1

Truth table
for 3-input
AND gate

IN	Out
AB	X
00	0
01	1
10	1
11	1

Truth table
for OR gate

IN	Out
AB	X
00	1
01	0
10	0
11	0

Truth table
for NOR gate

IN	Out
AB	X
00	1
01	1
10	1
11	0

Truth table
for NAND gate

IN	Out
AB	X
00	0
01	1
10	1
11	0

Truth table
for Exclusive
OR gate

occur on the inputs. Notice that the truth table lists the four possible input conditions; together with each of their resultant outputs for each condition. As shown by the truth table, the only time that the output goes "high" is when the *A* input "and" the *B* input are high.

Logic gates can have more than two inputs. Figure 13-2 illustrates the truth table for a three-input AND gate. Common logic gates are available with up to eight inputs.

Referring again to Fig. 13-1, note the OR gate and its associated truth table in Fig. 13-2. As the name implies, its output goes high whenever a high appears on the *A* input, or on the *B* input (or both).

In digital terminology, a *not* function means that a logical condition is inverted, or reversed. A *NAND* gate (short for *not AND)* is an AND gate, with the output inverted. Notice that the outputs in the truth tables for the AND gate, and the *NAND* gate, are simply inverted. This same principle holds true for the OR and NOR gates.

It is common for the output of one logic gate to provide inputs for several other logic gates. The maximum number of inputs that can be driven by a particular logic gate is specified as its *fanout.* Typical logic gates have fanouts ranging from 5 to 20. If it becomes necessary to drive a greater number of inputs than the fanout of a particular gate, a buffer is used to increase the fanout capability. The symbol for a buffer is illustrated in Fig. 13-1.

The need often arises to invert a logic signal. The symbol for an inverter is shown in Fig. 13-1. An inverter is sometimes called a *not* gate. Note that it has a small circle on its output just like the *NOT AND (NAND)* and *NOT OR (NOR)* gates. Anytime a small circle appears on an input or output of a logic device, it is symbolizing the inversion of the logic signals (or data). Also notice the horizontal line above the *A* output of the inverter. It is called a *not* symbol. Whenever a horizontal line is placed above a logic expression, it means that it is inverted.

Another common type of logic gate is the *exclusive OR gate.* Refer to its symbol and the associated truth table in Figs. 13-1 and 13-2. As the truth table indicates, its output only goes high when its inputs are different from each other. The *exclusive NOR gate* provides the same logic function with an inverted output.

Combining Logic Gates

The most basic type of digital system is designed to provide a logical output based on a set of input conditions. This is referred to as *conditional logic.* Figure 13-3 illustrates a hypothetical copy machine using a simple conditional logic system to monitor a copier's operation, and to provide operator feedback or shutdown.

The four status signals are designed to go high if a problem arises. The status 1 and status 2 signals will go high if a "low paper" or "low toner" situation occurs. Neither of these are major problems, so it is desirable to simply light an annunciator panel lamp, which usually says something similar to "check machine." OR gate 1 output will go high and turn on the annunciator, if either (or both) of these problems arise.

However, if status 3 goes high ("copy failure"), it means that the machine is completely out of paper or toner, which is a little more serious. Therefore, AND gate 2 is used to indicate (in digital language), "I have run out of paper OR toner AND I can no longer make a copy. Wake up!!" So it rings a bell, and the annunciator is left on.

Meanwhile, AND gate 3 is watching for a serious problem. If a "low paper" OR "low toner" condition happens AND there is a "copy failure" AND a "power failure" all at the same time, it shuts down the copy machine. There are no copy machines out in the real world with status controls as basic as the one illustrated in Fig. 13-3, but a good example has been provided of how some of these logic gates can be combined.

The most commonly used types of conditional logic systems have been integrated. Integrated circuits such as binary-coded decimal (BCD)-to-decimal decoders, BCD-to-seven-segment decoders, data selectors, and data routers are just a few examples. Almost any original design will incorporate some unique conditional logic.

Figure 13-3

Basic example of conditional logic.

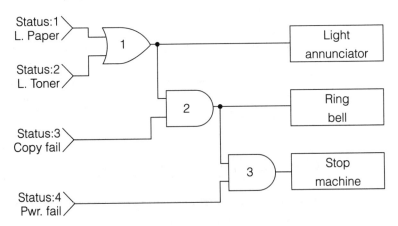

Multivibrators

Multivibrators are used extensively in digital electronics to provide clock signals (oscillators), count and store data, and control timing sequences. They can be divided into three major groups, or types: astable multivibrators (called *clocks* or *oscillators*), bistable multivibrators (flip-flops), and monostable multivibrators (one-shots).

You should already be familiar with *astable multivibrators* from our previous discussions regarding their use in sound circuits. The term *astable* means "not stable"; they cannot come to rest in either a high or low state. In other words, they *oscillate*. Because their outputs are in the form of a square wave, they are naturally suited to digital systems. IC forms of astable multivibrators are designed to operate at very high speeds.

Three common examples of *bistable multivibrators* are illustrated in Fig. 13-1. As the name suggests, bistable multivibrators have two stable states: "set" and "reset." They are usually called "flip-flops" (F-Fs).

Referring to Fig. 13-1, notice that each F-F has a "Q" and "NOT Q" output. The NOT *Q* output is always logically opposite of the Q output. When an F-F is "reset," the Q output will be a logic 0, thus meaning that the NOT Q is a logical 1. If an F-F is "set," the logical states of the Q and NOT Q will be reversed.

The first F-F illustrated in Fig. 13-1 is a "set-reset (RS) F-F," or *RS flip-flop*. If a logical 1 is applied to the R (reset) input, the Q output goes to logical 0. Similarly, if a logical 1 is applied to the S (set) input, the Q output will go to a logical 1. RS flip-flops have limited use in most digital systems. Their importance lies in their ability to latch, or remember, a logical status (if the logic levels to the RS inputs are not altered).

The second type of F-F illustrated is the *JK flip-flop*. The *JK* inputs are "clocked inputs." This means that the logical levels applied to the *JK* inputs have no effect without a coincidental pulse applied to the clock input. For example, if the *K* input is 1 and the *J* input is 0, the F-F will reset as soon as a clock pulse is applied to the clock input. If the *K* input changes to 0, and the *J* input goes to 1, the F-F will set as soon as (but not before) another pulse is received at the clock input.

If both the *J* and *K* inputs are held at logical 1, a *JK* flip-flop becomes a *toggle F-F*. The output, or Q status, of a toggle F-F will change state every time the correct *transitional change* occurs at the clock input. Toggle F-F's are designed to change state, or toggle, on either the "leading edge" or "trailing edge" of input clock pulses. For example, if a toggle F-F is specified to toggle on the trailing edge of the input clock pulses, a transitional change of the clock from a 0 to a 1 state will have no effect on the F-F. However, when the clock pulse changes from a 1 to a 0 (trailing edge), the F-F will toggle. Therefore, if a steady stream of clock pulses is applied to the clock input of a toggle F-F, the Q output of the F-F will be one "half" of the frequency of the input clock.

Referring to Fig. 13-4, three trailing-edge toggle F-Fs are connected together to form a *counter.* The lower part of the illustration shows a series of eight clock pulses that are applied to the clock input of the first toggle F-F. The righthand table lists the output status of the Q outputs after the trailing edge of each pulse. Remember, the leading edge is when the clock goes from a 0 (low) to 1 (high). In contrast, the trailing edge occurs when the clock goes from a 1 to a 0.

Continuing to refer to Fig. 13-4, note that a trailing edge occurs 8 times out of the eight pulses applied to the clock input. This means that

Figure 13-4
Basic modulo 8
counter.

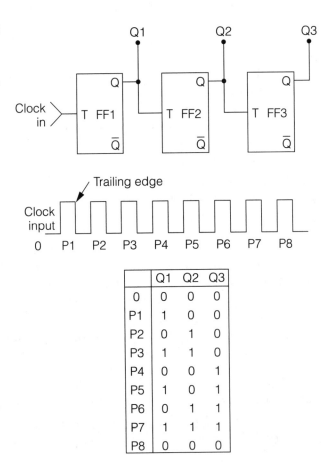

	Q1	Q2	Q3
0	0	0	0
P1	1	0	0
P2	0	1	0
P3	1	1	0
P4	0	0	1
P5	1	0	1
P6	0	1	1
P7	1	1	1
P8	0	0	0

F-F1 will toggle (or change state) 8 times, as shown in the "Q1" column of the chart. F-F1 has to toggle 8 times to produce four complete output pulses (a complete pulse requires two transitions: 0 to 1 and then 1 to 0). The result is a Q1 output of "four" pulses. These four pulses are applied to the clock input of F-F2, which also divides its clock input by 2, resulting in a Q2 output of two pulses. Similarly, these two pulses become the clock input of F-F3, and the resultant output of Q3 is one pulse. The overall response of all three F-Fs, as shown in the accompanying chart, is a *binary upcount,* where Q1 is the least significant digit and Q3 the most significant digit.

If continuous clock pulses are applied to the counter in Fig. 13-4, it will continue to cycle as shown in the chart. One full pulse will occur at the Q3 output for every eight pulses applied to the input. A counter of this configuration is known as a *modulo 8 counter.* There are two primary uses for such a circuit. The first use, quite obviously, is to count and accumulate totals (by adding additional F-F stages, virtually any size number can be counted, or accumulated). The second use is to *divide* a high frequency down to a lower frequency. For example, a 16-kHz square-wave frequency applied to the input of the Fig. 13-4 counter would be divided down to 2 kHz on the Q3 output. Logic gates can be incorporated with similar counter circuits to provide any *modulo divider* desired.

It is very common to use four toggle F-F stages with some associated logic gates to reset all four F-Fs when the counter attempts to increment up to a 10 count. This results in a 0 through 9 count, which is compatible with the decimal numbering system. The Q outputs of such a counter are called *binary-coded decimals* (BCDs). There are many integrated circuits available to decode BCD outputs into humanly recognizable forms. One common IC of this type is called a *seven-segment decoder/driver,* which decodes BCD outputs into the outputs needed to display decimal numbers on seven-segment LEDs. Many types of counters and decoders are available in IC form for a wide variety of applications.

If you would like to begin experimenting with some counter circuits, Fig. 13-5 illustrates a simple circuit for counting and displaying the count in BCD. The NTE4029B is a good example of a versatile counter in IC form. By applying logic levels to various pins, it can be made to count up or down, in BCD or binary. It can also be preprogrammed with a beginning number, reset, and set. Additional 4029B chips can be connected together (to the "carry in" or "carry out" pins) for a count as high as desired. The circuit illustrated in Fig. 13-5 will begin with a BCD 9, and count down with every pulse received.

Pin 8 = GND
Pin 16 = +5 V

The circuit illustrated in Fig. 13-6 adds a seven-segment decoder/driver IC (NTE4511) to display the count in decimal numbers. Any type of common-cathode seven-segment LED display can be used.

The circuits illustrated in Figs. 13-5 and 13-6 can be used as basic building blocks for a multitude of counting and dividing applications. A few cautions should be observed, however. First, these are CMOS devices, meaning that they are very susceptible to damage by static charges. Second, these circuits, like most other digital counting circuits, need a good-quality clock signal for proper operation.

Digital Clocks

Digital clocks can be in the form of continuous oscillations or gated pulses from other sources. They are always in the form of square waves, but they do not necessarily have to be symmetrical. In other words, the off time can be much longer than the on time, or vice versa.

For "solid" and repeatable operation of digital circuits, the clock pulses must have short transitional time periods. That is, the change from 0 to 1 and 1 to 0 must be very quick (these periods are called *rise time* and *fall time*, respectively).

Figure 13-6

Decade counter with
7-segment decoder
and display.

It is often desirable to use a mechanical switch to provide counting or control pulses to digital circuits. All mechanical switches, however, have an inherent characteristic called "bounce." When switch contacts come together to make contact, some mechanical vibration will result in the physical movement. For most applications, this vibration is unnoticed, but because digital circuits are designed to operate at extremely high speeds, they will respond to switch bounce. Therefore, if a mechanical switch is used in conjunction with digital circuits, it must be made "bounceless" for reliable operation.

Some examples of bounceless switches and digital clocks will be given in the circuit potpourri section of this chapter.

Shift Registers

A *shift register* is actually a modified form of digital counter. It consists of flip-flops, like a counter, but its purpose is to temporarily hold numbers for processing, display, or rerouting.

Shift register action is defined by how a number is put into a shift register, and how it is taken out. The possible combinations are serial in, serial out; parallel in, parallel out; serial in, parallel out; and parallel in, serial out.

A shift register can be designed so that the complete digital number is applied to the data inputs simultaneously and after one clock pulse, the number is loaded in. This is *parallel-in operation*. In contrast, if a number is loaded into a shift register one bit (short for *binary digit*) at a time (requiring one clock pulse per bit loaded), it is classified as *serial-in operation*. The same principle applies in removing a number from a shift register.

Shift registers are utilized within a digital system, in the same way a human uses a scratchpad, to hold various subtotals and interim calculations until they are needed for the final solution to a sequential problem. In this sense, they can be referred to as *latches* or *memory*.

Digital Memory Devices

The term *memory* is usually given to a large number of digital devices arranged specifically for the purpose of retaining large quantities of digital information, or *data*. It is technically accurate, however, to classify a single flip-flop as a 1-bit memory.

Digital memory is classified into two major categories: volatile and nonvolatile. *Volatile memory* will lose all accumulated data when the power is removed. For example, the most common type of volatile memory consists of thousands of flip-flops with their inputs and outputs arranged in a grid pattern. Data can be addressed (located) and read into, or read out of, the memory. However, when the circuit power is turned off and then back on, the thousands of flip-flops will "power up" in random fashion and all previously stored data will be lost. This, of course, is a major disadvantage.

There are many types of *nonvolatile memory devices* available on today's market. In integrated circuit form, nonvolatile memory is called *read-only memory* (ROM). IC ROMs operate on a variety of principles. Some retain data by converting them into electrostatic charges (which are retained after power is lost), and other types are permanently "programmed" by the physical destruction (burning in) of many diodes integrated into the chip.

In general, permanently programmed IC memory devices are called *read-only memory*, and IC volatile memory devices are referred to as *random-access memory* (RAM). Some types of RAM memory chips can be connected to a battery that supplies operating power for memory retention. This is called a *battery back-up* memory system. Long-life lithium batteries are typically used for this purpose.

Summary

Many logic devices are classified by their operational use, instead of by their internal design. In the same way that a transistor can be used for an amplifier, oscillator, buffer, or regulator; a flip-flop can be used as part of a counter, shift register, or memory.

The field of digital electronics has literally revolutionized the world. Although I would not classify it as difficult from a conceptual point of view, it is an extremely broad field that cannot be fully covered in any one book, and most certainly not a single chapter.

If you wish to learn more about digital electronics, experiment with some of the circuits included in this chapter, and spend considerable time researching available product information from the manufacturer's data books. If you anticipate a long-term hobby interest in digital electronic circuits, it will be necessary to acquire a good dual-trade oscilloscope (at least 40 MHz) for observance of the high-frequency circuit operations.

Circuit Potpourri

How Is Your Pulse?

Figure 13-7 incorporates the very popular NE555 timer IC to form a *digital pulser*, or clock. The output is a "clean" square wave, and it can be adjusted to a very slow rate; thus, the operation of a circuit under test

Figure 13-7
Digital pulser circuit.

IC1 = NE555 Timer

can be visually checked. If a flashing LED is desired to give visual indication of the output, the anode of any common type LED can be connected to the output in series with a 330-ohm resistor to ground. P1 and C1 are the primary frequency determining components. If higher output frequencies are required, their values can be reduced, and vice versa.

Improved Digital Pulser

Figure 13-8 illustrates an improvement over the simpler pulser circuit detailed in Fig. 13-7. The oscillator section is basically the same, and can be modified for different frequencies as described for the circuit in Fig. 13-7.

Switch S1 should be a momentary push-button switch with the "normally closed" contacts wired, as shown in the illustration. In this position, a constant train of logic pulses will be at the output (pin 4) of IC2. When S1 is depressed, it forces the output to a continuous high level. This circuit is very handy for checking the functional status of logic gates.

For convenient use, the circuit can be constructed on a narrow piece of universal breadboard and inserted in a round, plastic toothbrush holder, or in a similar hand-held package. The output is applied to a "probe tip" that can be purchased at an electronics parts store, or fabricated from a small nail. The power leads are typically channeled through the rear end of the holder.

Figure 13-8
Improved logic
pulser.

IC2 is a CMOS device, and normal static electricity precautions must be observed during construction. Only two of the four gates within this chip are used in the circuit. The remaining "inputs" (not outputs) of the unused gates should be grounded.

Digital Logic Probe

Figure 13-9 illustrates a logic probe circuit. This circuit differs from the logic pulser in several ways. It does not "inject" any pulses; it only reads the logic status of devices under test.

When the probe tip is placed on a test point with a logical high level, the red LED will light; a low level will light the green LED. If the test point is toggling, both LEDs will light.

Logic probes typically receive their operational power from the circuit under test. The $+V$ and GND leads should have small alligator clips on their ends so they can be attached to the power supply lines of the circuit under test. To protect this circuit against accidental polarity reversal, a common 1N4001 diode can be put in series with the $+V$ lead (cathode connected to R1 and anode connected to $+V$). IC1 will function properly on any DC voltage from 5 to 15 volts.

SW1 is a SPDT toggle switch. In the position illustrated, the logic probe is used to test CMOS circuits. Switch SW1 to the other position for testing TTL circuits.

This circuit should be constructed and enclosed in the same manner as that described for the circuit of Fig. 13-8. The NTE4049 is a

Figure 13-9
CMOS logic probe.

CMOS chip, so watch the static electricity, and connect all unused inputs to ground.

Bounceless Is Better

Figure 13-10 is a bounceless switch. As explained earlier in the text, any type of mechanical switch will produce very-high-frequency erratic pulses while "settling" from the vibrational shock of changing state. This is commonly known as "switch bounce." High-speed digital circuits will respond to switch bounce, causing erroneous operation. The circuit illustrated in Fig. 13-10 can be used to manually trigger digital circuits. SW1 can be any type of SPDT switch. The NTE4049 is a CMOS device, so take the necessary precautions against static electricity during construction, and connect the unused inputs to ground.

High-Stability Crystal Timebase

All of the types of oscillators that have been discussed thus far have one inherent problem—their operational frequencies will vary with temperature changes. For digital circuits requiring high-stability oscillators, a *crystal-based oscillator* is usually employed.

Crystals are "piezoelectric" elements with vibrational characteristics that are highly immune to temperature changes. Figure 13-11 is a crystal timebase oscillator. Depending on the crystal frequency chosen, start with a 10-Mohm resistor for R1, and increase its value as needed.

The NTE4011B is a CMOS device, so take precautions in assembly, as detailed previously.

Crystal oscillators operate at very high frequencies. To lower the output frequency, an appropriate *modulo* flip-flop divider circuit is required.

Pulse-Width Modulation Motor Control

Small DC motors are used extensively in the electrical and electronics fields, but they have one common shortcoming. As the applied voltage to a DC motor is reduced to reduce the speed (which is necessary for many applications), the motor loses its torque. At low voltages, DC motor operation becomes erratic, or it might fail to start at desired speeds. The circuit in Fig. 13-12 solves that problem.

Figure 13-10
CMOS bounceless
switch.

IC1 = NTE4049
PIN 1 = +V
PIN 8 = GND

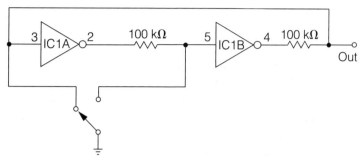

Figure 13-11
High-stability crystal
timebase.

IC1 = NTE4011B

Figure 13-12
PWM small-motor
speed control.

IC1 = NTE4049
D1,D2 = IN4148   C1 = 0.1μF, Mylar
D3 = IN4004   P1 = See text

The principle behind the operation of this type of motor control is called *pulse width modulation* (PWM). The *amplitude* of the pulses applied to the motor always stays the same, but this circuit varies the *duty cycle* (on time versus off time) of the pulses. The motor cannot respond to the rapid voltage changes, but it integrates the duty cycle into an equivalent "power" value. The speed of the motor will vary in accordance with any changes in duty cycle, but the torque characteristics are much improved.

Q1 is a Darlington transistor providing the additional current gain needed for this circuit. IC1C and IC1D are paralleled to provide more current drive for Q1. Any type of DC motor, drawing up to 2 amps, can be speed-controlled. Depending on certain variables, you might have to experiment with different values of P1 for best results. Start with a 10-Kohm potentiometer as a baseline.

The NTE4049 is a CMOS device, so use appropriate precaution during construction.

14

Computers

In today's world, the moment you hear the word *digital*, you tend to think of a computer. Computer advocates have sensationalized the minds and imaginations of many people with terms like "electronic brains," "artificial intelligence," and "cyberspace." Science fiction writers have had a field day with computers starting as far back as the 1950s. So, in (not virtual) reality, what is a computer?

A *computer* is a machine capable of performing analytical calculations, storing data, and controlling or monitoring redundant operations. Computers are very useful tools, because they can be programmed to do one thing unerringly at very high speeds. However, to even try to compare the largest computers, with the lowest forms of human intelligence, is ludicrous. At this very moment, as you are reading this text, your brain is receiving over 100 million bits of information per second. These "bits" of brain information, however, cannot be compared to the bits, or *binary digits*, of computer information. A bit of computer data is simply a 1 or a 0, but a bit of human brain information is complex electrochemical data, which has an almost infinite variation.

How a Computer Works

As Fig. 14-1 illustrates, a computer can be subdivided into four basic sections:

- Input
- Output
- CPU (central processing unit)
- Memory

For any logical system to be called a computer, it must meet five essential criteria:

- It must have input capability.
- It must have data storage capability.
- It must be capable of performing analytical calculations.
- It must be capable of making logical decisions.
- It must have output capability.

Referring to Fig. 14-1, the *input section* accepts information from a selected input device, and converts it into digital information, which can be understood by the CPU. The CPU controls the timing and data selection points involved with accepting inputs and providing outputs by means of the *input/output bus*. The CPU also performs all of the arithmetical calculations and memory storage/retrieval operations. The *memory address bus* defines a specific area in the memory to be worked on, and the *memory data bus* either stores or retrieves data from that specific

Figure 14-1
Block diagram illustrating the basic sections of a computer.

location. The *output section* accepts the digital information from the CPU, converts the information into a usable form, and routes it to the appropriate output device.

The analytical part of a computer is called the *central processing unit* (CPU). In the not-too-distant past, CPUs were relatively large printed circuit boards containing many integrated circuits. In modern computers, CPU printed circuit boards have been replaced with a single VLSI (very large-scale integration) chip called a *microprocessor.*

A microprocessor can be broken down into two main parts: the arithmetic logic unit (ALU) and the read-only memory.

The ALU controls the logical steps and orders for performing arithmetical functions. It interacts with the ROM to obtain instructions for performing redundant operations. The ROM also contains instructions pertaining to start-up, and to power-loss conditions.

The digital information sent to the microprocessor can also be broken down into two main types: *data* and *instructions.* These two types of digital information work with the two main parts of the microprocessor to perform all analytical operations.

Digital information is received at the microprocessor in the form of digital words, called *bytes.* In modern computers, a byte consists of 16 or 32 bits of data. A *bit* (binary digit) is simply a logical level; it can have only two states, either high or low. All 16 bits, constituting the byte, are applied to the microprocessor simultaneously. If the byte of information is an *instruction* word, it will be interpreted by the instruction code residing in the ROM section of the microprocessor. Instruction words tell the microprocessor what mathematical operations to perform on the data also being received.

Digital information is transferred throughout a computer system by means of "buses" (see Fig. 14-1). You might think of a *bus* as a communication pathway. Because bytes of information usually contain 16 bits, modern computer buses are either 16 or 32 electrical lines, with each line carrying one bit. This results in the bus being capable of transferring complete bytes of instruction and data information simultaneously.

Modern microprocessors contain numerous *registers,* or temporary memory cells. Mathematical operations are performed by shifting data words back and forth through the registers. Other registers serve a variety of purposes, such as storing instructions and keeping track of the current place in the program.

Basically, a computer is only capable of adding, subtracting, and accumulating data. Because it is capable of performing these simple operations at amazingly high speeds, complex mathematical calculations can be broken down into simple steps which the computer can then calculate. For example, a computer actually multiplies by *redundant addition* of the same number. In solving a multiplication problem such as 5×10, the digital form of the number 10 is placed in an accumulator register. The computer continues to accumulate (or add) tens in the same register until 5 tens have been added together. Of course, adding 5 tens together, or multiplying 10 by 5, provides the same answer. Division is performed by *redundant subtraction* in the same manner. By shifting and manipulating digital information in the various registers, which you might think of as "scratchpad" memory, a computer can perform virtually any mathematical calculation. The important point to recognize is that a computer does not perform calculations in the same manner as a human being does. In reality, its operation is much more similar to an ancient calculating instrument called an *abacus.*

With computers, timing is crucial. In order to maintain precise timing of all functions, a computer must have an *internal clock,* or oscillator, to provide timing pulses to all of the sections simultaneously. In this way, exact synchronization can be maintained. Because the frequency of this clock controls the ultimate speed of the computer, it is desirable to set this frequency as high as possible, while still maintaining reliable operation. Modern computers have speeds as high as 1.5 GHz.

The smallest single operation performed by a computer is the *machine cycle.* It consists of two stages: the fetch cycle and the execute cycle. During the *fetch cycle,* the processor fetches an instruction from memory. Then, during the *execute cycle,* the computer performs some action based on the content of that instruction. The processor knows which instruction to go to next from the address stored in the program counter. The *program counter* always contains the address of the next instruction. When computer programs (called *software*) are written, they are arranged in a sequential order. The program counter keeps track of the next instruction to be acted upon by simply incrementing (by 1) for each machine cycle. So, in essence, a computer follows a set of sequential instructions (called a *program*) by counting from beginning to end. When the end of the program is reached, it starts over again. When the computer is continually repeating the sequential steps in the program, the program is said to be "running."

Hardware, Software, and Firmware

The physical pieces making up a computer system are called *hardware.* Common examples of computer hardware include keyboards, videomonitors, floppy-disk drives, compact-disk (CD) drives, hard-disk drives, modems, power supplies, printed circuit boards, and electronic components. In other words, hardware is almost everything you can physically touch.

In contrast, *software* is all of the application-oriented programs used with or contained within a computer system. *Software programs* can be changed, manipulated, or customized at will. Depending on how the software programs are stored, they can be lost when operational power is removed.

Internal computer programs are also stored in the form of firmware. *Firmware programs,* usually referred to as *read-only memory,* are general-purpose programs that cannot be modified during normal computer operation. Nor when the operational power is removed. Examples of firmware include power-up programs, instruction code sets, and other computer routines which are common to all main, functional computer applications.

Memory and Data Storage

There are many different mediums for storing digital information. In this section, a few of the more common storage methods will be discussed.

Most RAM memories are "volatile," meaning that the information is lost when operational power is removed. Basically, RAM memories consist of many thousands, or millions, of flip-flops. Each individual flip-flop can "remember" one bit (i.e., latch into a high or low status). Programs are pulled out of a "nonvolatile" memory system (such as a hard-disk drive, a compact-disk drive, or a floppy disk) and loaded into the RAM memory, from which the computer actually runs the programs.

Technically speaking, hard and floppy disks are randomly accessible, and they are *nonvolatile* (meaning that the digital information is not lost at power down). However, colloquially speaking, they are not classified as RAM memory.

ROM can take on many forms. Some IC ROM chips can be programmed only one time. These types of chips are simply classified as ROMs. Other types of ROM chips can be physically removed from a computer mainframe and reprogrammed, using special ROM programmers. They are called *programmable read-only memory* (PROMs), *electrically alterable read-only memory* (EAROMs), and *erasable programmable read-only memory* (EPROMs), and are also known by the acronym *UVROMs* (ROM chips that can be erased by exposure to strong ultraviolet light and reprogrammed). The newest type of ROM memory is called *CD-ROM*. CD-ROMs can store tremendous quantities of digital information for their size, and they are randomly accessible.

Older computer systems used magnetic tape for operational data storage. Although some modern computer systems use magnetic tape for backup storage (permanent storage of important programs or data, in case of accidental destruction of the original), magnetic tape is no longer used as an operational storage medium because the data are not randomly accessible.

Input/Output Devices

Common examples of I/O (input/output) devices are keyboards, video monitors, printers, and modems (modulator-demodulators). All I/O devices are connected to the computer by means of ports. A *port* is just another name for a plug or a connector.

There are two main types of I/O ports used to connect the computer to the outside world: parallel and serial. *Parallel ports* have many parallel lines to enable data to be sent (or received) at a rate of 16 (or 8) bits at a time. *Serial ports* send (or receive) data on only one line, one bit at a time. Generally speaking, printers and data acquisition systems typically use parallel ports, but keyboards and modems use serial ports.

Computers communicate with I/O ports by means of *addressing*. Each I/O port has a specific code, or address, that is applicable to it alone.

When an I/O port "sees" its specific address on the I/O bus, it will activate for communication with the microprocessor. The microprocessor will "look" at the various I/O ports as it is instructed to do so by the running program. The functions of the I/O ports are defined by firmware or software.

Programming Computers

Generally speaking, computers can be programmed at three different levels. If you program the microprocessor directly using binary number commands, the process is called *machine-language programming* (MLP). This method is difficult and cumbersome, because the 16-bit digital commands are not easily recognizable to a human.

The next programming level above MLP is called *assembly-language programming*. When using assembly language, each binary number command is replaced with an easier to understand *mnemonic* (pronounced "nih-monic"). For example, instead of programming with commands like 0011010010011011, which is MLP, programming is performed with mnemonics like ADD A,B, which instructs the microprocessor to add the contents of the A and the B registers.

For long, complex programs, various high-level languages are used. Common examples are Pascal, FORTRAN, APT, and many others. High-level program languages are easy to use because they are very much like human language. However, after the program is written, it must be converted into MLP by a special computer program called a *compiler*.

Computer Processing of Analog Signals

When a computer needs to examine (or to output) a continuously variable signal (referred to as *analog* or *linear*), the analog signal must be converted into digital "words" before the microprocessor can understand it. Special *data acquisition I/O modules* used for this purpose are called *digital-to-analog* (D/A) *converters* or *analog-to-digital* (A/D) *converters*. A/D converters change analog input levels into equivalent digital numbers, understandable to the microprocessor. In contrast, D/A converters change digital words, from the microprocessor, into equivalent voltage or current outputs.

The process of converting analog levels to digital words, and digital words to analog levels, is called *digitizing*. To digitize an analog level, the A/D converter actually divides the level into thousands of incremental "steps," or pieces. The incremental level of the individual steps, which ultimately defines the accuracy of the conversion, is called the *resolution*. D/A converters simply reverse the process.

Computers as Electronic Tools

If you are considering the electrical and electronics fields as a long-term, ongoing hobby or occupation, what would you classify as being the most important and essential tool in achieving your individual goals (aside from your natural tools and talents, such as your mind, eyes, hands, etc.)? Okay, if you have trouble with that question, let's get specific. What is the most important electronic tool for designing circuits? Or, what is the most important tool for constructing circuits and projects? Or, what is the most important tool for advancing your skills and knowledge? Personally, I would have to answer "a computer" to all of these questions.

Today, there exists a new generation of design and analysis programs that have revolutionized the conventional methods of building, testing, analyzing, and designing electronic circuitry. I'm referring to a family of powerful design and simulation programs that provide the user with the ability to draw high-quality schematics, model the individual components in software, and then "test" the completed circuit using software simulation. In addition to the speed and cost savings provided by such CAD programs, I can't fathom a more enjoyable method of progressing in knowledge and experience. It's almost like learning electronics while playing a videogame!

After loading the circuit design program into a computer, the user is provided access to parts bins containing almost any electronic component needed to construct almost any circuit. The user simply "clicks" on the desired part and "drags" the part down into a schematic construction area. Once the part is in place, the user connects the various components together in the form of a typical schematic. The individual components are assigned values or part numbers, which places an accurate "model" of the component's characteristics in software. When the circuit is completed, the user can connect any combination of power supplies or signal sources to make the circuit operational. The circuit can then be simulated by dragging down a variety of "virtual instruments" (voltmeters, oscilloscopes, etc.) and analyzing the circuit's operation just as if it had been constructed on a real workbench and tested with real analysis equipment. In addition, the user can run a series of high-level analysis functions on the test circuit, such as noise, distortion, Fourier analysis, AC analysis, and many others.

I advocate circuit design programs for beginners in electronics, as well as experienced professionals (however, all of the professionals I know are already using such programs, and have been for a considerable time). For the beginner or hobbyist, such programs provide an efficient means of experimenting with simple circuits without the time and expense involved in literally building them. In addition, circuit design programs contain a parts inventory that is typically much more extensive than a hobbyist or amateur could afford (or have the space for). Depending on the desired applications, circuit design programs can cost anywhere from about $200 to $1000, and, of course, you must have a computer system to run them on. When you compare this to the cost of a good, professional-quality DVM (about $200) or a good 100-MHz oscilloscope ($1000 to $3000), the price doesn't seem comparatively high.

The software circuit design system I use is the new Multisim and Ultiboard systems marketed by Electronics Workbench. I have discovered their SPICE (Simulation Program for IC Emphasis) version simulation engine to be very accurate, and I appreciate the user-friendliness of the programs. They also have a knowledgeable and friendly support group, which, from my perspective, is probably the most important attribute of any software program. (Contact information for Electronics Workbench is provided in Appendix B.)

15

More About Capacitors and Inductors

You may wonder why I placed this chapter regarding capacitor and inductor fundamentals so far along in this book. The reason is simple. My experience in teaching electronics has shown me that it is much easier for students to perceive the information covered in this chapter *after* they have undergone some practical experience in other related areas of electronics. If you have progressed through this textbook sequentially, you have already developed a good feel for a few of the more abstract concepts relating to reactive components, including such concepts as the use of transformers in power supplies, implementing capacitors for coupling and bypass functions, and recognizing the effects of reactive components on the phase of AC voltages and currents. At this point, you will probably find it relatively easy to solidify some of the more vague concepts into definite mathematical terms and relationships.

I suggest that you read and study the information contained in this chapter, and then go back to reexamine many of the circuits incorporating capacitors and inductors in the previous chapters. This process will help to not only clarify a few of the operational principles contained in previous circuits but also drive home the points covered herein.

Inductive Reactance

As stated previously, capacitors and inductors are reactive components. This means that they "react," or oppose, changes in electrical variables. For example, a capacitor opposes, or reacts, to a change in voltage. Inductors, on the other hand, "react" to a change in current flow. This reactive effect has a profound relationship to frequency.

As the frequency of the applied voltage to an inductor is increased, the inductor's opposition to AC current flow increases. This is because the amount of energy capable of being stored in the inductor's electromagnetic field (its inductance value) remains constant, but the time period of the applied AC voltage decreases. As the AC time period decreases, less energy is required from the inductor's electromagnetic field to oppose voltage alternations. For example, it would take 10 times the energy to oppose 100 volts for 10 seconds than it would to oppose the same voltage for one second. The same principle applies with an increase in the frequency (decrease in time period) of the applied AC. Another way of stating this basic principle would be to say that *the reactance of an inductor increases with an increase in frequency.*

This frequency-dependent opposition to AC current flow through an inductor is called *inductive reactance.* Inductive reactance (X_L), as are impedance (Z) and resistance (R), is measured in ohms. The equation for calculating inductive reactance is

$$X_L = 2\pi fL$$

This equation states that inductive reactance (in ohms) is equal to 6.28 (the approximate value of 2π) times frequency times the inductance value (in henrys). For example, the inductive reactance of a 1-henry coil with 60-hertz AC applied to it would be

$$X_L = (6.28)(60)(1) = 376.8 \text{ ohms}$$

If the frequency of the applied ac voltage were increased to 100 Hz, the inductive reactance of the same inductor would be

$$X_L = (6.28)(100)(1) = 628 \text{ ohms}$$

Notice that the inductive reactance increases as the frequency of the applied AC increases.

Capacitive Reactance

Capacitors, like inductors, have a frequency-dependent opposition to AC current flow called *capacitive reactance* (X_C). When an AC voltage is applied to a capacitor, it will charge and discharge in an effort to maintain a constant voltage.

As the frequency of the applied voltage to a capacitor is increased, the capacitor's opposition to AC current flow decreases. This is because the amount of energy capable of being stored in the capacitor's electrostatic field (capacitance value) remains constant, but the time period of the applied AC voltage decreases. As the AC time period decreases, it becomes easier for the capacitor to fully absorb the charge of each half-cycle. In other words, from a relative point of view, it would require ten times the capacity for a capacitor to charge for 10 milliseconds than it would for 1 millisecond. Therefore, capacitive reactance decreases as the frequency of an applied voltage increases.

The equation for calculating capacitive reactance (X_C) is

$$X_C = \frac{1}{2\pi f C}$$

This equation states that capacitive reactance (in ohms) is equal to the reciprocal of 6.28 (2π), times the frequency, times the capacitance value. For example, at 60 hertz, the capacitive reactance of a 10-μF capacitor would be

$$X_C = \frac{1}{(6.28)(60)(0.00001)} = \frac{1}{0.003768} = 265 \text{ ohms}$$

(Note that 0.00001 $F = 10$ μF.) If the frequency of the applied AC voltage were increased to 100 hertz, the capacitive reactance of the same capacitor would be

$$X_C = \frac{1}{(6.28)(100)(0.00001)} = \frac{1}{0.00628} = 159 \text{ ohms}$$

Notice that the capacitive reactance decreases as the frequency increases.

After examining the previous examples of inductive and capacitive reactance, it should become apparent that if an electrical circuit contained a combination of resistors, capacitors, and inductors, the overall *impedance* of the circuit would be frequency dependent also.

Calculating the impedance of a resistive/reactive circuit is not as easy as simply adding reactance and resistance values together. True impedance requires compensation for the phase difference between voltage and current. The higher-level geometric equations used for calculating impedance are included in Appendix A; but, depending on your personal interests and goals, you might never need to use them.

Capacitive and Inductive Comparison

This section is devoted to studying the comparative nature of inductors and capacitors. The particular qualities of each component can be more fully appreciated in this way.

Inductance	Capacitance
Voltage leads the current.	Voltage lags the current.
Tries to maintain a constant current.	Tries to maintain a constant voltage.
Time constant = L/R.	*Time constant* = RC.
As frequency increases, reactance increases.	As frequency increases, reactance decreases.
AC power dissipation is zero in a purely inductive circuit.	AC power dissipation is zero in a purely capacitive circuit.
Exhibits minimum reactance at DC.	Exhibits maximum reactance at DC.
Stores energy in an electromagnetic field.	Stores energy in an electrostatic field.

Combining Inductors and Capacitors

Figure 15-1 illustrates the effects of combining inductors in series and parallel configurations. Notice that the method used for calculating total inductance is the same method used for calculating total resistance. In a *series* arrangement, the individual inductance values are simply added together. In a *parallel* configuration, the individual reciprocal values are calculated, then added together, then the reciprocal of the total is solved.

Figure 15-2 illustrates how the opposite is true for capacitors. Capacitance values are summed in parallel; but in series, reciprocals must be used to calculate total capacitance. The most common purpose for placing capacitors in series is to increase the overall voltage rating. For example, two capacitors of the same capacity, with a voltage rating of 100 volts, will have a 200-volt rating when placed in series. However, as the equation in Fig. 15-2 indicates, their capacitance value will be halved.

Reflected Impedance

If 120-volt AC, 60-hertz power is applied to the primary of a power transformer with its secondary open, a very small current will flow through the primary winding. Only a small primary current will flow because the inductive reactance of the primary winding is typically very high. However, if a load is then placed on the secondary winding,

Figure 15-1
The effects of connecting inductors in series and parallel.

$$L_T = \cfrac{1}{\cfrac{1}{L_1}+\cfrac{1}{L_2}+\cfrac{1}{L_3}+...\cfrac{1}{L_n}}$$

$$L_T = L_1 + L_2 + L_3 + ... L_n$$

Figure 15-2
The effects of con-
necting capacitors in
series and parallel
configurations.

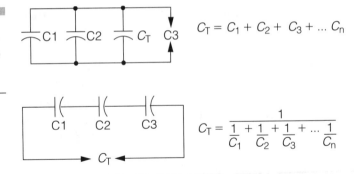

$$C_T = C_1 + C_2 + C_3 + \dots C_n$$

$$C_T = \cfrac{1}{\dfrac{1}{C_1} + \dfrac{1}{C_2} + \dfrac{1}{C_3} + \dots \dfrac{1}{C_n}}$$

the load impedance of the secondary will be reflected back to the primary winding. This effect is referred to as *reflected impedance.*

Reflected impedance is the effect that causes the primary current to automatically increase when the secondary current increases. Without it, a transformer could not function. Reflected impedance is calculated by multiplying the square of the turns ratio, by the secondary load. For example, assume that you have a transformer with a 10:1 turns ratio. Mathematically, a 10:1 ratio is the same as dividing 10 by 1, or simply 10. Therefore, the square of the turns ratio is 10 times 10, or 100. If a 5-ohm load is placed on the secondary of this transformer, the reflected impedance seen at the primary will be 5 × 100, or 500 ohms. This 500-ohm reflected impedance, seen at the primary, will be much less than the inductive reactance presented by the primary with the secondary open. Consequently, the primary current flow will automatically increase to compensate for the increased current flow in the secondary.

In the previous example, a "step-down transformer" was discussed. Reflected impedance works the same way in a "step-up transformer." For example, assume that you have a step-up transformer with a turns ratio of 1:10. Mathematically, this is the same as 1 divided by 10, or 0.1. The square of 0.1 is 0.01. If a 5-ohm load is placed on the secondary of this hypothetical transformer, the reflected impedance seen at the primary will be 0.01 times 5, or 0.5 ohm.

Resonance

You have learned that capacitors and inductors are reactive devices. As the applied frequency changes, the reactance values of inductors and capacitors will also change. However, their reactions to frequency changes are exactly opposite. This leads to some interesting effects when inductors and capacitors are combined in electrical circuits.

Consider the *series-resonant circuit* illustrated in Fig. 15-3. The source is a variable-frequency AC source allowing the user to vary the applied frequency throughout a wide range. This variable-frequency AC is applied to a series circuit containing some value of inductance and capacitance. Regardless of the AC frequency applied, both the inductor and the capacitor will pose some value of reactance.

If the applied AC source is of relatively low frequency, the inductor will present little opposition, but the capacitive reactance will be high. On the other hand, if the AC frequency is increased to a high value, the capacitor will now present less opposition, but the inductive reactance will increase significantly. In either of these two cases, very little circuit current will flow because one reactive device, or the other, will pose a high opposition.

As the frequency of the variable AC source is adjusted throughout its entire range, a specific frequency will be reached that will cause the inductive reactance to equal the capacitive reactance. At this specific point in the frequency spectrum, the circuit current will be the highest because it will not be blocked by either reactive component. From a mathematical viewpoint, the inverse characteristics of the inductive reactance and the capacitive reactance cancel one another, causing only the resistive element to remain in the circuit. This is called the *point of resonance.* Resonance occurs when capacitive reactance is equal to inductive reactance.

Figure 15-4 illustrates the effect on circuit current as the frequency of the applied AC is varied in the circuit shown in Fig. 15-3. Notice how the circuit current peaks at a specific frequency, but is substantially reduced at other frequencies above or below this point. The specific frequency causing the reactance values to equal each other is called the *resonant frequency.* Figure 15-3 shows the equation for calculating the resonant frequency when the inductance and capacitance values are known.

Figure 15-3
Series-resonant circuit.

f_r = Resonant frequency

$Z_{min} = f_r$

$$f_r = \frac{1}{2\pi\sqrt{LC}}$$

Figure 15-4
Current versus fre-
quency response of
Fig. 15-3.

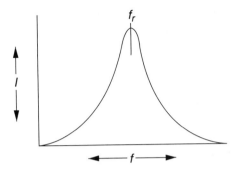

When a capacitor and an inductor are placed in series, the circuit current will always be highest at the point of resonance. Conversely, the circuit impedance at resonance is at its lowest point.

Consider the parallel-resonant circuit shown in Fig. 15-5. As can be
seen in this circuit, the inductor and capacitor are now in parallel.
Once again, the source is a variable-frequency AC source. Assume that
the variable-frequency AC source is adjusted to a low frequency. The
circuit current will be high because the inductor will present very little inductive reactance to the low frequency. This will allow the circuit
current to freely pass through the inductive leg of the parallel circuit.
If the AC source is adjusted to a high frequency, there will still be a
high circuit current because the capacitor will present very little capacitive reactance to the high frequency, allowing the circuit current to
freely pass through this parallel leg. In either of these two cases, the circuit current will be high because one reactive leg or the other will have
a low reactance.

As the frequency of the AC source is varied over a wide range, a specific frequency will be reached, causing the inductive reactance to equal
the capacitive reactance. At this frequency, the overall circuit current
will be at its minimum. This is the resonant frequency for the parallel
circuit. Any frequency above or below this point will cause the circuit
current to increase dramatically.

Figure 15-6 illustrates the effect on circuit current as the frequency is
varied over a wide range. Note that the circuit current is at its lowest
point at the resonant frequency, but it climbs on either side of this frequency. In a parallel-resonant circuit, the circuit impedance reaches its
highest point at the resonant frequency, thus causing the circuit current
to be at its lowest value. The equation for calculating the resonant frequency is shown in Fig. 15-5. Notice that it is the same equation as that
used for a series resonant circuit.

Figure 15-5
Parallel-resonant
circuit.

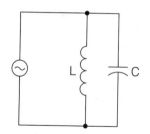

f_r = Resonant frequency

$Z_{max} = f_r$

$$f_r = \frac{1}{2\pi\sqrt{LC}}$$

Figure 15-6
Current versus fre-
quency response of
Fig. 15-5.

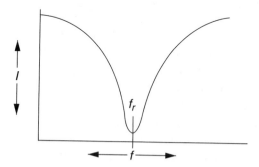

If a desired resonant frequency is known and you wish to calculate the needed inductance and capacitance values, the following *resonance equations* can be used:

$$L = \frac{1}{4\pi^2 f_r^2 C} \quad C = \frac{1}{4\pi^2 f_r^2 L}$$

where C = capacitance (in farads)
π = 3.14
f_r = resonant frequency (in hertz)
L = inductance value (in henrys)

Passive Filters

Previously, operational amplifiers were discussed as the primary "active" element in the design of active filters. In the vast majority of cases, active filters are preferred because of their small size, low cost (inductors can be large, expensive, and difficult to find), and ease of design and construction. However, active filters do have one drawback; they require operational power to function. You might run into some situations where a passive filter would be preferable. A passive filter will always attenuate the applied signal to some extent, but they do not require a battery or DC power supply for operation.

There are many different types of filters, depending on the needs and applications of the designer. However, on a fundamental level, there are only four basic filter types: low-pass filters, high-pass filters, bandpass filters, and band-reject filters. Essentially, the terms describe the basic action of the filters. *Low-pass filters* pass low frequencies, but reject (or block) high frequencies. *High-pass filters* are just the opposite, passing high frequencies while blocking low frequencies. *Bandpass filters* pass frequencies within a certain band, or range, blocking all frequencies either above or below the specified "passband" (i.e., the range of frequencies it is designed to pass). In contrast to bandpass filters, *band-reject filters* (often called "notch filters") block a specified range, or band, of frequencies, allowing all frequencies above or below the "notch" frequencies to freely pass.

With any of the previously described filter types, "ideal" performance is not achievable within practical means. Ideal operation of a low-pass filter, for example, would mean that all frequencies falling below the specified *cutoff frequency* (i.e., the frequency point that is the "dividing line" between low and high frequencies, relative to the design of the filter) would be passed without any attenuation, while all frequencies slightly above the cutoff frequency would be totally blocked (i.e., their amplitude would drop to zero). Such an ideal filter would exhibit a *crossover slope* of zero. In practical filter circuits, the crossover slope (i.e., the rolloff of signal amplitude around the cutoff frequency) is defined in terms of *decibels per octave*, which defines how rapidly the signal amplitude is boosted or attenuated relative to frequency. (Decibels were discussed in Chapter 8.) Typical filter circuits will exhibit crossover slopes in the range of 3 dB/octave or 6 dB/octave; some active filters achieve much higher crossover slopes, in the range of 12 dB/octave and higher.

Low-Pass Filters

The simplest type of low-pass filter consists of a single capacitor placed across, or in parallel with, a signal to be filtered. An example of this type of single capacitor filtering is illustrated in Fig. 12-12 (Chapter 12) with the 100-pF capacitor placed across the "audio input signal" connections. A single capacitor, however, is seldom classified as a filter.

In terms of their technical operation, power supply capacitors are a type of low-pass filters. In a typical raw DC power supply, the AC voltage from the power transformer secondary is rectified, converting it to a 120-hertz *pulsating DC* waveform (provided the rectifier is a full-wave rectifier). This pulsating DC rectifier output is applied to a filter capacitor, which charges to the peak of the DC voltage and converts the pulsating DC voltage to a smooth, continuous level of DC voltage. Assuming that you were utilizing a 1000-μF filter capacitor in a hypothetical power supply, consider the capacitive reactance (X_C) of the filter capacitor to the *AC component* of the pulsating DC:

$$X_C = \frac{1}{6.28 \ (120 \ \text{Hz}) \ (1000 \ \mu\text{F})} = \frac{1}{6.28 \ (120) \ (0.001)} = \frac{1}{0.7536} = 1.32 \ \text{ohms}$$

The previous calculation shows that the power supply filter capacitor exhibits 1.32 ohms of "reactive opposition" (i.e., capacitive reactance) to the 120-hertz *AC component* of the pulsating DC applied across it. Of course, the pulsating DC coming from the rectifier also contains a *DC component*. The frequency of DC is zero. Therefore, calculating the capacitive reactance of the *DC component*, we have

$$X_C = \frac{1}{6.28 \ (0 \ \text{hertz}) \ (1000 \ \mu\text{F})} = \frac{1}{6.28 \ (0) \ (0.001)} = \frac{1}{0} = \infty$$

(where ∞ is the symbol for infinity). The previous two calculations show that a 1000-μF capacitor exhibits only 1.32 ohms of capacitive reactance to a 120-hertz AC frequency, but exhibits an infinite resistance to DC (remember, a good capacitor cannot pass DC, so its opposition to DC would have to be infinite). Another way of stating this same phenomenon would be to say that *the filter capacitor allows the low-frequency component* (DC) *to pass to the power supply load, but blocks the AC component* (120-hertz ripple) *from passing on to the load*. The AC component is "blocked" in the sense that it is shorted to circuit common through the filter capacitor, which, for all practical purposes, attenuates it to such a low level that it can be considered negligible. (If any of this is confusing, you should review the material contained in Chapter 5.)

It can be accurately stated that filter capacitors in DC power supplies are low-pass filters; passing the zero-frequency DC on to a load while blocking the higher-frequency AC ripple component. As a matter of convention, however, they are not normally regarded from that perspective.

Figure 15-7a illustrates a common method of designing a low-pass filter. A resistance (R) is placed in series with the input signal and output signal. A capacitor (C) is placed in parallel with the output signal,

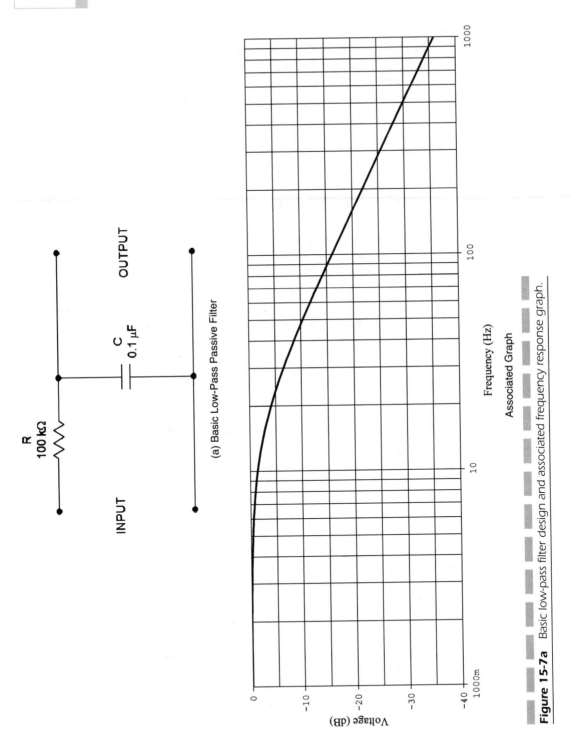

Figure 15-7a   Basic low-pass filter design and associated frequency response graph.

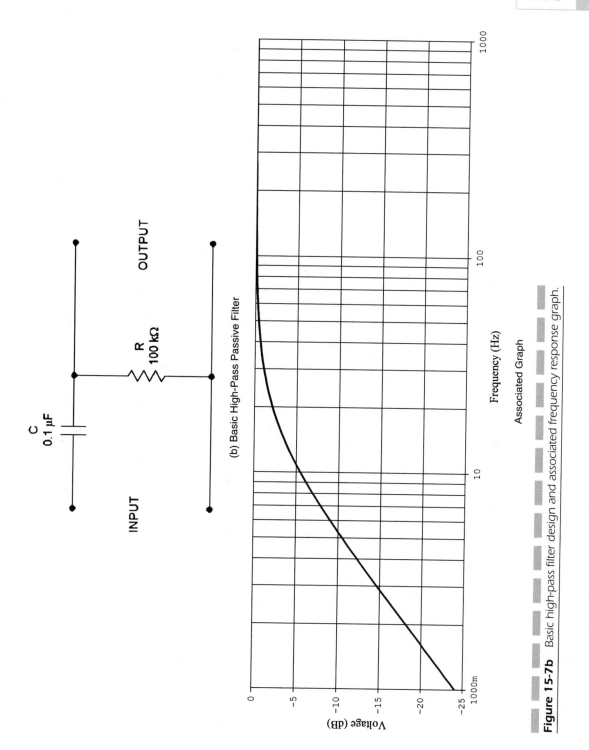

(b) Basic High-Pass Passive Filter

Associated Graph

Figure 15-7b   Basic high-pass filter design and associated frequency response graph.

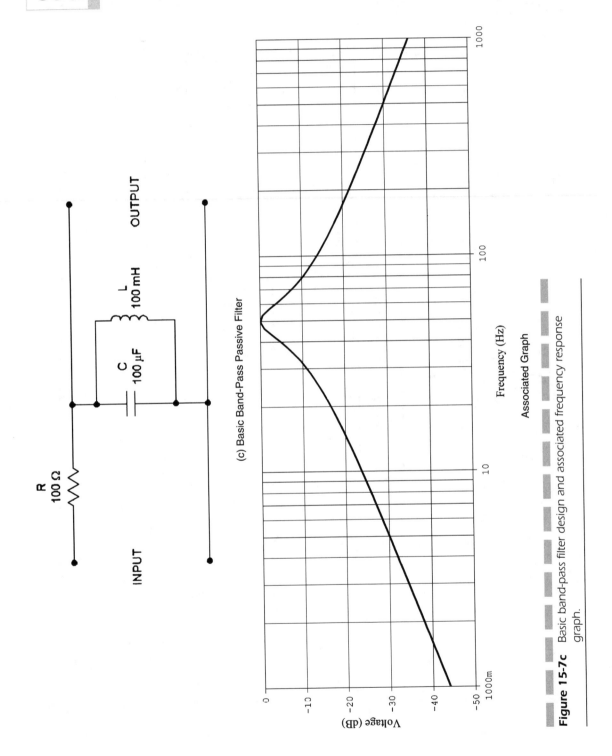

(c) Basic Band-Pass Passive Filter

Associated Graph

Figure 15-7c Basic band-pass filter design and associated frequency response graph.

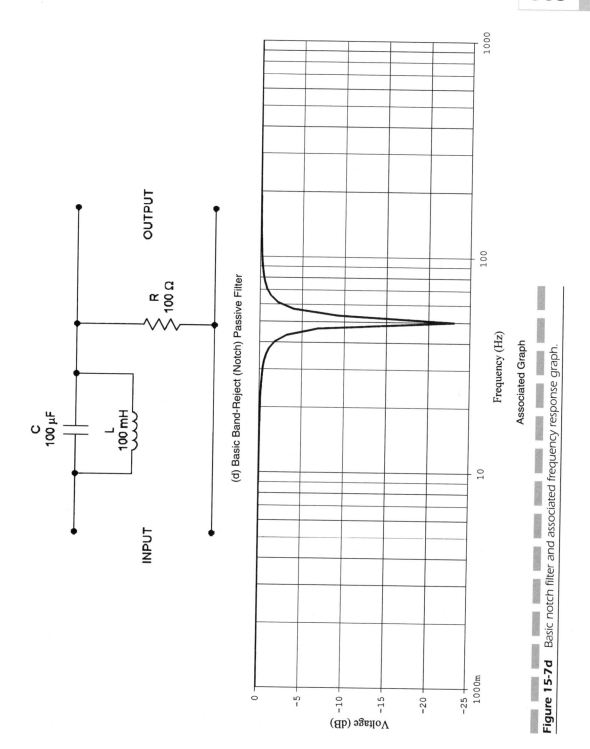

(d) Basic Band-Reject (Notch) Passive Filter

Associated Graph

Figure 15-7d Basic notch filter and associated frequency response graph.

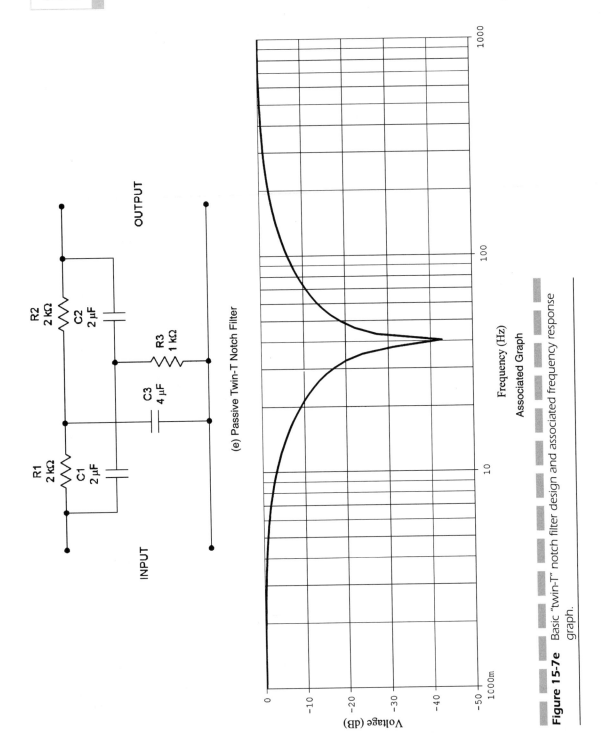

(e) Passive Twin-T Notch Filter

Associated Graph

Figure 15-7e Basic "twin-T" notch filter design and associated frequency response graph.

so that the output signal is actually the voltage drop across the capacitor. In essence, these two components form an *AC voltage divider*. At lower frequencies, the capacitive reactance of the capacitor will be very high in comparison to the resistance value of the resistor. Consequently, low frequencies will be "dropped" across the capacitor, and the output signal will be almost equal to the input signal in amplitude. However, at higher frequencies, the capacitive reactance will decrease, causing an increasingly greater percentage of the input signal to be dropped across the series resistance, attenuating the output signal greatly. In other words, low frequencies will be passed, but higher frequencies will be blocked.

As stated previously, the crossover slope of a passive filter is not abrupt, but rather a smooth rolloff transition from high amplitude to low amplitude. The associated graph for Fig. 15-7a illustrates the actual response of the circuit illustrated in Fig. 15-7a with the component values shown. The horizontal axis (abscissa) of the graph represents frequency, starting at 1 hertz (designated as 1000 mHz) and increasing to 1000 hertz. The vertical axis (ordinate) shows the voltage output of the filter in decibels, where "0" indicates the same amplitude as the input signal. (Incidentally, the response of the filter circuits illustrated in Fig. 15-7 was obtained by using computer simulation. Computer simulation is probably the best overall method of accurately designing more complex filter networks.)

Low-pass filter circuits are designed on the basis of their *cutoff frequency*. For the simple low-pass circuit of Fig. 15-7a, the equation for determining the cutoff frequency is

$$F_C = \frac{0.159}{RC}$$

For example, if you wanted to calculate the cutoff frequency of the filter illustrated in Fig. 15-7a, the calculation would be

$$F_C = \frac{0.159}{(100 \text{ Kohm})(0.1 \text{ } \mu\text{F})} = \frac{0.159}{(100,000)(0.0000001)} = \frac{0.159}{0.01} = 15.9 \text{ hertz}$$

If you observe the response graph for the Fig. 15-7a low-pass filter, you'll note that at approximately 15.9 hertz, the output signal has been attenuated by about 3 dB. This is considered the cutoff frequency of the Fig. 15-7a low-pass filter.

High-Pass Filters

Figure 15-7 *b* illustrates a common high-pass filter. As you will note, the high-pass filter is identical to the low-pass filter of Fig. 15-7 *a*, except the components have been reversed, with the capacitor (C) in series with the input and output signals, and the resistor in parallel with the output signal. Note that in this case, the output signal is the voltage drop across the resistor. Once again, the capacitor and the resistor form an AC voltage-divider network.

As might be expected, the filter circuit of Fig. 15-7 *b* provides a response that is opposite to the low-pass filter detailed earlier. At low frequencies, the capacitive reactance will be very high, causing most of the AC input voltage to be dropped across the capacitor, with very little appearing across the resistor or at the output. As the input frequency increases, the capacitive reactance will decrease, causing a greater portion of the AC input signal to appear across the resistor. Thus, this circuit will easily pass high frequencies, but block low frequencies.

Again, the design of a high-pass filter is based on the filter's cutoff frequency. The equation for calculating the cutoff frequency for the high-pass filter of Fig. 15-7 *b* is the same as that for calculating the cutoff frequency for the low-pass filter of Fig. 15-7 *a*. Since the component values are the same for both circuits, the calculated cutoff frequency will be the same, or about 15.9 hertz.

The response graph for Fig. 15-7 *b* illustrates the action of a typical passive high-pass filter. If you follow the horizontal frequency line until you reach the approximate location of 15.9 hertz, and then take this point vertically until it intersects the filter response line, you will note that the attenuation of the output signal is approximately 3 dB. Again, this is the cutoff point. If you compare the response graph for Fig. 15-7 *a* with the graph for 15-7 *b*, you will note that both output signals are attenuated by approximately 3 dB at about 15.9 hertz. However, their response is opposite to each other.

At this point, a few new terms can be interjected. Referring to the filter response graphs again, note that around the cutoff frequency the response is very curved and somewhat gradual. However, as the input frequency of the low-pass filter is increased above the cutoff point, the response line eventually straightens out, forming a flat attenuation slope line. The same response occurs with the high-pass filter as the input frequency is decreased below the cutoff frequency. In both cases, the flattened attenuation response represents a constant 6 dB/octave of output signal attenuation. As stated previously, the curved 3-dB/octave attenuation around the cutoff frequency is referred to as the *crossover slope*. The 6-dB/octave flattened attenuation response beyond the filter's cutoff frequency is called the *cutoff slope*. Filter circuits are classified and categorized according to the rate of attenuation of the cutoff slope. A *first-order* filter will have a cutoff slope of −6 dB/octave. A *second-order* filter will have a cutoff slope of −12 dB/octave, and this system continues on in −6-dB/octave increments. For example, a *third-order* filter will have a cutoff slope of −18 dB/octave, and so on.

Referring again to Fig. 15-7 *b*, imagine that R equaled the input circuit to a transistor amplifier stage. If that were the case, C would appear to be connected as the input coupling capacitor. In effect, that is exactly right! An input or output coupling capacitor to any circuit is actually a form of high-pass filter. In other words, the goal of a coupling capacitor is to block the DC bias (representing zero frequency), but pass the higher-frequency AC signals. Along this same line of thought, what other applications of capacitors have you examined throughout the course of this textbook that could be classified as a low-pass or high-pass filter? A *bypass* capacitor, placed in parallel with an emitter resistor, shorts the AC component to circuit common but retains the DC component of the emitter circuit. Therefore, it really functions as a type of low-pass filter, maintaining the DC while shorting the AC to circuit common. What about *decoupling capacitors*, detailed in Chapter 8? Again, they are a type of low-pass filter, shorting AC signals on the power supply lines to circuit common, but maintaining the DC component.

Understanding the action of low-pass and high-pass filters provides you with the informational tools to tailor the performance of various circuits according to your needs. For example, suppose that you would like to place a coupling capacitor at the base of a common-emitter amplifier stage, but you don't know what value of capacitor would be appropriate. Referring to Fig. 15-7 *b*, you would consider resistor R to be the input impedance of the amplifier stage, and then use the equation previously provided to determine the value of capacitance needed for a cutoff frequency slightly lower than the lowest frequency that you wanted to amplify. Thus, you'd be assured of having the capability of amplifying even the lowest frequency that you anticipated applying to the circuit.

Bandpass Filters

As the term implies, a bandpass filter allows a narrow band (or range) of frequencies to pass, while attenuating all frequencies either above or below the *passband.* There are many different ways and methods to construct bandpass filters, but the only two most common methods will be detailed in this section.

The most obvious method of creating a bandpass filter is to combine the actions of a low-pass filter and a high-pass filter into a single-filter network. For example, let's say that you connected a low-pass filter having a cutoff frequency of 200 hertz in series with a high-pass filter having a cutoff frequency of 100 hertz. If the input voltage were applied to the low-pass filter, all frequencies up to 200 hertz would be passed on to the high-pass filter. However, the high-pass filter would only pass frequencies above 100 hertz, so the final output of the filter network would be a passband from 100 to 200 hertz. All other frequencies (in theory) would be blocked.

The methodology of combining low-pass and high-pass filters to create bandpass filters is a common technique with operational amplifier-based "active" filters. Active filters using op amps as the active device exhibit very little interaction with each other, so low-pass and high-pass filters can be easily combined with predictable results. However, because of problems and complexities involving the interaction of internal impedances, this method is not very common when working with passive filter designs.

Figure 15-7c illustrates a passive bandpass filter that is relatively simple and highly practical. The capacitor and inductor form a parallel-resonant circuit. As you recall from the earlier discussion of resonant circuits in this chapter, a *parallel-resonant circuit* will exhibit maximum impedance at its frequency of resonance (theoretically, the impedance rises to infinity). Since the resistor and the resonant circuit are in series, the maximum voltage drop across the resonant circuit will occur at its point of resonance. At frequencies above or below the point of resonance, much of the input signal will be dropped across the resistor, since the resistor and the resonant circuit form another type of AC voltage divider. Obviously, since the output voltage is the voltage across the resonant circuit, the maximum output signal will occur at the frequency of resonance.

Regarding bandpass filters, there are two primary performance variables to consider: the *bandwidth* (BW) and *center frequency* (F_c). Referring to the response graph for Fig. 15-7c, note the typical response of a bandpass filter. The center frequency (F_c) is the frequency where the amplitude response "peaks." This frequency, of course, is the resonant frequency of the resonant circuit. *Bandwidth* defines the total range of frequencies above and below the center frequency that are not attenuated by more than −3 dB.

One of the nice attributes of the Fig. 15-7c bandpass circuit is that it is easy to design, and you already have the equations needed to calculate the variables. Interestingly, it works out that the equation for calculating the bandwidth (BW) of this circuit is the same equation for calculating the cutoff frequency (F_C) of the low-pass filter (i.e., the inductor is not included in the BW calculation). Therefore, the bandwidth of the circuit illustrated in Fig. 15-7 is

$$BW = \frac{0.159}{RC} = \frac{0.159}{(100 \text{ ohms}) (100 \text{ } \mu F)} = \frac{0.159}{0.01} = 15.9 \text{ Hz}$$

The center frequency of the Fig. 15-7c bandpass filter is determined by calculating the resonant frequency of the resonant circuit (i.e., the resistor is not included in the F_C calculation). Therefore

$$F_C = \frac{1}{(6.28) \sqrt{LC}} = \frac{1}{(6.28) \sqrt{(100 \text{ mH}) (100 \text{ } \mu F)}} = \frac{1}{(6.28) (0.00316)} =$$

$$\frac{1}{0.0198} = 50.39 \text{ Hz}$$

The previous two calculations tell you that the center frequency of the bandpass filter will be at about 50 hertz, and the passband (the total range of frequencies considered to be passed) will be about 15.9 hertz wide. If you observe the response graph for Fig. 15-7c, you will note that the peak amplitude (F_C) occurs at approximately 50 hertz, and if you go approximately 8 hertz above or below the center frequency (i.e., half of 15.9 hertz is approximately 8 hertz) the output signal amplitude drops off by about −3 dB.

Band-Reject Filters

As the name implies, band-reject filters are basically the opposite of band-pass filters. They block a narrow band, or range, of frequencies while passing all other frequencies above or below the center frequency point. Band-reject filters are often called *band-stop filters* or "notch filters."

As you may recall, a bandpass filter can be created by combining a low-pass filter and a high-pass filter in a series combination. Band-reject filters can be created by combining a low-pass and high-pass filter in a "parallel" configuration. For example, suppose that you connected a low-pass filter with a cutoff frequency of 100 hertz in parallel with a high-pass filter with a cutoff frequency of 200 hertz. Since the two filters are in parallel, the input signal would be applied to the input of both simultaneously. All of the input signal frequencies up to 100 hertz would be freely passed by the low-pass filter and appear at the output. Likewise, all input frequencies above 200 hertz would be freely passed by the high-pass filter and appear at the output. However, there would be a "notch" formed in the frequency response, occurring between the frequencies of 100 and 200 hertz. For example, a frequency of 150 hertz would be too high to be passed by the low-pass filter and too low to be passed by the high-pass filter.

As in the case of bandpass filters, active op-amp-based low-pass and high-pass filters are often combined to create practical band-reject filters, but this method is seldom used with common passive filter designs. As you may have guessed, one method of making a band-reject filter would be to convert the parallel-resonant circuit of Fig. 15-7c into a series-resonant circuit. Since a series-resonant circuit exhibits maximum impedance at all frequencies except the resonant frequency, the output of Fig. 15-7c would be maximum at all frequencies except those close to the resonant frequency. At this point, the impedance across the resonant circuit (i.e., the output) would drop to a low value, causing most of the input signal to be dropped across R. Consequently, a "notch" in the output frequency response would occur.

Another method of causing the same type of response is illustrated in Fig. 15-7d. This band-reject filter is identical to the bandpass filter of Fig. 15-7c, except the output is taken across the resistor instead of the parallel resonant circuit. The output frequency response graph for this circuit illustrates how a narrow band of frequencies is attenuated at the resonance frequency of the parallel circuit, while all other frequencies either above or below this frequency are passed.

As you may have guessed, the performance variables and the calculations for the Fig. 15-7d band-reject filter are identical to the bandpass filter of Fig. 15-7c. The bandwidth is calculated in the same way, but it refers to the band of frequencies that are blocked rather than passed. All frequencies attenuated by more than -3 dB are considered to be blocked. Likewise, the center frequency will be the resonant frequency of the parallel resonant circuit (calculated the same way as for the bandpass filter of Fig. 15-7c).

Twin-T Filters

The twin-T filter of Fig. 15-7e is another type of passive band-reject filter, but it is much more commonly used than the type of passive band-reject filter illustrated in Fig. 15-7*d*. The reasons for its popularity are its easy construction and the absence of any inductors (inductors are often large, bulky, and difficult to obtain with the inductance values required for specific applications).

The performance variables for twin-T filters are the same as those for any other band-reject filter (i.e., bandwidth and center frequency have the same meanings and definitions). However, the design calculations are different.

The calculations for a twin-T filter are as follows (refer to the Fig. 15-7*e* illustration):

$$R1 = R2$$

$$C1 = C2$$

$$R3 = \frac{R1}{2}$$

$$C3 = 2\,(C1)$$

$$F_C = \frac{1}{6.28\,(R1)\,(C1)}$$

Referring to Fig. 15-7*e*, with the component values shown, the center frequency would be

$$F_C = \frac{1}{6.28\,(2\,\text{Kohm})\,(2\,\mu\text{F})} = \frac{1}{6.28\,(2000)\,(0.000002)} = \frac{1}{0.02512} = 39.8 \text{ hertz}$$

As illustrated in the actual response graph for Fig. 15-7*e*, the center frequency "notch" occurs at approximately 40 hertz (i.e., very close to 39.8 hertz). Unfortunately, with passive types of twin-T filters, the designer has no control over bandwidth. However, the typical response is illustrated in the response graph.

The *Q* Factor

An important specification when dealing with filter circuits, resonant circuits, capacitors, inductors, and various other types of reactive circuitry is called the *Q factor*, which is short for "quality." The concept of "quality" gives the impression that a circuit with a high *Q* is somehow "better" than a circuit with a low *Q*. In reality, this isn't necessarily true. Depending on the application, low-*Q* circuits are sometimes better suited than high-*Q* circuits.

The quality factor (Q) is simply a performance parameter, and it can be associated with a wide variety of circuits and components in a variety of ways. For example, in Chapter 3, you were provided an equation for determining the quality of an inductor:

$$Q = \frac{L}{R}$$

where L is the inductance value (in henrys) and R is the DC resistance of the coil winding (in ohms).

Likewise, with reactive components (i.e., capacitors and inductors), Q is often expressed as the ratio of the *reactance* (at a specified frequency) compared to the component's DC resistance. With inductors, the equation is

$$Q = \frac{X_L}{R}$$

where X_L is the inductive reactance (in ohms) and R is the DC resistance (in ohms).

For capacitors *in a series circuit,* the equation is

$$Q = \frac{X_C}{R}$$

where X_C is the capacitive reactance (in ohms) and R is the DC resistance (in ohms).

The quality factor Q is also a common performance parameter associated with bandpass and band-reject filters (both active and passive). The equation for determining the Q of such a filter circuit is

$$Q = \frac{F_C}{BW}$$

As you may recall, the bandwidth for the bandpass filter circuit of Fig. 15-7c was 15.9 hertz and the center frequency was 50.39 hertz. Therefore, the Q of the Fig. 15-7c filter circuit is

$$Q = \frac{F_C}{BW} = \frac{50.39 \text{ hertz}}{15.9 \text{ hertz}} = 3.169$$

Generally speaking, Q provides a relative description of the sharpness of the frequency rolloff and the consequent range of the bandwidth. In other words, high-Q filters have a narrow bandwidth and very abrupt rolloffs, while low-Q filters have wide bandwidths and slow, gradual rolloffs on either side of the center frequency.

Circuit Potpourri

Having fun with filters and oscillators.

Proportional, Integral, and Differential Action

In the days of computers' infancy, "analog" computers performed mathematical functions using thousands of dual-vacuum-tube operational amplifiers. The "action" of these op amp circuits was defined in mathematical terms which are still commonly used today. Figure 15-8 illustrates a modern IC op amp, configured as a *proportional amplifier.* If you recognize this circuit as a simple noninverting amplifier, you're correct. The term *proportional* is simply a mathematical definition of a "linear" function.

Figure 15-9 illustrates an *integrator circuit.* Integrators produce an output that is proportional to the amplitude of the input multiplied by its duration. In mathematical terms, it is an "integral extractor." Although the mathematical function is probably of little use, the practical aspects of this circuit are very useful. First, it is a type of low-pass filter, and it works very well for many low-pass applications. But its most common use is for converting square-wave signals into "linearized" triangular waves, for audio synthesizers and tone generators. Triangular waves have a pleasing, mellow tone when used for audio applications and are easily converted to sinusoidal waves for other applications. Also, triangular

Figure 15-8
Proportional (linear) amplification.

Input

Output

$$A_e = 1 + \left(\frac{R_2}{R_1}\right)$$

Figure 15-9
Integrator.

$$R_3 = \frac{R_1 \cdot R_2}{R_1 + R_2}$$

$$R_2 = \frac{1}{F \cdot C_1}$$

F = Input fundamental frequency

$$C_1 = \frac{1}{F \cdot R_2}$$

$$R_2 \approx 10 \cdot R_1$$

waves are composed predominantly of "even harmonics," which makes them useful for function generator applications.

Note that in Fig. 15-9, the symbol F is defined as the fundamental frequency. The term *fundamental frequency* defines the frequency of a waveshape according to how often it repeats a full 360-degree cycle; exactly the same as you have learned for sine-wave AC voltages. However, every type of AC waveshape other than sinusoidal waveshapes contain "harmonics." *Harmonics* are multiples of the fundamental frequency added into the overall waveshape.

For example, theoretically speaking, a *square wave* contains all possible harmonics; however, the predominant harmonics are "odd" multiples of the fundamental frequency. By applying a square-wave input to an integrator, the odd harmonics are removed, leaving the "even" harmonics in the form of a *triangular waveshape*. By removing the even harmonics, a triangular waveshape becomes a *sinusoidal waveshape*, which is "pure," or without harmonic content. Waveshapes containing harmonics are referred to as *complex AC waveshapes*. Therefore, to be technically accurate, the frequency of a complex AC waveshape should be defined in terms of its fundamental frequency.

A simple *RC* network will provide a waveshape similar to a triangular waveshape from a square-wave input, but it will not be a high-quality triangular wave because a capacitor does not charge in a linear fashion. In contrast, the circuit shown in Fig. 15-9 will provide a highly linear triangular-wave output from a square-wave input. You can use this circuit as a low-pass filter, a square-to-triangle wave converter, or to "mellow out" the tone from many of the audio oscillator circuits illustrated in this book. R2 might (or might not) be needed, depending on the application and the operational amplifier used.

Figure 15-10 illustrates a *differentiator circuit*. Mathematically speaking, differentiators produce an output that is a derivative of the input. For most practical applications, you can think of a differentiator as a type of high-pass filter. Notice that the output of the differentiator in Fig. 15-10 with a square-wave input. The high-frequency "shifts" of the square wave are extracted, resulting in a series of dual-polarity pulses. Differentiators are useful in many types of pulse and timing circuits as high-pass filters, and they will also convert a triangle wave back into a square wave.

Square Waves Galore

IC operational amplifiers are easy to configure into *square-wave oscillators*. Figures 15-11 and 15-12 illustrate two common methods of accomplishing this end. Figure 15-11 has fewer components, but it requires a dual-polarity power supply. Figure 15-12 operates from a single supply. With the components illustrated, both will oscillate at about 2 to 3 hertz. By replacing the 2-μF capacitor with a 0.1-μF "cap," the frequency will increase to about 200 hertz; a 0.001-μF "cap" will set the frequency to about 10 kHz. The two feedback resistors and the capacitor constitute the frequency determining network components.

Depending on the op amp used, these circuits might not produce square waves with a sufficiently fast rise and fall times for use in digital circuits. If you want to use these circuits for digital applications, apply the output of the oscillator to the input of an inverter gate. Then, use the output of the inverter to interface with other digital circuits.

Figure 15-10
Differentiator.

$F =$ Input fundamental frequency

$$C_1 = \frac{1}{F \cdot R_2}$$

$$R_1 = \frac{1}{F \cdot C_1}$$

Figure 15-11
Op amp square-wave
generator.

Figure 15-12
Single-supply square-
wave generator.

Basic Tone Generator

Figure 15-13 illustrates a very versatile, general-purpose *tone generator.* Virtually any kind of transistor, resistor, or capacitor combination will work. P1 controls the frequency.

Where's the Fire?

Figure 15-14 is a variation of the circuit shown in Fig. 15-13. S1 should be a momentary push-button switch. When S1 is initially depressed, C1 begins to charge through R1. This applies an increasing bias to Q1, thus

Figure 15-13
General-purpose
tone generator.

Figure 15-14
Electronic siren.

causing the pitch of the oscillations to increase. When S1 is released, C1 begins to slowly discharge through R2 and the Q1 base, causing the pitch to decrease. The overall effect is that of a siren.

To make the operation automatic, S1 can be removed, and the top of R1 can be connected to one of the low-frequency, op amp square-wave generators illustrated in Figs. 15-11 and 15-12. The frequency of the square-wave generators might need to be reduced for optimum operation.

Building a Lab Function Generator

Figure 15-15 illustrates a very versatile square- and triangle-wave *function generator* for lab use or many other applications. It requires only two dual op amps, and it can be assembled on a small universal grid board. The square and triangle waves are independently amplitude-adjustable with low output impedances. The circuit is very forgiving of power supply variations and component tolerances. It is also easily modified for specific applications.

In Fig. 15-15, IC1A forms a comparator that is continually switched by the output of the IC1B integrator. Because the wiper of P1 provides the reference voltage, P1 acts as a symmetry control. The frequency of oscillation is controlled by P3, R1, and C1. IC2B serves as a buffer/attenuator amplifier with an adjustable DC offset controlled by P4 and IC2A.

When building the circuit, don't forget to connect the 12-V and −12-V power supply leads to their respective power input pins on the op amps (this is not shown on the schematic).

Figure 15-15
Square- and triangle-wave function generator.

After building the circuit and applying power, adjust p1 (trim-pot) for about 0.1 volts DC at pin 2 of IC1A. P2 controls the amplitude of the square-wave output. P3 controls the frequency, and P5 controls the output level of the triangle wave. P4 (trim pot) should be adjusted for a 0-volt DC output at pin 7 of IC2B.

Building a Percussion Synthesizer

The circuit illustrated in Fig. 15-16 is an extremely versatile circuit that can be used for a variety of applications. In its basic form, it is a twin-T *filter/oscillator,* with adjustable Q and frequency. If the Q is adjusted high enough, it breaks into oscillation; and thus it becomes a low-distortion *sine-wave oscillator.* If the Q is adjusted to "far below" the oscillation point, it is a *high-Q filter.* If the Q is adjusted to "just below" the point of oscillation, it is a *dampened waveform generator,* or percussion synthesizer.

To use the circuit as an oscillator: D1, R2, R1, and SW1 are removed. P2, the Q-adjustment potentiometer, is adjusted to the point of oscillation, and the frequency is adjusted with P1 and P3.

To use the same circuit as a high-Q filter, D1, R1, and SW1 are removed, and R2 is connected directly to pin 2 of IC1. The input signal is applied to the other end of R2. P2 is adjusted until all "ringing" (dampened waveforms) disappears when an input signal is applied. P1 and P3 are then used to "tune" the filter.

This circuit also forms the heart of a *percussion synthesizer* (electronic drums) in the dampened mode. There are so many possible variations to this basic circuit, I cannot possibly list all of them, but I trust your

Figure 15-16
Twin-T resonant frequency sine-wave oscillator (percussion synthesizer).

imagination will take over before you can try out the examples given in this context.

With the component values given, the circuit of Fig. 15-16 is the most basic form of the percussion synthesizer. The dual power supply voltages are not critical. They can be any level that is nominally used with operational amplifiers. SW1 should be a momentary, push-button switch. IC1 can be almost any type of internally frequency-compensated op amp (such as the common 741, NTE778A, and many others).

After building the circuit, connect the output to any type of audio amplifier. Set P1 and P3 in their (approximate) middle position, and adjust P2 until a constant tone is heard. Then, reverse the adjustment very slightly until the tone just dies out. Now, by pressing SW1, you should hear an electronic drum sound. By playing with the P1 and P3 values, the pitch of the drum should change, and even bell sounds can be produced.

You might also want to experiment with different triggering methods, depending on the type of op amp used. For example, try connecting the bottom side of SW1 to circuit common (instead of $-V$); or by disconnecting R1 from $+V$; or use a double-pole switch, and switch both. Use the method that provides the most "solid" initiating pulse.

To make a set of electronic bongo drums, refer to Fig. 15-17. One bongo is made by starting with two pieces of single-sided PC board material. They can be about the size illustrated in the diagram, or they might be cut into round pieces of suitable size, if so desired. Place one PC board on top of the other with the foil sides together (facing each other). Temporarily tape the boards together, and drill five small holes completely through both boards. Separate the boards, lay them foil side up, and form five blobs of solder over each hole on both of the boards (the holes are needed only to line up the solder contact points). Silver solder might be used to reduce the problem of oxidation on the surface of these switch "points." After soldering, lay the boards on a table, blob side up; lay a wide file on top of the blobs on one of the boards, and lightly file the tops until they are all *planar* (no one blob higher than another); then, repeat this process on the other board.

The final steps are to solder a piece of insulated hook-up wire to the foil side of each board to be able to substitute this assembly for SW1 in Fig. 15-17. A more professional-looking cable might be affected by substituting microphone cable, or audio coax (coaxial cable), for the hook-up wires. Using some thin pieces of foam rubber as spacers, position one board above the other (with the contact points facing each other) as illustrated. The contact points should not touch each other, but they

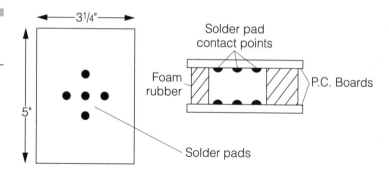

Figure 15-17
Physical construction
of a drum pad.

should be close enough so that a light pressure, or "tap," will cause them to make contact (the separation distance is exaggerated in the illustration for the sake of clarity).

This assembly can be held together by gluing both sides of the foam rubber to the PC boards (an epoxy "putty" does a good job). If a more rugged assembly is desired, holes can be drilled in the corners of the PC boards, and small bolts with locking nuts can be used to hold the boards together. This "large momentary switch" has a response similar to that of commercially available drum machine "pads." Obviously, two such assemblies (and two percussion synthesizer circuits) are needed for a "set" of bongos. One synthesizer circuit is tuned slightly higher than the other for a realistic bongo sound.

Figure 15-18 shows some other variations that can be incorporated into the basic percussion synthesizer. By removing R1 and SW1 (Fig. 15-16), a low-frequency, square-wave oscillator can be used to automatically operate the synthesizer. The oscillator output should be connected to the unconnected side of R2. Either one of the circuits illustrated in Figs. 15-11 and 15-12 will work very well for this application. Adjusting the frequency of the square-wave oscillator will set the "beat." A toggle switch could be installed to switch from "manual" (using SW1, or the drum pad assembly) to "automatic" (oscillator control).

A very impressive modification can be added to automatically tune the synthesizer while it is operating automatically. The effect is similar to that of a kettle drum; as the sound is decaying, it is also changing pitch. Referring to Fig. 15-16, P1 and P3 are replaced with two JFETs. The source and drain leads are connected to the same points as the two ends of the variable resistors, leaving the gate leads open. The orientation of the source and drain is not critical in this circuit. Referring back to Fig. 15-18, the output of the square-wave oscillator, which is automatically operating the synthesizer, is also applied to the input

Figure 15-18
Block diagram of
automatic operation,
automatic tuning per-
cussion synthesizer.

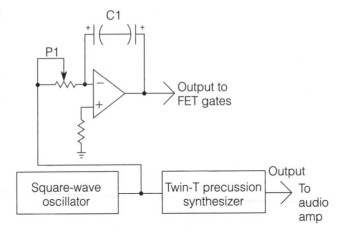

of an integrator. The triangle-wave output of the integrator is applied
to the gates of both JFETs (be sure that the voltage polarity is correct
for the type of JFET used). The two JFETs operate as voltage-controlled resistors, changing the pitch in perfect synchronization with
the beat. Synchronization is maintained because both the beat, and
the pitch shift, are being controlled by the same oscillator.

Continuing to refer to Fig. 15-18, you should experiment with a few
different values of P1 and/or C1 to explore the great variety of complex
sounds that can be generated from this circuit. A good starting point is
to use a 100-Kohm potentiometer for P1, and two back-to-back 10-μF
capacitors for C1 (as discussed previously, placing two electrolytic capacitors back to back creates a nonpolarized electrolytic capacitor).

There are some additional variations that could be incorporated into
the Fig. 15-18 circuit for some really spectacular effects, depending on
how far you want to go with it. The output of the integrator could be
applied to an inverting amplifier to change the direction of the pitch
shift. Another possibility is to increase the frequency of the square-wave
oscillator, and to use its output as the input to a digital counter. Two,
three, four (or more) Q outputs from the digital counter could be used to
initiate drum sounds from multiple synthesizer circuits. These multiple
synthesizer circuits would not beat at the same rate, but they would
beat "in time" because each would be beating at some "harmonic" division of the oscillator frequency.

I leave the remaining infinite number of possibilities to you, and to
your imagination and ingenuity.

CHAPTER 16

Radio and Television

Radios and television sets are electronic systems. Although it is a little out of context with the goals of this book to become involved with systems discussions, I believe that a well-rounded foundational knowledge of the electrical and electronic fields should include the basic operational theory behind radio and television. I also believe you will find this chapter interesting because radio and television have always been the backbone of the modern electronics industry.

The majority of the building blocks incorporated into radio and TV systems have already been discussed. This chapter will focus on the three main areas of theory needed to "fill the gaps" so that a final systems discussion can be more easily understood. These three topics are the modulation of signal intelligence, the transmission and reception of RF signals, and the cathode-ray tube.

Modulation

Modulation is the use of one electrical signal to "control" a primary variable of another. For example, if an audio signal voltage is used to control the "amplitude" of a carrier signal, the result is *amplitude modulation.*

It is important that you do not confuse "mixing" with "modulation." *Mixing* occurs when two (or more) signals are simply combined in a linear network. *Modulation,* however, requires one signal to "control" a variable of another; variables such as the amplitude of an RF signal [amplitude modulation (AM)], the frequency [frequency modulation (FM)], the pulse width [pulse width modulation (PWM)], the phase [phase modulation (PM)], or the pulse code [pulse code modulation (PCM)]. Unfortunately, the electronics industry has traditionally retained many circuit names that are incorrect in this regard. For instance, when you examine the actual circuit operation of many circuits labeled as a "mixer/oscillator," you will discover that it is really a "modulator/oscillator."

Strictly speaking, when two signals are mixed, they combine without the creation of any additional frequencies. When two signals are modulated, they are said to "beat" with each other, creating additional frequencies called "beat frequencies." If the two modulated signals are sinusoidal, the beat frequencies will be the sum and the difference of the original frequencies.

AM radio broadcast transmissions contain two signals of primary importance to the user: the carrier signal and the audio signal, or the program signal. The *carrier frequency* is the frequency to which the radio receiver is tuned for station selection. For example, the AM radio band (also referred to as the *medium-wave broadcast band,* or simply the *broadcast band*) is legally designated from 535 to 1605 kHz. If your favorite local radio station broadcasts on 830 kHz, this means that the carrier frequency being used for transmission is 830 kHz. The audio signal, or program, is riding on this carrier frequency.

Figure 16-1 illustrates an amplitude-modulated waveshape as it would appear when picked up by a radio antenna. Notice that the carrier frequency is much higher than the program signal riding on it. In actuality, there is not a literal program signal "on top of" the carrier. When the AM signal was broadcast, the program signal modulated the amplitude, or the level, of the carrier; this process formed an "envelope" of carrier amplitude, having the same shape as the program signal.

The "beat" frequencies, contained within the AM waveform of Fig. 16-1, will be the sum and the difference of the carrier and its program signal. For example, if the program signal were a constant 5-kHz tone, with a carrier frequency of 600 kHz, the beat frequencies would be 595 kHz (*difference frequency*) and 605 kHz (*sum frequency*). In a typical AM broadcast, the program signal will contain vocal and music information, making up a very wide range of frequencies. The highest frequency, of this range of frequencies, will determine the maximum separation of the beat frequencies from the carrier. For example, if the highest frequency in the program signal was limited to 1 kHz, then the beat frequencies would be 599 and 601 kHz. However, as you can see from the earlier example, when the highest frequency is limited to 5 kHz, the width (or distance from the carrier frequency) increases. The range of beat frequencies above (and below) the carrier frequency are called *sidebands*. The width of the sidebands is closely monitored at AM broadcast stations because, if they become too wide, they can interfere with adjacent stations.

Frequency modulation (FM) radio signals also have a carrier signal and a program signal. However, the program signal does not ride "on" the carrier frequency; it is contained "within" frequency variances modulated into the carrier signal. Because the program signal is not dependent on carrier amplitudes (as are AM transmissions), FM radio is largely immune to many forms of interference. Figure 16-2 illustrates an exaggeration of an FM-modulated waveform.

Figure 16-1
AM-modulated waveform.

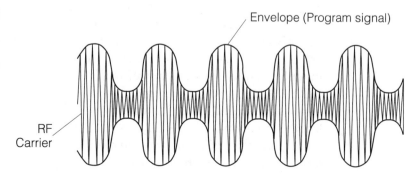

Envelope (Program signal)

RF Carrier

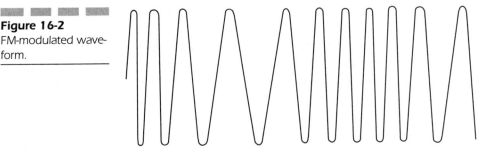

Figure 16-2
FM-modulated wave-
form.

Transmission and Reception of RF Signals

When AC current flows through a conductor, a moving electric field is developed around the conductor. This field, consisting of both electromagnetic energy (current-related) and electrostatic energy (voltage-related), is radiated perpendicular to the wire, similarly to visible light radiating outward from a light source. If the frequency of the AC current is within the radio-frequency spectrum (approximately 10 kHz to 30,000 MHz), the radiated energy is called *RF energy.* The most efficient transmission of RF energy occurs when the AC conductor is resonant at the frequency of the AC current. Special conductors designed to be resonant at specific RF frequencies are called *antennas* (either transmitting or receiving).

Think of the antenna as a transducer. A *transducer* is a bidirectional device that will take an electronic signal and convert it into another form of signal energy. For instance, piezoelectric elements can be used as speakers, translating the electronic signal into an audio emission. Conversely, they can also receive an audio signal, and then convert that signal into its electronic equivalent (a microphone). An antenna can receive an electronic signal, and then convert that signal into an electromagnetic wave that will radiate through space. It can also receive an electromagnetic signal, and then convert that signal into a series of electronic pulses. This conversion of energies can then be utilized, in a receiver, to extract the signal intelligence that had originally been impressed, or modulated, into it at the transmitter.

RF energy travels at the speed of light (186,280 miles, or 299,800,000 meters, per second). Every frequency of RF energy has a specific "wavelength" in meters, calculated by dividing 300,000 by the frequency in kilohertz. The most efficient transmission (or reception) of RF energy will take place when the antenna length is the same as the RF energy wavelength. At this length, the antenna becomes *resonant* to the RF energy. Slightly less efficient omnidirectional transmission and reception will also occur at even-harmonic subdivisions of the wavelength, such as at half-wave or quarter-wave lengths. Conversely, somewhat greater efficiencies are found in even-harmonic lengths that are greater than a wavelength. These variations of the efficiencies in an antenna occur because of the changes in the directional properties of the antenna at the various lengths. Because the wavelengths of the lower RF frequencies can be very long, half-wave or quarter-wave antennas are usually much more practical.

Now that some of the major definitions are out of the way, consider how an actual amplitude-modulated RF transmission takes place. To begin with, the RF transmitter must have an *oscillator* to produce the desired carrier frequency. The resonant frequency of the transmission antenna will be chosen to match the carrier frequency produced by this master oscillator. The signal from the oscillator proceeds through the *intermediate power amplifier* (IPA), sometimes referred to as the *driver.* The carrier frequency from the IPA is then applied to a *final power amplifier* circuit, for boosting to an efficient level for transmission.

The audio signal, having been amplified through the *speech amplifier* sections, then drives the high-powered modulator circuit. This modulator section might be connected into the plate supply (for tubular units), or into the collector or drain supply (for solid-state amplifiers). By varying the plate current, or the drain/collector current, the modulator impresses (or modulates) the program signal onto the carrier signal. This composite signal is then passed on to a final amplifier *tank circuit* for impedance matching, and then conveyed to the antenna. This system is often referred to as being a *high-level modulation.*

If the modulating signal is applied to the final amplifier through its cathode, or through one of its grids (for tubes), or through the source or emitter circuits (for solid-state amplifiers), the amount of modulating power is much less than would be required for high-level modulation. These methods are referred to as *low-level modulation.* As a general rule, tube-type amplifiers usually use high-level modulation, because of its greater efficiency. Solid-state power amplifiers usually are configured for low-level modulation and, in fact, often modulate several stages in phase. Any of these composite signals can then be applied to the input of a "linear" RF power amplifier, if further output power is needed.

RF oscillators and RF power amplifiers are essentially the same types of circuits that you have already studied. Some of the peculiarities of RF circuits involve the critical nature of parts placement. At very high frequencies, even slight amounts of stray capacitance or lead inductance can provide unwanted signal coupling or attenuation. High-frequency RF signals have a tendency to "ride on the surface" of conductors causing unexpected resistances on solid conductors (a phenomenon known as the *skin effect*). For these, and other reasons, RF circuits must be designed with more attention toward the physical details of a circuit, much more so than for lower frequency circuits.

Most high-power RF amplifiers are designed for class C operation in order to improve their efficiency. Class C amplifiers have a resonant *LC* circuit (called a *tank circuit*) between the amplifier and the load. A tank circuit can be thought of as a "flywheel" that smooths out the short pulses produced by a class C amplifier. The tank fills in the missing portions of the pulsed input signal. It produces a dampened sinusoidal waveform, at its resonant frequency, every time it is pulsed (just like the percussion synthesizer discussed in the previous chapter). Therefore, even though the class C amplifier only pulses the tank circuit, the carrier frequency output is a clean sine wave.

After the modulated RF signal is output from the RF power amplifier, it is coupled to the transmission antenna for broadcast into the atmosphere. The other job of the tank circuit is to match the output impedance of the amplifier to the characteristic impedance of the transmission cable and the antenna.

Lower-frequency radio waves ($\leq$300 kHz), called *longwaves*, have more of a tendency to travel along the curvature of the earth. Radio waves that propagate in this manner are called *ground waves*. Midfrequency radio waves (from 300 kHz to 3 MHz), called *mediumwaves*, travel by a combination of ground waves and *skywaves* (radio waves radiating upward, with a percentage of them bouncing off of the ionosphere layer of the upper atmosphere and returning to the earth). High-frequency (HF) radio waves (3 to 30 MHz), called *shortwaves*, are propagated mostly via skywaves, and they are commonly used for global communications. Very-high-frequency (VHF) and ultra-high-frequency (UHF) radio waves (30 to 300 MHz and 300 to 3000 MHz, respectively) behave more like a beam of light, traveling in a line-of-sight manner. For this reason, they are called *direct waves*.

Radio Receivers

The first stage of a radio receiver is, of course, the antenna. Receiver antennas are usually designed to be resonant at about the midpoint of the RF spectrum for the band being received. For lower frequency BCB reception (broadcast band; commonly and errantly referred to as AM), even a quarter-wave antenna is usually impractical to use (because of the long wavelength). Therefore, many mediumwave BCB antennas are "loaded" (reactive components are added) to make the antenna resonant at lower frequencies. This loading does somewhat decrease efficiency, however.

Most modern BCB antennas are actually part of the front-end tuned circuit (a tunable tank circuit). In a *loopstick antenna,* the antenna wire is wound on a ferrite rod, and placed in parallel with the tuning capacitor. The primary advantages of this system are that the antenna is very compact, and its *null* (direction of minimal signal reception) is very directional. This directional characteristic of the reception can be used to "null out" an unwanted station on the same frequency, while allowing the desired station to be clearly received. The end of the ferrite rod must point toward the offending transmitting station for maximum signal attenuation.

Virtually all FM receivers need an external antenna. Because of the higher frequency of the FM band, a half-wave FM dipole antenna is about 4.8 feet in length; a quarter-wave whip antenna is only about 29 inches.

After the radio signal is received by the antenna, this extremely small RF signal voltage needs to be amplified by an *RF amplifier stage* (some smaller, inexpensive radios omit this stage). Referring to Fig. 16-3, the RF amplifier stage contains a tunable capacitor (or a varactor) to allow it to amplify only the desired station to which it is tuned.

The amplified RF signal is then applied to a mixer stage. The mixer stage (technically, a modulator) modulates the amplified RF with a signal from the local oscillator. The frequency of this local oscillator is

Figure 16-3
Block diagram of a
superheterodyne
receiver.

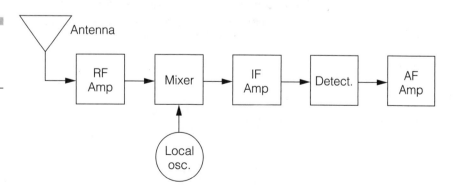

controlled by the radio's tuner to continuously produce a lower "mixing product" of (usually) 455 kHz called the *intermediate frequency* (IF). In other words, as you tune in a station on a radio, you are also changing the frequency of the local oscillator, causing it to "track" with the amplified RF signal, and to produce a constant 455-kHz IF.

This method of processing an RF signal provides higher gain than, and improved selectivity over, individually tuned RF amplifiers. By using a lower frequency, greater separation of adjacent stations might be afforded, thus resulting in greater selectivity. And, by concentrating the major part of the signal amplification process into one (IF) frequency, fewer components and stages are required, thus resulting in greater efficiency, simplicity, and savings. This system of mixing two signals in order to produce a third frequency is called *heterodyning*. Most modern radios use this process, and they are referred to as *superheterodynes*, or "superhets" for short. The processing of FM signals is identical to that described for AM, except that the IF frequency used in FM receivers is 10.7 MHz.

Continuing to refer to Fig. 16-3, the intermediate frequency is then fed into one (or more) stages of IF amplifiers, and it is eventually applied to the "detector" stage. The *detector* is the stage that "extracts" the original program signal from the IF carrier.

AM and FM Detectors

Figure 16-4 illustrates an AM detection circuit. Notice that it is hardly more than a diode. In the early days of radio, diodes were "crystals" of galena or pyrite. Many of these early "crystal" radios were merely a long, tunable antenna connected to a diode detector. Sensitive high-impedance headphones were used to listen to the program material.

To understand how a diode functions as an AM detector, refer back to Fig. 16-1. Notice that the program material is contained within the ampli-

Figure 16-4
Basic AM detector circuit.

tude variations of the carrier wave. In actuality, it is carried on the top and the bottom of the carrier wave, where the bottom is a "mirror image" of the top. However, at any given instant, the audio component voltages (but not the RF voltages) of the upper and lower halves tend to cancel each other out; they are 180 degrees out-of-phase with each other. Carefully examine where the positive and negative peaks of the two signals exist at several instants. The RF signal alternates its peaks. The audio does not.

If a diode were used to "half-wave-rectify" this signal, the result would be the full program signal (either the top, or the bottom) and half of the carrier wave, with no phase cancellation. In other words, it would be just like cutting the waveform horizontally, through the middle, and removing half of this energy to pass along to the audio stages. Referring back to Fig. 16-4, the half-wave rectified IF signal is applied across the *RC* network of C1 and P1.

The value of C1 is chosen to filter out the high-frequency IF carrier. This capacitance value has a low reactance to frequencies in the IF range, but it has a very high reactance to the lower audio frequencies. In this manner, the IF carrier is shunted to ground. The audio component, however, seeks the lower-impedance path through the pot (P1), to the wiper, and on to the audio amplifier stages. P1 is the audio volume control.

FM detection is a little more complicated. FM detection circuits are called *discriminators* or autodetectors. Figure 16-5 is technically an autodetector. The primary and secondary of the input transformer, together with C1 and C2, are both designed to be resonant at the FM IF of 10.7 MHz. The frequency variations making up the program signal cause the resonant circuit to look more inductive (or more capacitive, depending on the direction of the frequency variation). This, in turn, creates unequal voltage levels across D1 and D2, in proportion to the original program signal. C3 and C4 filter out the unwanted IF (in a fashion similar to blocking capacitor in the AM detector), and the remaining program signal is applied to an audio amplifier for reproduction. The discriminator in a quality FM receiver is much more complicated than the simplified circuit of Fig. 16-5, but the operational principle is the same.

Television

The majority of the principles that apply to radio also apply to television. The television RF signal (called the *composite video signal*) is transmitted from a broadcast station. It is received by a TV antenna, and the various components of the video signal (both AM and FM) are amplified and processed by superheterodyne action (in the same basic way as they are for radio). Of course, the composite video signal contains audio, video, and *chroma* (coloring) information. It is formatted with both FM (sound) and AM (picture) components. Even though the complexity of this signal is greatly increased, it requires only the addition of several resonant circuits to separate, or *split*, the various components: audio, video, synchronization, and colorburst signals.

The primary difference between radio and TV lies in the conversion of the video signal back into a visible picture. This conversion takes place in a special type of vacuum tube called a *cathode-ray tube* (CRT).

A basic side-view diagram of a CRT is illustrated in Fig. 16-6. A CRT has many things in common with most other vacuum tubes: a *filament* to heat the cathode, a heated *cathode* which emits electrons, an *anode* to which electrons are attracted, and a *grid* to control the flow of electrons. In a loose way, you can consider these elements to be similar to a bipolar transistor's emitter, collector, and base, respectively.

A unique aspect of a CRT is how the anodes are used as "electron accelerators." The electrons are attracted toward the anodes when they leave the cathode. But the ringed shape, and the voltage potentials (about 300 to 400 volts DC) of the beam-forming anodes, cause the electrons to overshoot them. This effect culminates into a narrow beam. After this beam is formed, the electrons are further attracted toward the screen area by a very high-voltage (10 to 25 kV DC), positive potential applied toward the front of the screen by the second anode. The forward velocity of the electrons forces them to hit a phosphor coating on the front of the screen, producing a tiny spot of light. The phosphor will absorb energy from the electron beam, and it will continue to glow for a short period, even after the beam is removed. This effect is known as *persistence*.

In order to light up the entire screen area, the tiny electron beam must be deflected in a linear pattern, called a *raster*, at a very high rate of speed.

Figure 16-6
Basic diagram of a
cathode-ray tube.

Cathode

Phosphor

Electron
beam

Anodes

Grid

Deflection plates

The movement of the beam is fast enough to allow it to cover the entire
screen area, and to return to its starting point, before the beginning spot
of phosphor totally loses its glow. In this manner, the whole screen surface
can stay continually bright. This process is repeated at a 60-hertz rate.

The CRT illustration in Fig. 16-6 is actually the type of CRT used in
oscilloscopes. CRTs used in TVs do not use deflection plates, as shown in
the illustration. They incorporate *deflection coils* placed around the narrow part of the CRT, in an assembly called the "yoke." However, for illustration purposes, it's much easier to show the action of deflection plates.
The basic operational principle is the same for both.

For beam deflection purposes, two oscillators are utilized. The horizontal oscillator output is applied to the horizontal deflection plates. As
the voltage varies on these plates, the electron beam will be attracted or
repelled, depending on the voltage polarity, causing the electron beam
to be deflected from right to left or left to right across the screen. The
vertical oscillator output is applied to the vertical deflection plates, and
this moves the beam up and down.

The horizontal and vertical oscillators are synchronized to cause the
beam to start at the top of the screen and "draw" 263 horizontal lines of
light from top to bottom; then the beam returns to the top, and draws
another 262 lines in between the original 263 lines. This whole process
occurs 30 times per second, with the two rasters of 263 and 262 lines
adding up to form a total raster of 525 lines. The total number of individual phosphor dots that must be excited for each full screen is about
315,000. This process is known as *interlaced scanning.*

By modulating the intelligence information onto the electron beam
(causing it to vary in intensity, as it scans the screen), a black-and-white
picture will be produced.

The conventional color CRT is made with three *electron guns* for the three primary colors: red, green, and blue. Each phosphor dot on the CRT screen is formed by separate red, blue, and green phosphors arranged in a triangle. The picture is thus scanned by a triangle of beams converging on each triangle of dots at the same rate as in a black-and-white picture. Individual intelligence information is provided to each gun, by the chroma board, telling it when (and how strongly) to fire. The various combinations of these guns' emissions form the resulting color picture on the screen.

APPENDIX A

SYMBOLS AND EQUATIONS

The following symbols and equations are used often in the electrical and electronic fields. In addition to equations discussed throughout the text, this section also contains many other equations for future reference. I have tried to illustrate the "case" of the symbols, and the abbreviations as they will appear most often; but many sources may use uppercase letters, where this section lists them as lowercase, and vice versa.

Letter Symbols, Abbreviations, and Acronyms

A	Ampere; amp
A	Length of the "adjacent" side; usually the longer; in a right triangle, in the same units as the other sides
AC (ac)	Alternating current
Ah	Ampere-hour
AM	Amplitude modulation
amp	Ampere, or amplifier
ant	Antenna
ASCII	American Standard Code for Information Interchange
assy	Assembly
aud	Audio
aux	Auxiliary
AWG	American Wire Gauge
B	Susceptance (measured in siemens or mhos); reciprocal of reactance
BW	Bandwidth
C	Capacitance (measured in farads)
C	Collector; coulomb

D	Dissipation factor; reciprocal of storage factor Q
d	Thickness of the dielectric material in a capacitor (measured in centimeters); depth
dB	Decibel (One-tenth of a bel); the logarithmic ratio between two levels of power, voltage, or current
DC (dc)	Direct current
DPDT	Double-pole, double-throw (switch)
DPST	Double-pole, single-throw (switch)
E (emf)	Electromotive force (measured in volts)
E	Emitter
EMT	Electrical metallic tubing
ERP	Effective radiated power
eV	Electronvolt
F	Farad (a measure of capacitance)
F	Temperature (measured in degrees Fahrenheit)
FM	Frequency modulation
f	Frequency (measured in hertz)
G	Conductance (measured in siemens or mhos)
GHz	Gigahertz
H	Henry (a measure of inductance)
H	Length of the hypotenuse of a right triangle, in the same units as the other (adjacent and opposite) sides
h	Height
HF	High frequency
Hp	Horsepower
HT	High tension (high current and high voltage)
HV	High voltage
Hz	Hertz
I	Electrical current (measured in amperes)
IC	Integrated circuit
ID	Inside diameter

IF	Intermediate frequency
IN (in)	Input
I/O	Input/output
IR	Infrared
J	Energy, work, or quantity of heat (measured in joules)
K	Coupling coefficient
k	Dielectric constant
K	Temperature in kelvins
k	Kilo- (thousand)
kHz	Kilohertz
kV	Kilovolt
kW	Kilowatt
kWh	Kilowatt-hour
L	Inductance (measured in henrys)
l	Length
LF	Low frequency
LP	Low-pass
LSB	Lower sideband
LSI	Large-scale integration
LW	Longwave
M	Mega- (million)
M	Mutual inductance (measured in henrys)
m	Milli- (one-thousandth)
mA	Milliampere
MeV	Megaelectrovolt
mH	Millihenry
mic	Microphone
mom	Momentary (switch)
Mohm	Megohm
ms	Millisecond
mV	Millivolt
mW	Milliwatt
MW	Mediumwave; megawatt

N or *n*	General symbol for numbers
n	Nano- (one-billionth)
NC	No connection
neg	Negative
neut	Neutral
nF	Nanofarad
NO	Normally open
nom	Nominal
norm	Normal
ns	Nanosecond
O	Length of the side opposite to the adjacent side (A) in a right triangle in the same units as the other sides
osc	Oscillator
out	Output
P	Power (measured in watts)
PC	Printed circuit board
PCM	Pulse-code modulation
pF	Picofarad
pf	Power factor
pk	Peak
pos	Positive
pot	Potentiometer
preamp	Preamplifier
pri	Primary
PS	Power supply
PU	Pickup
PWM	Pulse-width modulation
pwr	Power
Q	Quality factor; the ratio between reactance and resistance; ratio between inductance and resistance, when specifying inductor quality; quantity
R	Resistance (measured in ohms)
RC	Resistive/capacitive
rcv	Receive

rcvr	Receiver
rect	Rectifier
ref	Reference
res	Resistor
RF	Radio frequency
RFI	Radio-frequency interference
RLY	Relay
RL	Resistive/inductive
rms	Root mean square
rmt	Remote
rot	Rotate
rpm	Revolutions per minute
rps	Revolutions per second
rpt	Repeat
S	Siemens (conductance)
S	Area of one plate of a capacitor (measured in square centimeters)
SB	Sideband
SNR	Signal-to-noise ratio
sec	Secondary
sel	Selector
sft	Shaft
sig	Signal
sol	Solenoid
SPDT	Single-pole, double-throw (switch)
spk	Speaker
spkr	Speaker
SPST	Single-pole, single-throw (switch)
sq	Square
stby	Standby
subassy	Subassembly
SW	Shortwave
sw	switch
T	Temperature

t	Time
tel	Telephone
UF	Ultrasonic frequency
UHF	Ultrahigh frequency
V	Volt
V/A, VA	Voltampere
vid	Video
VLF	Very low frequency
VOM	Volt-ohmmeter or volt-ohm-milliammeter
VTVM	Vacuum-tube voltmeter
W	Watt
w	Width
W_C	Cutoff frequency
W_0	Midband frequency
wdg	Winding
WVDC	Working volts DC
X	Reactance (measured in ohms); the opposition to alternating current exhibited by reactive components
Y	Admittance (measured in siemens or mhos); the reciprocal of impedance
Z	Impedance (measured in ohms); the reactive and resistive opposition to the flow of alternating current
θ	90 degrees
λ	Wavelength (measured in meters)
μ	Micro- (one-millionth)
μA	Microampere
μF	Microfarad
μH	Microhenry
μS	Microsecond
π	π; 3.1416...

Formulas

Admittance

$$Y = \frac{1}{\sqrt{R^2 + X^2}} \quad Y = \frac{1}{Z} \quad Y = \sqrt{G^2 + B^2}$$

Average value

$$\text{Average value} = 0.637 \text{ (peak value)}$$
$$= 0.900 \text{ (rms value)}$$

Capacitance

Capacitors in parallel:

$$C_{total} = C_1 + C_2 + C_3 + \ldots C_n$$

Capacitors in series:

$$C_{total} = \frac{1}{\dfrac{1}{C_1} + \dfrac{1}{C_2} + \dfrac{1}{C_3} + \ldots + \dfrac{1}{C_n}}$$

Two capacitors in series:

$$C_{total} = \frac{C_1 C_2}{C_1 + C_2}$$

Capacitance value of a capacitor:

$$C = 0.0885 \, \frac{KS(N-1)}{d}$$

Quantity of charge stored:

$$Q = CE$$

where Q = charge (in coulombs)
 C = capacitance (in farads)
 E = voltage across capacitor (in volts)

Amount of stored energy:

$$W = \frac{CE^2}{2}$$

where W = energy (in joules)
  C = capacitance (in farads)
  E = voltage across capacitor (in volts)

Conductance

$$G = \frac{1}{R} = \frac{I}{E}$$

$$G_{total} = G_1 + G_2 + G_3 + \ldots G_n \text{ (resistors in parallel)}$$

Cosine

$$\cos \theta = \frac{A}{H} = \sin (90 - \theta)$$

$$= 1/\sec \theta$$

Cotangent

$$\cot \theta = \frac{A}{0} = \tan (90 - \theta)$$

$$= \frac{1}{\tan \theta}$$

Decibel

$$dB = 10 \log \frac{P_1}{P_2} \quad \text{(POWER)}$$

$$= 20 \log \frac{E_1}{E_2} \quad \text{(VOLTAGE)}$$

$$= 20 \log \frac{I_1}{I_2} \quad \text{(current)}$$

$$= 20 \log \frac{E_1 \sqrt{Z_2}}{E_2 \sqrt{Z_1}} \quad \text{(source and load impedance are unequal)}$$

$$= 20 \log \frac{I_1 \sqrt{Z_1}}{I_2 \sqrt{Z_2}} \quad \text{(source and load impedances are unequal)}$$

Figure of merit

$$Q = \frac{X}{R} = \tan \theta = \frac{L}{R}$$

Frequency

$$f = \frac{3.0 \times 10^5}{\lambda} \text{ (meters)} = \frac{9.84 \times 10^5}{\lambda} \text{ (feet)}$$

Impedance

$$Z = \sqrt{R^2 + X^2} = \sqrt{G^2 + B^2}$$

$$= \frac{R}{\cos \theta} = \frac{X}{\sin \theta} = \frac{E}{I}$$

$$= \frac{P}{I^2 \cos \theta} = \frac{E^2 \cos \theta}{P}$$

Inductance

Inductors in series:

$$L_{total} = L_1 + L_2 + L_3 + \ldots + L_n$$

Inductors in parallel:

$$L_{total} = \cfrac{1}{\cfrac{1}{L_1} + \cfrac{1}{L_2} + \cfrac{1}{L_3} + \ldots + \cfrac{1}{L_n}}$$

Two inductors in parallel:

$$L_{total} = \frac{L_1 L_2}{L_1 + L_2}$$

Coupled inductances in series with fields aiding:

$$L_{total} = L_1 + L_2 + 2M$$

Coupled inductances in series with fields opposing:

$$L_{total} = L_1 + L_2 - 2M$$

Coupled inductances in parallel with fields aiding:

$$L_{total} = \cfrac{1}{\cfrac{1}{L_1 + M} + \cfrac{1}{L_2 + M}}$$

Coupled inductances in parallel with fields opposing:

$$L_{total} = \cfrac{1}{\cfrac{1}{L_1 - M} + \cfrac{1}{L_2 - M}}$$

Mutual inductance of two coils with fields interacting:

$$M = \frac{L_A - L_O}{4}$$

where L_A = total inductance of both coils with fields aiding
L_O = total inductance of both coils with fields opposing

Coupling coefficient of two RF coils inductively coupled so as to give transformer action:

$$K = \frac{M}{L_1 L_2}$$

Meter formulas

$$\text{Ohms/volt} = \frac{1}{I} \quad \text{(meter sensitivity)}$$

where I = full-scale current in amperes

Meter resistance:

$$R_{meter} = \frac{E_{full\text{-}scale}}{I_{full\text{-}scale}}$$

Current shunt:

$$R_{shunt} = \frac{R_{meter}}{N-1}$$

where N = new full-scale reading divided by the original full-scale reading (both in the same units)

Voltage multiplier:

$$R = \frac{\text{full-scale reading required}}{\text{full-scale current of meter}} - R_{meter}$$

where reading is in volts, and current is in amperes.

Ohm's law for DC circuits

$$I = \frac{E}{R} = \sqrt{\frac{P}{R}} = \frac{P}{E}; \quad R = \frac{E}{I}$$

$$R = \frac{P}{I^2} = \frac{E^2}{P}; \quad E = IR = \frac{P}{I}$$

$$E = \sqrt{PR}; \quad P = I^2R = EI = \frac{E^2}{R}$$

Ohm's law for AC circuits

$$I = \frac{E}{Z} = \sqrt{\frac{P}{Z \cos \theta}} = \frac{P}{E \cos \theta}$$

$$Z = \frac{E}{I} = \frac{P}{I^2 \cos \theta} = \frac{E^2 \cos \theta}{P}$$

$$E = IZ = \frac{P}{I \cos \theta} = \sqrt{\frac{PZ}{\cos \theta}}$$

$$P = I^2 Z \cos \theta = IE \cos \theta = \frac{E^2 \cos \theta}{Z}$$

Peak value

$$E_{peak} = 1.414 \ (\text{rms value}) = 1.57 \ (\text{average value})$$

Peak-to-peak value

$$E_{P\text{-}P} = 2.828 \ (\text{rms value}) = 3.14 \ (\text{average value})$$

Phase angle

$$\theta = \text{arc tan} \ \frac{X}{R}$$

Power factor

$$pf = \cos \theta; \quad D = \cot \theta \ (\text{dissipation})$$

Reactance

$$X_L = 2\pi f L; \quad X_C = \frac{1}{2\pi f C}$$

Resistance

Resistors in series:

$$R_{total} = R_1 + R_2 + R_3 + \dots + R_n$$

Resistors in parallel:

$$R_{total} = \cfrac{1}{\cfrac{1}{R_1} + \cfrac{1}{R_2} + \cfrac{1}{R_3} + \dots + \cfrac{1}{R_n}}$$

Two resistors in parallel:

$$R_{total} = \frac{R_1 R_2}{R_1 + R_2}$$

Resonance

$$f_{res} = \frac{1}{2\pi \sqrt{LC}} \; ; \quad L = \frac{1}{4\pi^2 f^2 C} \; ; \quad C = \frac{1}{4\pi^2 f^2 L}$$

Right triangle

$$\sin \theta = \frac{O}{H} ; \quad \cos \theta = \frac{A}{H} ; \tan \theta = \frac{O}{A}$$

$$\sec \theta = \frac{H}{A} ; \quad \cot \theta = \frac{A}{\cap}$$

Root mean square (sinusoidal waveshapes only)

$$rms = 0.707 \text{ (peak value)}$$
$$= 1.111 \text{ (average value)}$$

Secant

$$\sec \theta = \frac{H}{A} = \frac{1}{\cos \theta}$$

$$\sec \theta = \text{cosecant } (90 - \theta)$$

Sine

$$\sin \theta = \frac{O}{H} = 1/\text{cosecant}\theta$$

$$= \cos (90 - \theta)$$

Susceptance

$$B = \frac{X}{R^2 + X^2} = \frac{1}{X}$$

$$B_{\text{total}} = B_1 + B_2 + B_3 + \ldots + B_n$$

Tangent

$$\tan\theta = \frac{O}{A} = \frac{1}{\cot\theta} = \cot (90 - \theta)$$

Transistors, bipolar

$$I_c = I_b (\beta_{\text{DC}}) \; ; \quad I_e \approx I_c \; ; \quad E_e = E_b - 0.7 \text{ volts DC (silicon)}$$

$$E_e = E_b - 0.3 \text{ volts DC (germanium)} \; ; \quad Z_b = R_e (\beta_{\text{DC}})$$

$$I_c = \frac{E_e}{R_e} \; ; \quad A_p = \frac{P_{\text{output}}}{P_{\text{input}}}$$

Transistors, field-effect

$$G_{\text{fs}} = \frac{\Delta I_D}{\Delta E_{GS}}$$

where G_{fs} = transconductance value
ΔE_{GS} = a change in gate to source voltage
ΔI_D = a subsequent change in drain current

Temperature

$$°C = (0.556°F) - 17.8; \quad °F = (1.8°C) + 32; \quad K = °C + 273$$

where K = kelvins.

Transformer ratio

$$\frac{N_p}{N_s} = \frac{E_p}{E_s} = \frac{I_s}{I_p} = \sqrt{\frac{Z_p}{Z_s}}$$

Units of energy

$$\text{Watt-hours} = PT$$

where P = power   (in watts)
T =   (in hours)

Wavelength

$$\lambda = \frac{300,000}{f} \quad \text{(in meters)}$$

APPENDIX B

SOURCES FOR ELECTRONIC MATERIALS

The sources listed in this section are classified under the following sub-headings:

- Data Book Sources
- Electronic Kit Suppliers
- Full-line Suppliers
- Surplus Dealers
- Tools and Hardware Suppliers
- Electronic Test Equipment Dealers
- Electronic Periodicals
- Recommended Reading

Please keep in mind that there is much overlap within these categorizations. For example, few surplus dealers sell "only" surplus items, and many are quite diversified in their product lines. The goal here is to provide some means of organization, based on "primary" commercial efforts.

Data Book Sources

NTE Electronics
44 Farrand Street
Bloomfield, NJ 07003
1-800-631-1250

MCM Electronics
650 Congress Park Drive
Centerville, Ohio 45459-4072
1-800-543-4330

Jameco Electronic Components
1355 Shoreway Road
Belmont, CA 94002-4100
1-800-831-4242

Electronic Kit Suppliers

(SEAL Electronics offers many of the projects contained within this book in kit form.)

SEAL Electronics
3898 Kentucky Route 466
P.O. Box 268
Weeksbury, KY 41667
1-606-452-4135

Ramsey Electronics, Inc.
793 Canning Parkway
Victor, NY 14564
1-716-924-4560

Active Kits
345 Queen Street West
Toronto, Ontario M5V 2A4
1-800-465-5487

Full-Line Suppliers

Jameco Electronics
1355 Shoreway Road
Belmont, CA 94002-4100
1-800-831-4242
(Jameco is a great source for electronic data books.)

Parts Express
725 Pleasant Valley Drive
Springboro, OH 45066
1-800-338-0531

MCM Electronics
650 Congress Park Drive
Centerville, OH 45459-4072

Mouser Electronics
958 North Main Street
Mansfield, TX 76063-4827
1-800-346-6873

RadioShack
P.O. Box 1981
Fort Worth, TX 76101-1981
1-800-442-7221

Digi-Key Corporation
701 Brooks Avenue South
P.O. Box 677
Thief River Falls, MN 56701-0677
1-800-344-4539

Surplus Dealers

Fair Radio Sales
1016 E. Eureka Street
P.O. Box 1105
Lima, Ohio 45802
1-419-227-6573

B. G. Micro, Inc.
P.O. Box 280298
Dallas, TX 75228
1-800-276-2206

Brigar Electronics
7—9 Alice Street
Binghamton, NY 13904
1-607-723-3111

Herbach and Rademan
16 Roland Avenue
Mt. Laurel, NJ 08054
1-800-848-8001

C and H Sales
2176 East Colorado Boulevard
Pasadena, CA 91107
1-800-325-9465

All Electronics Corporation
14928 Oxnard Street
P.O. Box 567
Van Nuys, CA 91408-0567
1-800-826-5432

Marlin P. Jones & Associates, Inc.
P.O. Box 12685
Lake Park, FL 33403-0685
1-800-652-6733

Hosfelt Electronics, Inc.
2700 Sunset Boulevard
Steubenville, OH 43952-1158
1-800-524-6464

Tools and Hardware Suppliers

Techni-Tool
5 Apollo Road
Box 368
Plymouth Meeting, PA 19462-0368
1-800-832-4866

Contact East, Inc.
335 Willow Street
North Andover, MA 01845-5995
1-800-225-5334

Jensen Tools, Inc.
7815 S. 46th Street
Phoenix, AZ 85044-5399
1-800-426-1194

Electronics Workbench
908 Niagara Falls Boulevard
Suite 068
North Tonawanda, New York 14120-2060
1-800-263-5552

(Electronics Workbench distributes the MultiSim and UltiBoard electronic design and simulation software.)

Electronic Test Equipment Dealers

Tucker Electronics
P.O. Box 551419
Dallas, TX 75355-1419
1-800-527-4642

Global Specialties
70 Fulton Terrace
New Haven, CT 06512
1-800-572-1028

Electronic Periodicals

Poptronics Magazine

Gernsback Publications, Inc.
500 Bi-County Boulevard
Farmingdale, NY 11735
(631)-293-3000

Recommended Reading

Stan Gibilisco, *Electronics Portable Handbook,*
McGraw-Hill, New York, 1999 (ISBN 0-07-134415-2).

G. Randy Slone, *High-Power Audio Amplifier Construction Manual,* McGraw-Hill, New York, 1999 (ISBN 0-07-134119-6).

Neil Sclater and John Markus, *McGraw-Hill Electronics Dictionary,* 6th ed., McGraw-Hill, New York, 1997 (ISBN 0-07-057837-0).

APPENDIX C

The following figures are to be used for making printed circuit boards and constructing the audio power amplifier in Chapter 8.:

Figure C-1
12-watt audio amplifier: (a) and (b) top views silkscreen layout; (c) copper artwork.

(a)

(b)

(c)

Figure C-2 Bottom-view reflected artwork for the 50-watt professional-quality amplifier.

INDEX

Note: **Boldface** numbers indicate illustrations.

ABOUT THE AUTHOR

G. Randy Slone is an electrical engineer with nearly two decades of experience in data processing and industrial electronics. A private engineering consultant, his clients have included DuPont, Champion International, and Ralston Purina. Also a former college instructor, Slone spends much of his time working in his state-of-the-art home electronics laboratory.